Building Vulnerability Assessments

Industrial Hygiene and Engineering Concepts

Building Vulnerability Assessments

Industrial Hygiene and Engineering Concepts

Edited by
Martha J. Boss • Dennis W. Day

CRC Press
Taylor & Francis Group
Boca Raton London New York

CRC Press is an imprint of the
Taylor & Francis Group, an **informa** business

CRC Press
Taylor & Francis Group
6000 Broken Sound Parkway NW, Suite 300
Boca Raton, FL 33487-2742

First issued in paperback 2019

ISBN-13: 978-1-4200-7834-3 (hbk)
ISBN-13: 978-0-367-38547-7 (pbk)

Library of Congress Cataloging-in-Publication Data

Building vulnerability assessments : industrial hygiene and engineering concepts
/ editors, Martha J. Boss, Dennis W. Day.
p. cm.
Includes bibliographical references and index.
ISBN 978-1-4200-7834-3 (alk. paper)
1. Buildings--Security measures. 2. Industrial hygiene. 3. Sanitary engineering.
I. Boss, Martha J. II. Day, Dennis W. III. Title.

TH9705.B85 2008
658.4'73--dc22   2008049151

Visit the Taylor & Francis Web site at
http://www.taylorandfrancis.com

and the CRC Press Web site at
http://www.crcpress.com

About the Cover Photo

Nel Bernard Larson, the maternal grandfather of Martha Boss, took this picture on a tour of upper Michigan, circa 1950s.

Birch trees grow throughout the world, spreading through forests and establishing birch tree colonies. The interaction between birch trees and their unseen symbiotes is an example of nature reducing vulnerability. The symbiotes are mycorrhizal fungi (i.e., *Laccaria bicolor*) that establish a partnership with plant roots creating a mycorrhizal root. Mycorrhizal fungi lack the enzymes involved in degradation of the carbohydrate polymers of plant cell walls but maintain the ability to degrade nonplant cell walls. Consequently, the fungi transition between two distinct lifestyles—the benign saprotroph, able to use decaying matter of animal and bacterial origins, versus the symbiont, living in mutually profitable harmony with plant roots. The fungi translocate nutrients and water captured in soil pores inaccessible to the host plant roots significantly improve photosynthetic carbon assimilation by plants. The plant, in turn, protects the fungi within its roots from competition with other soil microbes and the fungi gain preferential access to carbohydrates within the plant.

The interweaving of the mycorrhizal roots from various plants, including trees, beneath the soil surface creates a rhizosphere. This rhizosphere benefits trees, plants, and their mutualistic fungi that enter into this unseen web beneath the soil surface. The rhizosphere may also provide a primordial chemical communication link. So, on a biological, albeit chemically, mediated level, the rhizosphere uses assessment of the environment, realization of interactions between entities, and communication to limit risk. These plant systems through time have limited their vulnerabilities!

Contents

Preface

BUILDING VULNERABILITY ASSESSMENTS: INDUSTRIAL HYGIENE AND ENGINEERING CONCEPTS

Often vulnerability is defined only with respect to security issues leading to vulnerability. Consequently, emphasis may be placed solely on security upgrades, guards, and entrance barriers (both physical and electronic). However, in addition to these vulnerabilities, the status of the building as designed and ultimately constructed and used should be considered. This book focuses on the range of vulnerabilities that can and should be addressed—from initial security from intrusion to security given the building's industrial hygiene status. The intent is to aid building owners and operators in risk management, both during and after design implementation.

Initial Chapters 1 and 2 deal with a suggested outline for vulnerability assessments conducted either in-house or in coordination with a third party. Modifications of the outline presented will make the assessments site specific for your particular needs. Decision logic should also be included during planning stages to elucidate why the vulnerability assessment items were included or not included.

Chapter 3 presents summaries of both applicable and relevant regulations used to determine if chemicals create a risk to off-site locations or constitute a homeland security vulnerability. While the emphasis is on United States regulations, many other international regulations meet or exceed the U.S. requirements and should also be considered.

Since physical security may be the first line of defense from intrusion, Chapter 4 discusses physical security in terms of both the building's geographic and virtual (electronic) footprint. Then Chapters 5 through 7 discuss the chemical, biological, and radioactive (CBR) threat potentials. The concepts presented in these chapters can be expanded upon to deal with other CBR threats not implicitly discussed.

The remainder of this handbook deals with control systems to reduce vulnerabilities. Emphasis is placed on ventilation system controls to ensure initial safe evacuation or shelter-in-place success. The reader must always keep in mind that a building or facility already contaminated will be easier to contaminate further. This onus is particularly true for buildings with latent biological contamination, thus, Chapters 15 through 17 address issues associated with new, latent, and residual biological contamination within building infrastructures.

Finally, Chapter 18 presents basic emergency planning provisos. This planning can be augmented as further vulnerabilities are determined through successful assessment iterations and use of emergency drills or table-top exercises. The intent is to assist the normal operations and emergency planners, as well as future designers, in making decisions that may lessen the impact of emergencies.

By using this handbook, building managers, planners, and the emergency response community will increasingly see buildings and their sites in a new way.

This handbook is both a reference and a wake-up call provided by scientists and engineers who have dealt in the past with building and siting failures.

In the words of an ancient Navaho chant, "It was the wind that gave them life. It is the wind that comes out of our mouths now that gives us life. When this ceases to blow we die. In the skin at the tips of our fingers we see the trail of the wind, it shows us the wind blew when our ancestors were created." We must all strive to see this wind and the effects on all of us—for our vulnerabilities when discovered are our opportunities for improvement and protection.

About the Authors

Martha and Dennis would like to thank their colleagues and clients who, over the past four years, have helped us test our ideas and hone this text. We have been honored to work with these contributors and the authors. We thank them all for their dedication to applied science as a tool to make this world a safer and healthier place.

Martha Boss is a practicing Certified Industrial Hygienist and Certified Safety Professional (Safety Engineer). Martha is a principal toxicologist for URS Corporation. She is a Diplomate of the American Board of Industrial Hygiene and participates on the Editorial Advisory Board for Stevens Publishing. She has over nineteen years experience in industrial hygiene, safety engineering, emergency response and risk management planning, indoor air quality and ventilation evaluation, due diligence assessments, biological risk assessment, pollution prevention and waste minimization studies, and industrial hygiene and safety engineering design analysis. Martha is co-editor and author of the *Biological Risk Engineering Handbook* (2003) and *Air Sampling and Industrial Hygiene Engineering* (2001) texts published by CRC Press, Boca Raton, Florida. Martha has audited international programs to determine client compliance with United States, European Union, and specific governmental requirements. She has worked in the United States, United Kingdom, Norway, Germany, Spain, Switzerland, Brazil, Indonesia, Poland, and other international locales. She holds a Bachelor of Science degree in biology with an engineering emphasis and a bachelor of arts degree in chemistry with a chemistry emphasis.

Dennis Day is a practicing Certified Industrial Hygienist and Certified Safety Professional (Safety Engineer). Dennis is a regional health and safety manager for URS Corporation. He is a Diplomate of the American Board of Industrial Hygiene, a member of the American Industrial Hygiene Association, and an instructor for Global Village Engineers. He has over thirty years' experience in industrial hygiene, safety engineering, emergency response and risk management planning, indoor air quality and ventilation evaluation, due diligence assessments, biological risk assessment, pollution prevention and waste minimization studies, and industrial hygiene and safety engineering design analysis. Dennis is co-editor and author of the *Biological Risk Engineering Handbook* (2003) and *Air Sampling and Industrial Hygiene Engineering* (2001) texts published by CRC Press, Boca Raton, Florida. Dennis has audited international programs to determine client compliance with United States, European Union, and specific governmental requirements. He has worked in the United States, United Kingdom, Germany, Spain, Switzerland, Brazil, Indonesia, China, Georgia (the country), Poland, and other international locales. He holds a Bachelor of Science in forest management, and chemistry and biology secondary teaching endorsements. Dennis received the Achievement Medal for Civilian Service for his emergency industrial hygiene support following Hurricane Andrew, December 1992.

Charles Allen is currently a vice president for URS Corporation in Atlanta, Georgia. He is a Federal Programs Manager and has been the responsible principal for environmental and engineering projects for the U.S. Army Corps of Engineers; the U.S. Air Force Center for Environmental Excellence; the U.S. Postal Service; the U.S. Department of Agriculture, Forest Service; the U.S. Department of Veterans Affairs; the U.S. Department of Energy, Savannah River Site; and the U.S. Navy, Naval Facilities Engineering Command. Prior assignments have included Chief Engineer, DAMISA, a URS Spanish subsidiary and principal engineer for projects on four continents. Charles is a practicing civil and environmental engineer with over forty years of experience on projects with URS and is a registered engineer in Georgia and Alabama. He holds a Bachelor of Science in civil engineering from the Georgia Institute of Technology and is a member of the Academy of Distinguished Engineering Alumni from the College of Engineering, Georgia Tech.

Randy Boss is currently the Military Program Manager for the U.S. Army Corps of Engineers (USACE) Northwestern Division in Omaha, Nebraska. His previous U.S. Air Force work engagements included Supervisory Engineering for Headquarters, Strategic Air Command (SAC); Deputy Base Civil Engineer for Moody Air Force Base (AFB) in Georgia; and Missile Civil Engineer at Norton AFB. Prior to that time, Randy was deployed by USACE as a lead engineer for Saudi Arabian projects at both Headquarters Rear Middle East Division and at various Middle East locations. Randy is a practicing Professional Engineer with over thirty years of experience providing engineering support and management for military, environmental, and civil works programs. He is a registered Professional Engineer in the Commonwealth of Virginia. Mr. Boss holds Bachelor of Civil Engineering and Masters of Environmental Engineering degrees from Iowa State University, and is a graduate of the U.S. Air War College.

Kendall G. Christenson is a mechanical engineer with the U.S. Corps of Engineers in Omaha, Nebraska. Mr. Christenson has over twenty years of design and project management experience on military projects. His chemical, biological, and radiological (CBR) experience includes performing building pressure tests, shelter-in-place design, filtration system design, system testing, system certification, and criteria development. He holds a Bachelor of Science degree in mechanical engineering from North Dakota State University.

Jon A. Cummins is the Chief Operating Officer of Amerimar Enterprises, Inc., a position he has held since 2001. Mr. Cummins joined Amerimar in 1989. His primary responsibilities are the general oversight of the company's operations as well as the sourcing, acquisition, asset management, property management, financing, and disposition of the company's multifamily and hotel properties. Notable redevelopment projects led by Mr. Cummins include the Hotel George in Washington, District of Columbia; the Sheraton Atlanta Hotel in Atlanta, Georgia; the Westin Governor Morris in Morristown, New Jersey; Pennsylvania Place in Harrisburg,

Pensylvania; and the Hotel Derek in Houston, Texas. Mr. Cummins is also a member of Amerimar's Investment and Executive Committees. Amerimar's focus is on the acquisition of "value-added" hotels, apartments, office, and retail properties in major urban and suburban markets throughout the United States. The company's expertise encompasses all facets of the acquisition, financing, renovation, redevelopment, repositioning, marketing, leasing, asset management, property management, and disposition of its assets. Prior to joining Amerimar, Jon worked in the Real Estate Investment Banking Division at Lehman Brothers in New York where he was involved in the placement of debt and equity financing vehicles and the acquisition of property for the firm's own account. Jon earned a Bachelor of Science degree in economics from the Wharton School of the University of Pennsylvania. Mr. Cummins has served on the board of directors of the New Jersey Apartment Association and is actively involved with the Cystic Fibrosis Foundation.

Donald C. Dittus is a mechanical engineer who has worked for the U.S. Army Corps of Engineers (Omaha, Nebraska) for twenty-five years. He has extensive experience in pressure testing numerous buildings to reduce air infiltration. The test results are used to help reduce building energy consumption or to minimize an external chemical, biological, and radiological (CBR) threat. He also has experience designing air filtration systems that minimize threats from industrial chemicals and weaponized CBR agents. He holds a bachelor's degree in mechanical engineering from the University of Nebraska and is a licensed Professional Engineer in that state.

Lawrence Fitzgerald is a senior security consultant and program manager with URS Corporation in their Hallowell, Maine office. In this role he assesses security performance and effectiveness on a facilitywide as well as enterprisewide basis throughout North America, and develops strategies to enhance performance of the security function. He has led numerous threat and vulnerability assessments of critical infrastructure (CI) and key resources (KR) for government entities and private sector clients (i.e., commercial, energy, educational, chemical, and financial facilities). Mr. Fitzgerald has directed and developed security policies and procedures and designed security solutions for numerous types of facilities. Previously he managed a 24/7 security operations center (SOC) and central station that provided security at hundreds of government, commercial, and industrial facilities throughout the northeastern United States. Due to his experience with hazardous materials and scientific background, he was intimately involved with the response to the anthrax bioterrorism events in 2001. Mr. Fitzgerald earned a bachelor of arts in geology from the University of Rhode Island, is a certified Physical Security Professional (PSP) by ASIS International, and is licensed as a Professional Geologist (PG) in several states.

Tony Host is a practicing Certified Safety Professional and Certified Fire Protection Specialist. Tony is a Senior Environmental Scientist for URS Corporation. He has over 36 years experience in air quality management and environmental chemistry, specializing in industrial hygiene, environmental, health and safety auditing, fire protection, indoor air quality, hazardous waste management, and air and waste emission measurements. Mr. Host has been responsible for the management of both non-commercial and commercial analytical laboratories and has worked internationally to determine client compliance with United States, EU, and specific governmental requirements.

Gayle Nicoll is an environmental scientist and associate project manager for URS Corporation in Omaha, Nebraska. Dr. Nicoll has conducted several years of research in radiochemistry at academic cyclotrons. She has also researched methods to degrade CFCs, using excimer lasers and FT-IR to probe the mechanism of decay. As a graduate student, Dr. Nicoll identified many undergraduate-level general chemistry misconceptions related to bonding and had her first sole-author paper published prior to receiving her master's degree. Dr. Nicoll spent fifteen years teaching and researching in academia, including Big Ten/Twelve universities, private colleges, and community colleges. She holds a Bachelor of Science degree in chemistry and physics from Indiana University, a master of science degree in analytical chemistry from Purdue University, and a doctorate in chemistry education from Purdue University.

Alisa Otteni is a Certified Professional Environmental Auditor and lead auditor with URS Corporation. Ms. Otteni sits as the Auditor Resource Chair for the Auditing Roundtable and also sits on the BEAC Certification Committee. She has over seventeen years of experience in the environmental compliance field. She is certified as an ISO 14000 Lead Auditor with regard to environmental management systems. Her work performance has been both throughout the United States and internationally.

Greg Tsouprake is a registered Professional Engineer with over twenty years of environmental engineering experience. He has a B.S. in civil engineering from Worcester Polytechnic Institute (WPI), 1988, and an M.S. in civil engineering from WPI, 1992. His relevant experience includes environmental due diligence and investigations, civil and environmental engineering, geotechnical investigations and engineering, wastewater and storm water permitting and engineering, air quality permitting and engineering, environmental, health and safety audits and compliance actions, RCRA and CERCLA closures, and land surveying. Mr. Tsouprake is a group manager and senior project manager with URS Corporation and has managed a wide variety of civil, environmental, health and safety projects for private clients around the world.

CHAPTER 1

Introduction to Vulnerability Assessments

Martha Boss, Dennis Day, Lawrence Fitzgerald, and Gayle Nicoll

CONTENTS

Whether through natural events or solely due to human factors, disasters do happen. This text is provided to alert owners, managers, engineers, scientists, workers, site visitors, and residents to the inherent vulnerabilities faced as property is managed. Many lessons have been learned in the United States and around the world as disasters and their aftermath have been faced. Some of these lessons indicate that current site, facility, and building vulnerabilities are a factor in disaster response and the rehabilitation that comes after the emergency events. Steps can be taken now to lessen these vulnerabilities provided the vulnerabilities can be identified.

The sample Vulnerability Assessment Program (VAP) elements and methodologies presented in Chapters 1 and 2 encompass the physical and cyber components and the interdependencies among these components with the other critical national infrastructures. Example components of the vulnerability assessments and vulnerability surveys and quick-turnaround assessments tools are presented. The example components can be modified to address other types of concerns. The chapters that follow provide information to be used in the vulnerability assessment process.

Many inspections currently conducted for due diligence, insurance, regulatory compliance, management, and code compliance can be adapted to provide information on vulnerabilities as well. One aspect of all such inspections should be availability of utility shutoffs, fire prevention equipment (either on-site or remote), water supply evaluations, and detection system potentials. The entire facility, including outdoor environs, rooftops, interiors, system conduits, subslab conduits, crawl spaces, and subgrade structures, should be inspected.

Building and facility vulnerability may be exposed through natural disasters or through man-made chemical, biological, or radiological (CBR) attack. Both the threat profile and a security assessment provide information on current vulnerability. The decisions concerning which protective measures should be implemented should be based on several factors, including the perceived risk associated with the building and its tenants, engineering and architectural feasibility, and cost. The physical security of the facility's infrastructure and site is the first feature to be analyzed—simply because the most vulnerable site is the one easily reached. Preventing terrorist access requires physical security of entry, storage, roof, and mechanical areas, as well as securing access to the outdoor air intakes of the building heating, ventilation, and air conditioning (HVAC) system. Building owners and managers should be familiar with their buildings. This familiarity must include an understanding of what assets require protection and what building or occupant characteristics make the facility a potential target. Since many of these systems and the facility or site operations are computer controlled, an assessment of the computer systems must be simultaneously conducted.

The basic hazards associated with CBR risks must be known and incorporated into all VAP and planning decision logic. Some of the risks may already be inherent in the controlled use of materials at a site. When a disaster strikes, these materials may take on a new identity as site contaminants, acute hazards, and long-term risks. Knowing current site vulnerabilities prior to a CBR risk event is a required element in vulnerability analysis. Basic concepts vital to chemical contaminant control, and biological and radiological (or radioactive) threats must be understood in order for vulnerability to be assessed and delimited.

While the identification and resolution of building vulnerabilities will be specific to each building, some physical security actions are applicable to many building types. A walk-through inspection of the building and building systems, including the heating, ventilation, and air conditioning (HVAC), fire protection, and life-safety systems is the first step in evaluating levels of security. If as-built drawings are available, these drawings and the decision logic used during design should be consulted during the walk-through inspection. Alterations from the original design that occurred either during construction or throughout the building use should be noted on updated drawings.

Often the easiest way to defeat a building security system is through the building ventilation system. HVAC system reliability during an emergency must be evaluated. Since ventilation systems are both a vector to introduce hazards into a facility and a potential risk mitigation technique, a basic understanding of the HVAC systems is required. This understanding will also include the potential for sheltering in place versus evacuation during an emergency event. In order for a site or facility to be considered for sheltering in place, adequate lighting, building ventilation, water control, power supply, and physical security must be assured. HVAC systems may play several roles depending on the needs to isolate areas of the facility from contaminants, remove contaminants from the airstream, and/or zone areas for sheltering in place.

The current disposition of debris, polluted waters, and contaminated materials within a facility's visible structure and normally inaccessible interstitial structures must be addressed. Latent residual contamination may become another inherent threat as CBR contaminants are introduced and if biological contamination amplifies. If the HVAC system is incorrectly used, the system itself may become contaminated and add to the dispersion and site loading of contaminants. Because an initial comprehensive evaluation of current HVAC status is crucial to evaluating facility vulnerabilities, a detailed analysis of HVAC systems is presented in this text. This information must be supplemented by site-specific, installation-specific, and manufacturer's information during a site-specific HVAC vulnerability assessment. Current HVAC installations can be compared to the United States Government Service Administration (GSA) design parameters presented in this text. This analysis can extend to plumbing and water supply systems both for the HVAC and building occupants' direct use.

Decontamination of personnel, materials, and equipment may be an ongoing need. Alternately, if evacuation is the first response, decontamination may be required before the building is safe for reentry. The use of decontamination solutions, biocides, vacuuming, steam cleaning, and mechanical demolition must all be considered given the potential for secondary transmittal of CBR hazards. The type of building materials and impacted surfaces will determine which decontamination methods have proven efficacy for the range of contaminants. Examples with imbedded decision logic of how building material decontamination needs may be assessed (prior to a disaster) and actual decontamination methods are presented.

This text introduces the concept of Vulnerability Assessments and then provides information about CBR threats, mitigation techniques, and essential building systems. Knowledge of the threats primes the assessors to expand on the general concepts of vulnerability assessments to specifically address the implications of a CBR incident. Such preparation is needed given the current threat to facilities and sites worldwide.

1.1 ASSESSMENT PROGRAMS AND METHODOLOGY

VAP development may take many forms. The VAP presented in this text illustrates a format originally presented by the Department of Energy's Office of Energy Assurance. VAPs are developed to better understand threats and vulnerabilities, determine acceptable levels of risk, and stimulate action to mitigate identified vulnerabilities.

The National Institute of Justice developed the Vulnerability Assessment Methodology (VAM) in collaboration with the Department of Energy's Sandia National Laboratories. The use of that vulnerability assessment methodology was limited to preventing or mitigating terrorist or criminal actions that could have significant national impact (e.g., loss of chemicals vital to the national defense or economy) or could seriously affect localities (e.g., release of hazardous chemicals that would compromise the integrity of the facility), contaminate adjoining areas, or injure or kill facility employees or adjoining populations. This VAM addresses physical security at fixed sites but not cyber and transportation security issues.

The combined VAP and VAM elements address physical, cyber, and chemical security issues. The VAP must be site specific and focused on the critical assets at hand.

- If a VAP has been fully implemented and a solid security infrastructure (staffing, plans/procedures, funding) is present, a cursory review of the program documents may suffice. Implementation of the VAP is the main focus of the assessment.
- If, however, an insufficient VAP and organization team to implement the VAP is determined to be present, the majority of time should be spent at the organizational level to identify the appropriate staffing and funding necessary to implement a VAP. Research into specific facility deficiencies should be limited to finding just enough examples to support VAP recommendations, since VAP implementation cannot be evaluated if a VAP is not in place.

Programmatic elements that in concert create an integrated VAP can be evaluated in lieu of a stand-alone VAP. For example, emergency planning documents, information technology (IT) security requirements, and site standard operating procedures in concert may be reviewed as a VAP.

1.2 BENEFITS

The direct benefits of performing a vulnerability assessment include:

- **Build and broaden awareness.** The assessment process directs senior management's attention to security. Awareness is one of the least expensive and most effective methods for improving the organization's overall security posture.
- **Establish or evaluate against a baseline.** If a baseline has been previously established, an assessment is an opportunity to gauge improvement or deterioration of an organization's security posture. If no previous baseline has been performed (or the work was not uniform or comprehensive), an assessment is an opportunity to integrate and unify previous efforts, define common metrics, and establish a definitive baseline. The baseline also can be compared against best practices to provide perspective on an organization's security posture.
- **Identify vulnerabilities and develop responses.** Generating lists of vulnerabilities and potential responses is a core activity. The assessment process documents the decision logic used and the rationale for choices made to address the vulnerabilities—even if the chosen response is nonaction.
- **Categorize key assets and drive the risk management process.** The assessment facilitates consensus through the ranking of key assets, combined with threat, vulnerability, and risk analysis.
- **Develop and build internal skills and expertise.** A security assessment promotes security skills and expertise while focusing various groups to communicate about issues that affect some or all groups. External assessors should be used as needed, with an emphasis on teaching and collaborating.
- **Promote action.** The assessment focuses management attention and resources on solving specific and systemic security problems. Legal, financial, and executive resources can then be mobilized. A well-designed and executed assessment identifies vulnerabilities, makes recommendations, gains executive buy-in, identifies key players, and establishes a set of crosscutting groups that can convert recommendations into action.
- **Kick off an ongoing security effort.** An assessment should be a catalyst to involve people throughout the organization in security issues, build crosscutting teams, establish permanent forums and councils, and harness the momentum generated by the assessment to build an ongoing institutional security effort. The assessment can lead to the creation of either an actual or a virtual (matrixed) security organization.

1.3 VULNERABILITY ASSESSMENT METHODOLOGY MODEL

The prototype VAM model is a systematic, risk-based approach in which risk is a function of the severity of consequences of an undesired event, the likelihood of adversary attack, and the likelihood of adversary success in causing the undesired event. For the purpose of the VAM analyses:

Risk is a function of S, L_A, and L_{AS}.

S = severity of consequences of an event
L_A = likelihood of adversary attack
LS = likelihood of adversary attack and severity of consequences of an event
L_{AS} = likelihood of adversary success in causing a catastrophic event

The VAM compares relative security risks. If the risks are deemed unacceptable, rec-ommendations can be developed for measures to reduce the risks. The severity of the consequences can be lowered in several ways, such as reducing the quantity of hazard-ous material present or siting chemical facilities (CFs) farther from populated areas.

Although adversary characteristics generally are outside the control of CFs, facilities can take steps to make themselves a less attractive target and reduce the likelihood of attack. Reducing the quantity of hazardous material present may also make a CF less attractive to attack. The most common approach, however, to reduc-ing the likelihood of adversary success in causing a catastrophic event is increasing protective measures against specific adversary attack scenarios.

Because each undesirable event is likely to have its own consequences, adversar-ies, likelihood of attack, attack scenario, and likelihood of adversary success deter-mine the risk for each combination of risk factors. Although the VAM is usually used for some or all CFs that are required to submit Risk Management Plans (RMPs), the VAM can also be used for undesired events of lesser consequence than those found in RMPs. The VAM has 12 basic steps:

Preassessment
1. Screening for the need for a vulnerability assessment
2. Defining the project
3. Characterizing the facility
4. Deriving severity levels
5. Assessing threats
6. Prioritizing threats
7. Preparing for the site analysis

Assessment
8. Surveying the site
9. Analyzing the system's effectiveness
10. Analyzing risks
11. Making recommendations for risk reduction
12. Preparing the final report

1.3.1   Likelihood of Adversary Attack

After the threat spectrum has been described, the information can be used together with statistics of past events and site-specific perceptions of threats to categorize threats in terms of the likelihood that each would result in an undesired event. The Department of Defense (DoD, 2000) standard definitions have been modified for use in categorizing the threats—likelihood of adversary attack (L_A) definition:

- 1 = Threat exists, is capable, has intent or history, and has targeted the facility.
- 2 = Threat exists, is capable, has intent or history, but has not targeted the facility.
- 3 = Threat exists and is capable, but has no intent or history and has not targeted the facility.
- 4 = Threat exists, but is not capable of causing an undesired event.

An initial threat analysis for all assets should be conducted as illustrated in the Threat Environment Methodology Table which must be made site or facility specific.

Threat Environment Methodology

The on-site analysis of the threat environment takes place in three phases:

Phase I: An initial screening of sources, prior to the arrival of the vulnerability assessment team, is conducted to identify individual(s) and/or group(s) who are potentially threatening and to establish contact with local, state, and federal law enforcement agencies (LEAs) to begin the analysis process.

Phase II: An on-site assessment is performed, beginning with interviews of facility security managers. The purpose of these interviews is to:
Determine current corporate security measures (vis-à-vis threats)
Identify LEAs with whom the *facility* security managers routinely liaison

Phase III: Office calls are made to federal, state, and local LEAs. The following questions are pursued to determine the threat environment.

The identities and modi operandi of known or suspected individual(s)/group(s) who have:

- Initiated hostile actions against any public utilities in the region
- Threatened hostile action against any public utility in the region
- Threatened or executed hostile action within the community or its surrounding area to further their "cause"
- Overall assessment of potential threats to any public utility in the area
- The availability of regularly scheduled LEA intelligence briefings to assist the *facility* Points of contacts *facility* security personnel can utilize for questions concerning incidents and/or threats of any nature by current or former employees of facility

List of Organizations to Contact for Threat Information

Organization	Contact (fill in)	Phone (fill in)
Federal Bureau of Investigation (FBI) Joint Terrorist Task Force (JTTF), Domestic Terrorism Division		
FBI Field Office		
Department of the Treasury, Bureau of Alcohol, Tobacco, and Firearms (ATF) Field Office, Domestic Extremist Section		
ATF Field Office, Intelligence Division		
Department of Justice, U.S. Marshal Service		
City Police Intelligence Division		
North American Electric Reliability Corporation (NERC)		
County/Sheriff's Office, Crime Analyst; Drug Enforcement Agency, Crime Analyst (organized crime); U.S. Marshal Service, Department of Justice; U.S. Secret Service; U.S. Customs; U.S. Border Patrol; U.S. Immigration and Naturalization Service (INS)		

1.3.2 Severity of Consequences

The severity of consequences for each undesired event must be derived. For facilities that have conducted process hazard analyses (PHAs) modified to account for the consequences of a malevolent (rather than an accidental) event, the PHAs will provide valuable information. Another source of data to help determine the severity of consequences is the analysis of the off-site consequences of the worst case and alternative-release scenarios.

The following example (for chemical facilities) should be made site specific because various facilities and communities may assign different severity levels to similar consequences. Each undesired event will be assigned a severity level based on the consequences defined by the severity level definition. This severity value (SV) will be used in the risk analysis.

- 1 = Potential for any of the following resulting from chemical release, detonation, or explosion: worker fatalities, public fatalities, extensive property damage, facility disabled for more than one month, major environmental impacts, evacuation of neighbors.
- 2 = Potential for any of the following resulting from a fire or major chemical release: nonfatal injuries, unit disabled for less than one month, shutdown of road or river traffic.
- 3 = Potential of any of the following resulting from chemical release: unit evacuation, minor injuries, or minor offshore impact (i.e., odor).
- 4 = An operational problem that does not have the potential to cause injury or a reportable chemical release with no off-site impact.

1.3.3 Likelihood of Adversary Success in Causing a Catastrophic Event

An effective protective physical system (PPS) should neutralize the adversary and prevent an undesired event with a high degree of confidence. The more effective the PPS, the less likely the adversary will succeed. Thus, the likelihood of adversary success in causing a catastrophic event, L_{AS}, is derived directly from estimates of the PPS effectiveness. The assessor should develop levels of likely adversary success definitions for the physical protection system that is specific to the site. An example set of L_{AS} definitions is (U. S. Department of Justice [DoJ], 2002):

- 1 = Ineffective or no protection measures; catastrophic event expected
- 2 = Few protection measures, catastrophic event probable
- 3 = Major protection measures; catastrophic event possible
- 4 = Compete protection measures; catastrophic event prevented

1.4 PHASES

The VAP methodology includes three basic phases: preassessment, assessment, and postassessment. The specific elements or tasks associated with each assessment phase meet specific assessment objectives. In general the steps are shown as the

General Steps in Vulnerability Assessments shown here (U.S. Department of Energy [DoE], 2002).

General Steps in Vulnerability Assessment

List critical company assets. The list should be prioritized.

Discuss with company personnel the strengths and weaknesses of security programs protecting the critical assets.

Identify assessment areas that provide the most benefit. These become the major focus of the assessment activities.

Review documentation associated with the physical security programs present for the critical assets.

Complete the worksheets initially during documentation reviews.

Confirm the information contained in the worksheets during interviews with personnel responsible for the physical security programs.

Verify the functioning of the security organization and implementation of companywide plans and procedures.

Identify appropriate staffing and funding necessary if further implementation of security programs is needed.

Conduct interviews of personnel responsible for the physical security programs.

Review documentation associated with the specific physical security elements.

Conduct tours of the critical assets. Verify the information recorded in the assessment worksheets.

Consider the value of company assets and potential threats to these assets when defining the relationship between appropriate level of physical security and cost associated with protecting the assets.

Assess low-cost security elements (locking doors, wearing identification badges, and escorting visitors) that develop a "security state of mind" for employees.

Consider more stringent security elements (access control points, cameras /alarms, guard force) using a cost-benefit approach.

Discuss initial judgments with company security personnel and other assessment team members to develop final recommendations.

1.5 PREASSESSMENT

The preassessment phase defines the assessment scope, establishes appropriate information protection procedures, identifies critical assets, and ranks assessment components.

1.5.1 Scope

Scope includes the type of assessment and scheduling decisions. Objectives may include:

- Critical vulnerabilities identification (physical, cyber, and interdependencies)
- Response option development
- Key assets identification and ranking

- Feasibility studies given the business case rationale to justify investments and organizational change
- Awareness and training, as needed, to foster ongoing risk management
- Operating procedure integration with security provisos and risk management concepts

Four basic strategies are used for conducting assessments:

1. **Internal.** In-house technical and organizational expertise are used to perform the assessment. Internal staff members have a resident understanding of the facility or site, organization, technology, and policies and practices currently in effect. In-house experts may also have a historical perspective and a sense of future plans.
2. **Facilitated.** In-house technical experts are guided by outside facilitators. The organizational and methodological aspects of the assessment are offloaded to the facilitators to efficiently leverage internal staff for their specific technical expertise.
3. **External.** An external assessment team conducts the assessment. This approach provides outside objectivity, intra- and interindustry perspectives, visibility into trends and benchmarks, and access to specialized staff with specific expertise.
4. **Hybrid.** Internal staff members perform some elements or tasks, and external experts conduct others.

1.5.2   Information Protection

A nondisclosure agreement is typically developed that defines the policies for the storage, transmission, handling, and disposition of all sensitive data gathered and generated during the assessment.

1.5.3   Critical Asset Identification

Asset analysis examines the workforce, physical, and operational assets. Included in this element is an examination of asset utilization, system redundancies, and emergency operating procedures. Vulnerability analyses include reviews of overall system operation controls, communications among operation staff, physical and IT security, general communication with infrastructure providers and utilities, and policies and procedures. Organizational structure is also critiqued. Trends in staffing (human factors analysis), maintenance expenditures, and infrastructure investments are defined in terms of potential vulnerabilities.

Critical asset identification and prioritization of assets focuses the vulnerability assessment. The risk characterization task focuses on the critical assets and consequences of loss. Identification of asset criticality enables:

- Consideration of factors that affect risk, including threats, vulnerabilities, and consequences of loss or consequences that compromise the asset.

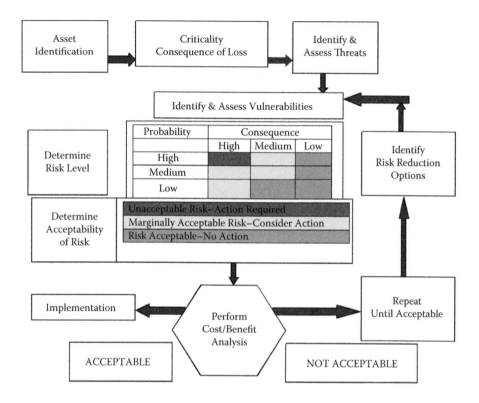

Figure 1.1 (See color insert following page 34.) Example risk management process (DoE, 2002).

- Focused and thorough consideration of risk mitigation options.
- Development of robust methods for managing asset loss consequences.

A typical security-based risk management process is depicted in Figure 1.1.

Sectors that have or control valuable assets must be consulted in the critical asset identification process. The sector representatives should have a reasonable understanding of the operational workings of the facility or site, as well as finance, auditing, risk management, and security. All representatives present their organizations' list of critical assets and discuss the ranking of the assets based in part on perceived consequence of asset loss.

The first step is to identify the criticality dimensions and attributes. The next step is to identify significant assets that might be potentially critical. This includes both information assets (data, networks, systems, processes) and physical assets. If the facility operates computer-based systems, the scope must include assessments of these facilities. Sample assessment items for network architecture are shown in the Network Architecture Methodology Table.

Network Architecture Methodology

Three techniques are used in conducting the network architecture assessment:

1. *Analysis* of network and system documentation during and after the site visit;
2. *Interviews* with the facility staff, managers, and Chief Information Officer (CIO); and
3. *Tours* and physical *inspections* of key facilities.

Documentation of network diagrams, sample system reports, results from previous assessments, and Request for Information (RFI) preliminary background information.
The primary site visit consists of interviews and interaction with facility employees, including technical staff, managers, and the CIO. Questions are asked involving such topics as:

- Network architecture
- Deployed security measures
- Remote access
- Intrusion detection
- Incident response
- Vulnerability assessment activities
- Configuration management
- Cyber security training
- Software development
- Specific questions regarding mission-critical systems such as supervisory control and data acquisition (SCADA)/energy management system (EMS)

The information generated by these discussions provides the core material for the findings and observations.
In addition, tours and physical inspections are made at key facilities. During these tours, the network architecture assessment focuses primarily on:

- *Networking equipment*—hubs, switches, routers, firewalls
- *Production equipment*—key servers, workstations
- *Visible connection points*—attached terminals, external modems
- *Physical location of equipment*—wiring closets, raised floors, control rooms, Halon or other fire-protected zones

These inspections provide additional information as well as verification of information previously obtained.

Network Architecture Tips

Conduct routine system-level security reviews or vulnerability assessments of internal or trusted systems.
Supplement outsourced security reviews and vulnerability assessments with frequent, system-level, self-assessment of internal systems across the entire network infrastructure.
Assure that appropriate external and internal intrusion detection are in place. Intrusion detection (the detection of malicious activities on the network, such as unauthorized packet sniffing, port and vulnerability scanning, user access and privilege escalation, or use of common system exploits) is an essential component of any cyber-security program. Facilities should enhance the capability to detect internal malicious activity.
Consistently implement or screen security measures for remote access, monitoring, and maintenance. Facilities should implement and screen all remote access points.

Risks of noncritical assets must also be considered. If the threats and vulnerabilities are high, even noncritical assets may collectively contribute to overall risk. In general, these conditional assets rank below those assets with unconditional criticality.

Additional assets may be identified that have dimensions and attributes that are not easily analyzed, perhaps because criticality information is not available. A separate listing of these special focus areas should be developed.

The final step is an associated ranking of all vulnerabilities to prioritize measures for reducing vulnerability and risk. Note: Measures used to protect assets that are more critical often facilitate risk reduction broadly for all assets. Conversely, an absence of broadly applied security measures directly increases vulnerability and has the indirect effect of increasing vulnerability.

1.5.4 Request for Information

After the initial criticality ranking is determined (see Appendix A, Table A-10), a Request for Information (RFI) should be provided to the facility or site. This RFI will identify the information needed, documents to be obtained for review, and personnel to be interviewed. The site should determine what information can be provided prior to the site visit and/or during the site visit and the method of information transmittal. Secure internet Web sites, transfer of electronic media (DVDs, CDs) or hard copy data transmittal may be used. Any information that cannot be made available should be noted. The information request may consist of specific information to be provided (e.g., maps, diagrams, documents) or be more like a survey with focused questions.

Example RFIs are provided in Tables A.1 through A.7 found in Appendix A.

1.5.4.1 *Adversarial Situations and Design Basis*
Threat Information Needs

Realistically, facility personnel will probably not have accurate knowledge of a specific adversarial situation beforehand. Therefore, judgments must be made in defining the threat. The more complete the available threat information is, the better those judgments will be. The written definition of the threat is called the Design Basis Threat. The type of information that is needed to describe a threat includes the adversary's:

- Type
- Potential actions
- Motivation
- Capabilities

Adversaries can be divided into these three types: outsiders, insiders, and outsiders in collusion with insiders. Outsiders might include terrorists, criminals, extremists, gangs, or vandals. Insiders might include hostile, psychotic, or criminal employees or employees forced into cooperating with criminals by blackmail or threats of violence against them or their families.

A discussion of the adversary's potential actions must include what sorts of crimes these adversaries are interested in and capable of carrying out and which of these crimes could be committed against the specific site. Examples are theft, destruction, violence, and bombing.

Knowing the adversary's possible motivation can provide valuable information. Potential adversaries may undertake criminal actions because of ideological, economic, or personal motivations. Ideological motivations are linked to a political or philosophical system and include those of political terrorists, extremists, and radical environmentalists. Economic motivations involve a desire for financial gain, such as theft of hazardous materials for ransom, sale, or extortion. Personal motivations for committing a crime range from those of the hostile employee with a grievance against an employer or coworker to those of the psychotic individual.

The capability of the potential adversary is an important concern to the designer of a physical protection system. Factors in determining the adversary's capability include the following:

- Number of attackers
- Their weapons and explosives
- Their tools and equipment
- Their means of transportation (i.e., truck, helicopter, ultralight, or radio-controlled vehicle)
- Their technical skills and experience
- Their knowledge of the facility and its operations
- Possible insider assistance

1.5.5 Facility Characterization Matrix

The facility characterization matrix organizes the security factors for each activity and provides a framework for determining and prioritizing the critical activities. The following example applies to a chemical facility. However, the same general principles also apply to any facility. For chemical facilities, the following matrix items would be used to qualify each activity:

1. **Process activity.** Describe the activity (for example, from flow diagram, piping and instrument diagram [P&ID], reactor, pipe, storage tank, transportation).
2. **Covered chemicals.** Enter the names of all chemicals used in this activity. Enter "Yes" if the chemical is listed in 40 CFR 68.130 or 29 CFR 1910.119.

To quantify each activity, the following matrix items would be used (DoJ, 2002, p. 9).

1. **Quantity of covered chemicals.** Enter
   - "1" if the quantity is more than 25 times the threshold quantity (TQ)
   - "2" if the quantity is 10–25 times TQ
   - "3" if the quantity is 1–10 times TQ
   - "4" if the quantity is TQ or less
2. **Process duration.** Enter:
   - "1" if the process is 100% continuous
   - "2" if the process is 50–99% continuous

- "3" if the process is 25–49% continuous
- "4" if the process is less than 25% continuous

3. **Recognizability.** Enter:
   - "1" if the target and importance are clearly recognizable with little or no prior knowledge
   - "2" if the target and importance are easily recognizable with a small amount of prior knowledge
   - "3" if the target and importance are difficult to recognize without some prior knowledge
   - "4" if the target and importance require extensive knowledge for recognition

4. **Accessibility.** Enter:
   - "1" if easily accessible
   - "2" if fairly accessible (target is located outside or in an unsecured area)
   - "3" if moderately accessible (target is located inside a building or enclosure)
   - "4" if not accessible or only accessible with extreme difficulty

1.6 INDUSTRY BEST PRACTICES: SECURITY PROGRAM

As the VAP is assessed, one of the most important considerations will be the status of the security program. RFIs should be formulated and ultimately reviewed given accepted best practices.

1.6.1 Security Governance

The most important and often most overlooked aspect of security is a corporate or enterprisewide mandate for security functions. For a security function to be effective, the various integral functions must be authorized and legitimized at the highest levels within the organization. Without this upper management support, security will be viewed as unimportant or something of a nuisance. Whether the mandate is from the chief executive officer (CEO) of a corporation, the mayor of a city, or the board of trustees for a transit agency, sending the security message throughout the organization is critical. The message must state that security is viewed as an important function and all employees are expected to aid in the security function. Figure 1.2 illustrates an exemplary security policy hierarchy.

This mandate can take several forms. One of the most effective is a high-level policy statement (typically 1–2 pages) that acknowledges:

- The safety and security of all employees, visitors, company property, and the public at large are of paramount importance to the organization, and
- All appropriate resources of the firm will be used to provide a safe and secure environment in which to work and conduct business.

Additional information should include:

- Compliance with laws and regulations
- Effective control access to sensitive areas

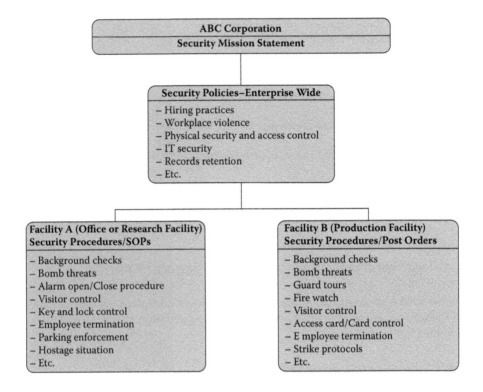

Figure 1.2 Example security policy hierarchy.

- Due diligence for all new hires as appropriate to the position to be filled
- Basic security awareness training for all employees, including how to conduct their duties to achieve the security goals

Once high-level commitment and focus is provided to address the security function within an organization, the building blocks of a security program can be formed and supporting documents (e.g., overarching policies, site-specific policies, and specific procedures) can be developed. This high-level commitment can also be used as a guide to recommend funding of security budgets and to define the position of a security director or manager.

If executive management is reluctant to provide a high-level policy statement or other form of mandate, vulnerability assessments should evaluate whether the organizational leadership does not value security, does not understand the need for security, or does not want an effective security program. On the other hand, organizations that have the vision to provide such a mandate, have legitimized the role security will have in their organization from the top down, underscored how the executive leadership needs and wants security to function, and prioritized security functions.

When assessing security at a facility or across an organization, understanding if a security mandate exists or not (whether verbal, in writing, or in practice) is a challenging task. Managers will at times verbally communicate what the managers perceive the assessor wants to hear. As an example, the plant manager states that he/ she is the security manager and that security policies are in effect and used throughout the facility. However, when the plant receptionist and switchboard operator (in that position for 5 years) is later asked to whom she would go with a security related issue, she replies: "I do not know since a security person is not on-site." These incongruities must be noted in a vulnerability assessment since security governance, if present, should be detectible at all (or at least most) levels of the organization. Observations of daily operations and discussions with line mangers, executives, maintenance staff, human resources staff, receptionists, purchasing, and other employees will provide data to assess the real mandate security has (or does not have) to operate.

1.6.2 Security Policies and Procedures

Security policies and procedures can be developed on an organizational or enterprisewide basis, or can be specific to a single facility or department. In more mature organizations, policies are developed at the organizational level and then site-specific procedures (and sometime policies) are developed to implement policies at the site-specific or department (or other local entity) level.

For example, the organization may develop a policy that workplace violence or threat of violence will not be tolerated. The policy will likely describe some of those unacceptable activities and various consequences for employees that conduct themselves in a violent or threatening manner. At each local facility area within the organization, the local facility will develop specific procedures to implement the workplace violence policy.

Not all facilities will be equipped to react in the same way to the same type of incident. The hierarchy of the enterprisewide policy and the site-specific procedure to implement that policy is one way to accommodate facility differences. For example:

- At Facility A, the procedure may be to immediately notify the plant manager, initiate a lockdown of the building, and call the police.
- At Facility B the procedure may be to page all supervisors, then call 9-1-1, close all vehicle gates except the main gate, send a security guard to meet the arriving police vehicles at the main gate, and have all employees lock the doors to the rooms the employees occupy.
- At Facility C the procedure may be to notify process control to initiate an emergency shutdown of a hazardous chemical process, notify site security personnel and police, and secure all rooms and buildings.

In an ideal situation, all employees would know all policies and procedures relevant to their work responsibilities. Since this rarely happens without routine training and drills, the frequency of drill and exercise occurrence and the type of training is important for the assessor to understand and evaluate. The old adage "practice makes perfect" is certainly relevant to security. If only certain managers and a few security personnel are familiar with policies and procedures that address security-related issues, the assessor should note this fact as a vulnerability.

A well-established and mature security organization policy compendium will usually address the following:

- Workplace violence
- Background checks and investigations
- Access control
- Bomb threats
- Confidential records
- IT/computer use
- Security awareness training
- Building evacuation
- Building shutdown
- Security staffing
- Executive succession
- Travel (international and domestic)
- Ethics

1.6.3 Post Orders

The security department's policies and procedures should specifically describe how the security function is expected to operate. Typically called *post orders*, security guard force polices and procedures are specifically developed to anticipate activities that security guards and officers may encounter. Policies and procedures should also be developed for other security activities, such as those at security operations centers. Emergency preparedness activities should also be developed.

1.6.4 Effectiveness and Assessments

A security assessment should consider the policies and procedures (or lack thereof) with regard to the overall effectiveness of the security function. Consequently, the assessor must understand not only what policies and procedures have been developed, but how well these procedures are understood and implemented. Often during the assessment process, the security assessor will determine that a disconnect exists between what management thinks happens with regard to security, and what the "on the ground reality" is. The assessor's responsibility is to independently verify that what management thinks security is doing, is reality.

Approximately 50% of the recommendations that typically result during a security assessment address the lack of good policies and procedures or the lack of implementation of the same. When conducting an assessment, policies and procedures should be reviewed to check content and relevancy, as well as indicators such as the date of the procedure. Procedures should be periodically updated (preferably every 2–3 years or as appropriate). If the current procedure being reviewed is 12 years old, this aged procedure could be an indication of how up to date (or more likely how out of date) the procedure is.

Here are a few examples of the lack of good policies or procedures.

Example 1:

Scenario: Part of the duties of the Mike Randall, the treasurer for a mid-sized city, was responsibility for the receipt of tax revenues, licensing fees, and similar income streams, and making sure that these monies were properly deposited to the municipal bank account. Mike made daily deposits of cash and checks ranging from several hundred thousand to a million dollars at the local bank. The procedure for the deposit run was that Mike would take the daily receipts, put them in a canvas bag with bank markings on the outside of the bag, walk out of City Hall, and walk two blocks down the street through a high crime neighborhood, pass a bus stop, and walk into the bank lobby. The procedure was used for over five years and did not change even when Mike carried over a million dollars.

Problem: Obviously, this procedure was flawed and had very little security built into the decision logic. Mike faced great personal risk of attack, and the city faced a risk for a significant financial loss.

Solution: The city asked CMI, an independent consulting firm, to conduct a vulnerability assessment. As part of the final report, CMI recommended that Mike have an armed escort (the city has a sizeable police force) and armored car pick-up. CMI also advised using bags that were not clearly labeled with the name of the local bank. These recommendations were procedural changes that did not involve large expenditures for new equipment or systems, or even a change of staffing—only a change in duties for the armed escort.

Example 2:

Scenario: OilCo, a major oil refinery, has a substantial security force on duty that patrols the interior of the facility as well as the perimeter fence line. OilCo's security plans depend in part upon visual observations of the perimeter by these security forces. A procedure was in place for security to report any malfunctioning lights or poorly lit areas to maintenance for repair.

Problem: During a vulnerability assessment of the facility, the assessment team noted that numerous exterior area light poles were present throughout the refinery. During the day, lighting was adequate. However, when the assessment team returned to observe OilCo's second shift at night, one entire side of the perimeter was dark and twenty light poles were not lit.

Solution: Upon further investigation, the assessor discovered that a whole lighting circuit had been disconnected six months prior due to a complaint from a nearby homeowner that a poorly aimed light was disrupting their home life. The lights were disconnected that day in response to the complaint and never reestablished (turned on) until the assessment revealed that security was compromised due to the lack of light. The security managers were never informed that the lights were disconnected and no issue was identified until the assessors brought the issue to the attention of OilCo's managers. The procedures that were developed in response to this problem were that the lights were reestablished and that the poorly aimed light was redirected away from the home.

1.6.5 Security Response Force

Actions taken by the security response force (e.g., on-site security personnel or local law enforcement officers) can prevent adversarial success. Response consists of interruption and neutralization. Interruption not only includes stopping the adversary's progress, but also the following:

- Communicating accurate information about adversarial actions to the response force and deploying the response force. The effectiveness measures for response communication are the probability of accurate communication and the time required to communicate with the response force.
- Neutralization of the adversary. Neutralization is the act of stopping the adversary before the goal is accomplished. The effectiveness measures for neutralization are security police force equipment, training, tactics, cover capabilities, and engagement ability.

The measure of overall response effectiveness is the time between the receipt of a communication of adversarial actions and the interruption and neutralization of the action.

1.7 PHYSICAL PROTECTION SYSTEM

In addition to the elements described above, an effective PPS has these specific characteristics:

- Protection in depth
- Minimum consequence of component failure
- Balanced protection

1.7.1 Protection in Depth

Protection in depth means that an adversary should be required to avoid or defeat several protective devices in sequence to accomplish the goal. For example, an adversary might have to penetrate three separate barriers before gaining entry to a process control room. The effectiveness of each barrier and the time required to penetrate them may differ, but each requires a separate and distinct act as the adversary moves along the planned path.

1.7.2 Minimum Consequence of Component Failure

Every complex system will have a component failure at some time. Causes of component failure in a PPS can range from environmental factors that may be expected to adversary actions beyond the scope of the threat considered in the system design. Contingency plans to ensure continued operation given operational failures must, therefore, consider outside influences that may not be standard design elements or decision logic.

1.7.3 Balanced Protection

Balanced protection means that no matter how adversaries attempt to accomplish their goals, the adversaries will encounter effective elements of the PPS. In a completely balanced system, all barriers would take the same time to penetrate and would have the same probability of detecting penetration. However, complete balance is probably not possible or desirable and would constitute overdesign.

All of the hardware elements of the system must be installed, maintained, and operated properly. The procedures of the PPS must be compatible with the procedures of the facility. Security, safety, and operational objectives must be coordinated.

1.8 INDUSTRY BEST PRACTICES: CHEMICAL MANAGEMENT

1.8.1 Flow Diagrams

If chemicals are present on-site, a process flow diagram should be in use that shows reportable chemicals that can be exploited to create an undesired event. The chemical manufacturing process is usually divided into the following five stages, each of which may contain one or more processing activities for the chemical:

1. Incoming: Ingredients are received
2. Staging In: Ingredients are temporarily staged or stored awaiting use in production
3. In Process: Product is in process
4. Staging Out: Product is temporarily staged or stored awaiting shipment
5. Outgoing: Product is being shipped out

A chemical may not present a security hazard during all processing activities. For example, a hazardous chemical may be converted to a nonhazardous material during production. One way to determine which processing activities provide a potential for an undesired event and are therefore critical activities, is to review the following attributes for each activity:

- Process or manufacturing activity underway
- Specific chemicals being used and whether or not those chemicals are listed and regulated in 40 CFR 68.130 or 29 CFR 1910.119

- Quantity, form, and concentration of the chemicals
- Accessibility and recognizability of the chemicals
- Location and duration at that location of chemical stores
- Potential for off-site release of the chemicals

Once these attributes have been analyzed, the following types of measures related to facility security or protection against a chemical release or spill should also be reviewed:

- Physical protection measures
- Process control protection measures
- Active and passive measures to mitigate the harm resulting from a chemical spill or release
- Plant safety measures

A flow diagram can be developed for the process control system for each critical activity.

1.8.2   Process Logic Controls

For computer-mediated process logic control, the process is normally a closed cycle in which a sensor provides information to a process control software application through a communications system. The application determines if the sensor information is within the predetermined (or calculated) data parameters and constraints. The results of this comparison are fed to an actuator, which controls the critical component. This feedback may control the component electronically or may indicate the need for a manual action. This closed-cycle process has many checks and balances.

The investigation of how process control can be subverted is likely to be extensive because all or part of the process control may be oral instructions to an individual monitoring the process. The process may be fully computer controlled and automated, or may be a hybrid in which only the sensor is automated and the action requires manual intervention. Further, some process control systems may use prior generations of hardware and software, while others are state of the art (DoJ, 2002).

REFERENCES

U.S. Department of Energy (DoE), Office of Energy Assurance, 2002, *Vulnerability Assessment Methodology*, Electric Power Infrastructure.

U.S. Department of Justice (DoJ), Office of Justice Programs, 2002, *A Method to Assess the Vulnerability of U.S. Chemical Facilities.*

APPENDIX A: REQUESTS FOR INFORMATION AND A CRITICALITY ASSESSMENT

Table A.1 Physical Security Request for Information

How are the assets/sites/facilities currently being protected?

Security Program

1 What is the level of management support for the security program?
2 Has top management established, and effectively disseminated, security policies?
3 Is the security policy a part of all managers' responsibilities?
4 Are adequate resources (budgetary, staffing) allocated to the security program?
5 What is the structure of the security organization within the company? (Attach organizational chart.)
6 How many staff members are assigned to the security function?
7 How are security staff responsibilities broken out, by functional area (e.g., management, personnel, physical, protective [guard] force, information, operations security)?
8 Are policies and procedures documented and in place for the security functional areas (i.e., physical and operations security, review of security policies)?
9 Are disciplinary procedures in place?
10 Is a security policy in place for handling disgruntled or at-risk employees?
11 Is a security policy in place for handling terminated employees?
12 Has an executive (senior management) protection plan been developed?
13 Has the issue of bomb threats been addressed in policy and communicated to personnel?
14 Are telephone "Bomb Threat Checklists" available to personnel?
15 Is a self-assessment program in place to evaluate the effectiveness of security programs?
16 Are security policies made available to company staff members?
17 How are company staff members made aware of security policies?
18 Are security staff members provided with adequate training to accomplish their functions?
19 Are company staff members provided with initial and refresher security education/ awareness training?
20 What is the frequency of this training? Are training records (attendance) maintained?
21 What does the training consist of (e.g., lecture, computer-based, flyers, posters, hand-out materials)?
22 What are the expected responsibilities of management and staff with regard to security?
23 How is company property/inventory accounted for (e.g., property tags, periodic inventories, change control) and by whom?
24 Is theft/damage of property investigated (and by whom)?
25 Is a personnel security (Employment/Human Resources) organization in place to conduct employment screening (background checks/criminal/financial)?

Sites/Facilities

1 What is the layout of the site(s)? (Attach map.)
2 Are barriers and postings (no trespassing signs) in place to clearly delineate site boundaries and advise the general public of access restrictions and control points?
3 Have the legal aspects and prosecution options been evaluated for trespassing?
4 What types of barriers are used at site boundaries (e.g., type [fencing, barricades, alarm zones])?
5 Are access control posts staffed and used for entrance to the site?
6 What types of entry controls are used for site access (pedestrian/vehicle gates)?

(Continued)

Table A.1   Physical Security Request for Information (Continued)

7   What types of barriers are used at facility boundaries (i.e., construction materials/walls and doors, windows/bars, one-way film)?
8   Do any delineated "security" areas have restricted access (i.e., sensitive storage, computing facilities)?

Access Control

1   What methods of access control are implemented for site access?
2   What methods of access control are implemented for facility access?
3   Are access control staff (e.g., receptionists) used for controlling access to any sites/facilities?
4   Is a lock-and-key program in place? Who administers this program?
5   Are automated access controls (magnetic stripe, proximity card, bar code) used at the sites/facilities? Who administers these systems?
6   What is the implementation strategy (policy) for lock-and-key and/or automated access controls?
7   How is the implementation strategy determined and who approves?
8   What is the specific type of access control device or configuration (control) a site/facility will use?
9   How do personnel request access (i.e., keys, automated access control credentials)?
10 Who approves those who are granted access, (by key or automated access control credential) to specific areas, and how is this determined?
11 Are any checks made before access is granted to an individual?
12 How, and where, do personnel obtain their approved keys/automated access control credentials?
13 How are keys/automated access control credentials tracked (e.g., entered into a database, paper trail)?
14 Are audits conducted of the keys/automated access control programs (i.e., for issued/lost/stolen keys)?
15 What happens to an individual's keys/automated access control credentials when he/she voluntarily leaves employment or is terminated?
16 What happens if key/automated access control credentials are determined to be lost or stolen (e.g., locks rekeyed, access removed from automated system)?
17 Is a policy in place for delineating under what circumstances locks are to be rekeyed?

Protective Force (PF) (i.e., guards, sentries)

1   Is there a PF dedicated to the site (if so, give the number)?
2   What is the command structure of the PF? (Who reports to whom?)
3   What are the PF's protection responsibilities?
4   Are the responsibilities delineated in policy and procedures?
5   Are PF personnel armed?
6   Are PF personnel commissioned (arrest authority, deadly force, credentials)?
7   What equipment is issued to PF personnel (vehicles, uniforms, vests, weapons, flashlights)?
8   What types of communications equipment are used (two-way radios, telephone, intercom, cellular phone)?
9   Is a training program in place for PF personnel?
10 What types of training are provided to PF personnel (physical, weapons, assessment)?
11 Who certifies and administers the training?
12 Are contingency plans in place for incidents that may require PF action?
13 Are practice exercises conducted for PF personnel?
14 Is an Emergency Preparedness Organization in place?
15 Is a Fire Department and/or other hazardous material response capability in place?
16 Does the PF coordinate with Emergency Preparedness and Fire Department personnel (including exercises and daily functions)?

Table A.1 Physical Security Request for Information (Continued)

Law Enforcement Agency (LEA)

1 Is there an LEA with site protection/incident responsibility?
2 If an LEA is the primary response agency, is a memorandum of understanding or other form of agreement in place identifying the arrangement?
3 What is the LEA's protection responsibility? (What are the expectations of the company?)
4 Are the responsibilities delineated in policy and procedures?
5 Is a site-specific training program in place for LEA personnel?
6 Who administers the training?
7 Are contingency plans in place for incidents that may require LEA action?
8 Are practice exercises conducted for LEA personnel?
9 Does the site coordinate with LEA, Emergency Preparedness, and Fire Department personnel on-site (including exercises and daily functions)?

Intrusion Detection/Alarm Systems

1 Are alarm systems used as part of the protection strategy?
2 What assets/locations are protected with alarms?
3 What types of alarm sensors are used?
4 What transmission method is used for alarm systems (hardwire, RF)?
5 Is line supervision used for alarm lines?
6 Are alarm transmissions encrypted?
7 Where are the alarms monitored? Who monitors the alarms?
8 What types of alarm monitoring equipment are used?
9 How is alarm information reported or displayed?
10 Do any alarm systems interface with other emergency systems (e.g., fire detection and suppression, HVAC, water, electric)?
11 Are assessment or surveillance devices (closed circuit television [CCTV]) used?
12 Is adequate lighting in place (internal and external) for alarm/intrusion assessment by CCTV and/or human means?
13 What power sources (primary and backup) are used for alarm equipment (line, generator, battery or uninterrupted power service [UPS])?
14 Is a performance testing/system maintenance program in place for alarm systems?
15 Who conducts the testing/maintenance?
16 What is the frequency of testing?
17 How is maintenance prioritized (e.g., routine, preventive, emergency)?
18 How are failed tests handled?
19 Are contingency plans in place to address system failures?
20 Does the PF/LEA respond to alarms?
21 What strategies and authorities are used by the PF/LEA to respond to alarms (delay, interdiction, containment)?

Personnel to Interview

Security Director
Physical Security Manager
Facilities Manager
Contractor Guard Liaison
Chief Information Officer
Chief Operations Officer
Program (i.e., control center) Operations Manager(s)
Network Security Manager
Human Resources Manager

(Continued)

Table A.1   Physical Security Request for Information (Continued)

Issues To Be Addressed

Security plan/procedures/organization
Current security strategy
Changes since 9/11
Proposed security changes
Historic/current physical security concerns

Table A.2   Computer System Request for Information

1   Design goals (target performance, throughput, reliability, availability target [99.99% uptime?])
2   Design process with an emphasis on generation and inclusion of security requirements
3   Current network maps (AutoCad, Visio, PowerPoint) of infrastructure supporting critical operations systems, business systems, and desktop computing
4   Authentication mechanisms for local (on-site), Internet, and dial-up access to infrastructure systems
5   Authorization and access control mechanisms for systems and data
6   Technical countermeasures (firewall, filtering, proxies, intrusion detection) currently deployed
7   Network equipment (routers, switches, firewalls, intrusion detection) vendor, model, and the version/versions of code running on the equipment
8   Network protocols in use?
9   Routing protocols in use?
10 Maintenance/spares for equipment kept on-site
11 Methods for scheduling preventive maintenance (downtime)
12 Methodology for disaster recovery, or fail-over to secondary sites
13 Configuration management, both in the core of the network
14 Network contact for clarification on infrastructure issues
15 Network operational policies
16 Network operational procedures
17 Network security plans

Personnel to Interview

Chief Information Officer
Network Security Officer
Network/Infrastructure designers
Network administrator(s) to walk through the network maps
Network administrator(s) to walk through firewall configurations on routers
Network administrator/administrators that manage wide area networking

Issues To Be Addressed

Single points of failure
Known vulnerabilities
History of failures, break-ins, or break-in attempts

Table A.3 Physical Asset Analysis Request for Information

1	Existing security plan for physical assets
2	List of facilities, control centers, etc., where equipment or personnel are stationed
3	List of primary assets (including function and location) for electric operations
4	System maps showing interconnectivity of assets and components, related capacities, critical (internal and external) operations
5	Maintenance procedures and standard practices
6	Emergency preparedness plan
7	List and location of support equipment (maintenance/repair, spare/replacement parts, communications, transportation)
8	Historic problems that have impacted system operations
9	Top (known) threats to safe and continuous operation of electric or gas operations

Personnel to Interview

Chief Operations Officer
Maintenance Director
Chief Financial Officer

Issues To Be Addressed

Maintenance practices relative to industry
Investment strategies toward physical assets/maintenance
Historic/current problems and concerns

Table A.4 Operations Security Request for Information

1	Site map
2	Simplified facility (facilities) floor plan diagram
3	List all information/asset categories that are considered sensitive to the operations/ marketing functions — in order of relative priority
4	Company policy/procedure documentation that would address:
5	Information security (i.e., information protection, storage, marking, transmission, recycling, disposal)
6	Operations security
7	Security training
8	Site badging
9	Requirements for review of information prior to distribution on company Web site
10	Code of Conduct statement

Training

1	How are company security-related policies distributed to site personnel (e.g., hard copy, e-mail, posters, Web site, staff meetings, computer-based training, group training)?
2	Are personnel provided with initial and/or refresher security awareness training?
3	Does the security education/awareness program address operations security issues (e.g., information exploitable by adversaries/competitors)?

Personnel Identification

1	Are badges (or other credentials) issued to site personnel?
2	Do site personnel wear badges (or other credentials)?
3	Is there a visual distinction between types of employee badges (e.g., to provide visual indication that access to some areas is limited)?

(Continued)

Table A.4   Operations Security Request for Information (Continued)

4   How is visitor/vendor access handled for sites/facilities (e.g., visitor badges, visitor logs, escorts, hosts)?

5   Are site personnel encouraged/instructed to challenge individuals who do not display badges?

6   Who handles janitorial services (contracted out)?

7   What are the janitorial service hours?

8   What level of access is granted to janitorial services?

Computing

1   Is any process in place to review the Internet/Intranet information content, for sensitive information, prior to placing it on the Web?

2   Who reviews information for release to the public (i.e., is security involved in the review process, as well as communications/public relations/legal)?

3   Can sensitive information inadvertently be placed in the public domain?

4   Can nonsensitive pieces of information be combined to produce or lead to sensitive information?

5   Are periodic reviews conducted of the company's public Web site for operations security concerns?

6   Is access to the Intranet controlled (e.g., password-protected)?

7   Is a policy in place for computer users to use passwords (for network access) and/or password-protected screen savers (for desktop computers)?

Information Handling

1   Is a policy in place for identifying and protecting sensitive information?

2   How is computer processing of sensitive information handled (designated locations, use of encryption)?

3   How is sensitive information marked (to identify it as sensitive)?

4   How is sensitive information protected (when in use, need-to-know policy)?

5   How is sensitive information stored (locked rooms, cabinets, security containers)?

6   Are special receptacles provided for employees to discard sensitive information?

7   Are procedures in place for destruction of sensitive information?

8   How is sensitive information destroyed (e.g., shredded, burned, pulped, buried)?

9   Who is responsible for the destruction of sensitive information? Is it done internally, or contracted out?

10   Are trash receptacles periodically inspected to determine if sensitive information has been improperly thrown out?

11   How is sensitive information transmitted and received (encryption)?

12   Is there any form of secure fax or phone set up for sensitive facsimile/voice/data transmissions?

Additional Operations Security (Indicators/Awareness Issues)

1   Would any of the following company practices inadvertently provide an adversary or competitor with access to sensitive information or activities of the company?

2   Publishing of certain events, such as schedules, test preparations, routine switches

3   Abrupt changes or cancellations of schedules

4   Purchasing of specialized equipment for sensitive activities

5   Purchasing paperwork that may include information that could identify the sensitivity of the work requiring purchased equipment

6   Sensitive facilities/areas that telegraph the existence of something special happening

7   Increased telephone calls, conferences, longer working hours (relating to sensitive upcoming events)

8   Exercises to test concepts of operations immediately prior to sensitive upcoming events

9   Unusual or increased levels of travel and/or conferences by senior personnel

Table A.4 Operations Security Request for Information (Continued)

10	"Talking around" sensitive subjects in locations where conversations could be heard by unintended ears
11	Discussing personnel, operations, logistics, and communications plans over nonsecure communications (phones, fax, e-mail, radio)
12	Company policies and procedures that may reveal sensitive information
13	Distinctive emblems or logos (e.g., markings on uniforms, equipment, or supplies) that may indicate association with sensitive activities
14	Memorandums/advance plans regarding sensitive activities or information
15	Access restrictions implemented prior to sensitive activities (telegraphing intentions)
16	Overt increases or changes in security operations prior to sensitive activities
17	Press releases, company brochures, annual reports concerning general company activities that provide more information than is necessary about staff, company capabilities
18	Telephone listing with job titles, organizations, and other personnel information identified

Personnel to Interview

Physical Security Director
Chief Information Officer
Chief Operations Officer
Chief Financial Officer
Network Security Director
Human Resources Manager

Issues To Be Addressed

Protecting sensitive information

Table A.5 Policies and Procedures Request for Information

1	Technical security and countermeasures
2	Communication (internal and external)
3	Computer security
4	Protected communication
5	Proprietary information
6	Emergency management
7	Physical security
8	Operations security
9	Human resources (all)
10	Contracting
11	Training
12	Employee manual
13	Supervisor/manager manual
14	Benefits manual
15	Annual report
16	Organizational charts
17	Emergency management plan
18	Code of Conduct or other ethics statements

Personnel to Interview

Security Officer
Emergency Management Officer
Corporate Communication Officer

(Continued)

Table A.5   Policies and Procedures Request for Information (Continued)

Public Relations Officer
Human Resource managers
Network/System administrators
Information Security Manager
Training Manager
Contracts Officer
Corporate Attorney
Employees

Issues To Be Addressed

Life-cycle management of policies/plans/procedures
Policies/plans/procedures in place
Education/training associated with policies/plans/procedures
Overall effectiveness of policies/plans/procedures
Alignment of policies/plans/procedures with corporate objectives and functions

Table A.6   Infrastructure Interdependencies: Request for Information

1   Building diagrams showing internal (i.e., HVAC, fire suppression) infrastructure locations
2   Building diagrams showing external (i.e., electric, telecommunications, water, natural gas feeds) infrastructure connections
3   External infrastructure connections to internal infrastructures and upstream routes (include maps showing routes if available)
4   List of facilities and facilities managers
5   Emergency and contingency plans
6   Risk management activities relating to interdependencies (e.g., investment criteria, risk exposure)
7   Contingency plans for telecommunications, transportation (road, rail, air), water, banking, and finance
8   List of critical infrastructures—electric power, natural gas, oil, emergency services, and government services
9   What function(s) are performed with the commodity/service used?
10   Who are the service providers and points of contact?
11   What types of contracts/service agreements are in place?
12   How would the utility be affected by disruptions to the critical infrastructures that serve the utility facilities?
13   What is the severity of disruptions in terms of the utility operations?
14   How do the impacts, and interdependencies concerns, change if the disruptions occur during the workday?
15   How do the impacts, and interdependencies concerns, change if the disruptions occur during peak load conditions?
16   How do the impacts, and interdependencies concerns, change if the disruptions occur during alerts?
17   How do the impacts, and interdependencies concerns, change if the disruptions occur at night?
18   How do the impacts, and interdependencies concerns, change if the disruptions occur on a weekend?

Table A.6   Infrastructure Interdependencies: Request for Information (Continued)

19 What types of backup systems or other mitigation mechanisms are in place to reduce the impacts to the utility operations from disruptions to supporting infrastructures?

20 What are the limitations of the backup systems as a function of outage duration?

21 How are the backup systems affected by the prolonged outage of other interdependent infrastructures?

22 Does the frequency of disruption affect the backup systems or the infrastructure reliability and response mechanisms?

23 Does the frequency of disruption introduce new interdependencies concerns?

24 What infrastructure services directly or indirectly impact restoration activities?

25 Do dependencies on other infrastructures exacerbate response and recovery efforts?

Personnel to Interview

Facilities Manager
Telecommunications Contractor Coordinator
Critical Systems Manager
Corporate Services Manager
Safety and Security Coordinator
Operations Manager
Emergency Response Coordinator
User Support Services Manager
Director of Financial Planning/Treasurer
Director of Utility Operations
Senior Network Analyst
Human Resources (policies and procedures)
Lead Strategic Contingency Planner

Issues To Be Addressed

Single-point infrastructure failures
Infrastructure backup
Commercial infrastructure reliance
Historic/current problems and concerns

Table A.7   Risk Characterization: Request for Information

1   Current risk management approach and activities
2   Internal and external investment decision criteria

Personnel to Interview

Chief Financial Officer
Capital Budgeting Officer
Physical Security Director
Chief Information Officer
Chief Operations Officer
Network Security Director
Human Resources Manager

Issues To Be Addressed

Current risk management approach and activities

Table A.8   Threat Environment: Request for Information

1	Do employees receive annual security briefings that contain information on potential threats?
2	What office is responsible for developing and updating security briefings?
3	Does the participant have a relationship with law enforcement agencies to maintain an understanding of potential threats?
4	If yes, what agencies and who are the points of contact for these relationships?
5	Can the VAP assist participants in establishing contact with law enforcement agencies to obtain current threat information?
6	Have physical or electronic intrusions (breaches in security) occurred?
7	Has a disgruntled/disenchanted employee ever caused, or threatened to cause, property damage?
8	If yes, please provide as much information as possible concerning each incident, including preventive measures taken

Personnel to Interview

Physical Security Director
Chief Information Officer
Human Resources Manager
City police department representative (crime analyst)
Local FBI representative (Domestic Terrorism Department)
Local ATF representative (Domestic Terrorism and Hate Groups Departments)
U.S. Marshal's representative
Local Drug Enforcement Agency representative
Local Secret Service representative
State Homeland Security representative (intelligence analyst)
County/Sheriff's representative (crime analyst)
State/Highway Patrol representative (crime analyst)

Issues To Be Addressed

Historical incidents
Known threats
Specific threats to the company
Threats to the industry other than the specific company
Methods/techniques commonly used by individuals/groups to cause damage in the
  community and to gain media attention
Extreme environmental and animal rights groups in the area and their modi operandi
Known hate groups and their modi operandi
Presence of possible international terrorist organizations within the community

Table A.9 Example: Criticality Assessment

Item	Consequence		
	High	**Medium**	**Low**
	Legal Liability		
Property damage	Not mitigated by insurance	Mitigated by insurance	Mitigated by insurance
Health and safety	Multiple loss of life	Loss of life	Minor injury, lost time
Customer relations	Regional loss of service, >48 hrs	Systemwide loss of service	Localized loss of service
Service interruption	Regional	Industrial/large	Small
	Environmental, Safety, Health		
Regulatory	Major environmental release	Minor environmental release	Environmental release (nonreportable)
	Serious and willful OSHA citation	Serious OSHA citation	De minimum OSHA citation
	Major accident — loss of life, disability	Minor accident — short-term disability	First aid medical treatment
Employee relations	Propertywide strike and walkout	Propertywide strike	Breach of trust
	Financial		
Shareholder value	Loss capital	Lesser profits than anticipated	Profits anticipated, no business growth for future
Community relations	Alienation of community	Irritation of community	Disgruntled activists
Competitive impact	Loss of major strategic asset	Loss of customer demographic entotal	Loss of major nonstrategic asset — customers, suppliers
Business interruption	Major loss computer files	Computer network major disruption	Loss of network terminal data (backed up elsewhere
Supply chain impact	Loss of XYZ supply	Critical equipment failure	Equipment failure
Political impact	Significant increase in regulations	Anticipated response to regulatory changes	Significant change in administration
	Operations		
Operations	Total loss of ABC	Loss of DEF facility	Disruption of operation
Product delivery	Simultaneous, coordinated attack on multiassets	Systemwide loss (service, communications)	Loss of one major asset
Operations	Loss of entire facility	Loss of one production unit	Loss of one piece of equipment

CHAPTER 2

Assessment
Threat Analysis

Martha Boss, Dennis Day, Lawrence Fitzgerald, Randy Boss, and Gayle Nicoll

CONTENTS

The characterization of a facility includes a description of building structures, traffic areas, infrastructure, terrain, weather conditions, and operational conditions. The first step is to gather information that will be helpful in identifying potential security vulnerabilities. An understanding of the threat environment is a fundamental element of risk management. This element should include:

- A characterization of threats
- Identification of trends in these threats
- Ways in which vulnerabilities are exploited

To the extent possible, characterization of the threat environment should be localized, that is, within the organization's service area. The facility should participate in State Homeland Security activities and other security-related working groups to broaden its depth and understanding of threats.

Both internal and external threats are considered in the threat environment assessment. The on-site analysis of the threat environment takes place in three phases:

Phase I: Initial screening of sources, prior to the arrival of the vulnerability assessment team, conducted to identify individual(s) and/or group(s) who are potentially threatening and to establish contact with local, state, and federal law enforcement agencies (LEAs) to begin the analysis process.

Phase II: An on-site assessment is performed, beginning with interviews of facility security managers. The purpose of these interviews is to:

- Determine current corporate security measures (vis-à-vis threats)
- Identify LEAs with whom the facility security managers routinely associate

Phase III: Office calls made to federal, state, and local LEAs. The LEAs are questioned about the identities and modus operandi of known or suspected individual(s)/group(s) who have:

- Initiated hostile actions against any facility in the region
- Threatened hostile action against any facility in the region
- Threatened or executed hostile action within the community or its surrounding area to further their "cause"

Facility security managers need to be proactive in seeking up-to-date threat advisories from LEAs, and all facility personnel should remain vigilant. An overall assessment of potential threats to any facility in the area is then developed. This dialogue includes whether regularly scheduled LEA intelligence briefings to assist facility security personnel will be provided. Points of contact are provided, which facility security personnel can use in the future. The threat assessment also includes a summary of any incidents and/or threats of any nature by current or former employees of the facility. Discharged employees are as much of a concern as disenchanted employees in this analysis (U.S. Department of Energy [DoE], 2002).

2.1 PENETRATION TESTING OR WAR GAMING

Penetration testing or war gaming does not require a Request for Information (RFI). The tester(s) acquire information through public sources combined with on-site analysis. Penetration testing utilizes active scanning and penetration tools to identify vulnerabilities that a determined adversary could easily exploit. In general, penetration testing should include a test plan and details on the rules of engagement (ROE). ROE are required because penetration testing is an active assault on current systems, rather than a passive assessment based on decision logic and programmatic elements.

The purpose of a penetration test is to detect potential vulnerabilities of a target environment and its connected resources using a range of known attack tools and techniques. A one-time penetration test can provide valuable feedback. However, testing on a regular basis is more effective. Repeated testing is recommended because new threats develop continuously. The testing can be incremental or comprehensive.

Penetration testing should include identified vulnerabilities and, in particular, whether access could be gained to systems or devices that have a critical role. Details regarding which attack tools and techniques should be used must be provided by the on-site testing agent prior to the onset of the penetration test to a *White Cell*. This *White Cell* has individuals with "insider" information. This approach, from a facility, human factors, and infrastructure perspective, is similar to orchestrated emergency plan drills. However, the focus is more global with normal operations being considered, too. The penetration testing process consists of four steps.

2.1.1 Step 1: The Rules of Engagement

The ROE are the ground rules for when and how the penetration test will be conducted. Defining the ROE includes developing an ROE authorization list of what should and should not be done and which facility or site components are eligible for the test. The ROE agreement must be in detail and in writing to serve as an indisputable consensus agreement between the facility or site and the testing agent. Issues to consider when developing the ROE include:

- When will the test start—day and time? What is the duration of the test—one day, multiple days, or ongoing?
- Is a priori knowledge of the environment available?
- Should the testing agent ask permission prior to attempting penetration on a potentially vulnerable system?

- Will the tester(s) be given any computer network or system information, or must this be obtained by using reconnaissance techniques? Are port scans of systems and network devices permitted? Are penetration attempts of firewall systems allowed? How far past the perimeter of the network can the testing agent go?
- What systems or system components are eligible for testing and which are off limits?
- Are physical penetration attempts and facility surveillance allowed?
- How often and when should the testing agent inform the facility being tested of testing activities and findings?
- Will employees be notified that a test is being performed? If so, which employees?

Depending on the comfort levels, more ground rules can be specified to ensure a successful and nondisruptive test. An experienced testing agent will help identify potential issues and provide guidance during the ROE planning phase of the testing process. Below is a list of issues to consider that illustrates questions that should be asked as the ROE planning proceeds.

Issues to consider when developing the ROE include:

Is a priori knowledge of the customer's environment available? Will the tester(s) be given any network or system information, or must it be obtained by using reconnaissance techniques?

When will the test start—day and time?

What is the duration of the test—one day, multiple days, or ongoing?

What networks and systems are eligible for testing and which are off limits?

Are social engineering techniques permitted?

Are physical penetration attempts and facility surveillance allowed?

Are port scans of systems and network devices permitted?

Are penetration attempts of firewall systems allowed?

How far past the perimeter of the network can the testing agent go?

How often and when should the testing agent inform the customer of his/her activities and findings?

Should the testing agent ask permission prior to attempting penetration on a potentially vulnerable system?

Are home computers of employees off limits? (A potential attack path into a network is from employee computers that dial into or connect via ISP to a remote access gateway on the corporate local area network.)

Are computers and networks of business partners off limits? (Another potential attack path into a corporate network is via trusted third parties, such as equipment vendors, business partners, and customers that may have direct network connections into the corporate network or otherwise have trusted access through the perimeter via the Internet.)

Unless the third parties are informed and consent to a penetration test, this method of testing should not be authorized.

Depending on the comfort level of the customer, more ground rules can be specified to ensure a successful and nondisruptive test. Customers know their environment best and understand what ground rules need to be established. An experienced testing agent will help the customer identify potential issues and provide guidance during the ROE phase of the testing process.

2.1.2 Step 2: The *White Cell* and Incident Command

The *White Cell* is the group of people who are aware of when a penetration test is being conducted and what the ground rules are. The White Cell consists of the personnel conducting the test and the personnel identified as "in the know" at the site. Members of the White Cell must provide and exchange contact information and maintain communication during the testing period. The White Cell is a medium for the tester(s) to continually update the progress of the test and clarify or resolve potential issues evolving during the test.

While the test is being conducted, the tester(s) maintain constant communication within the White Cell, especially during the exploitation phase and to a lesser degree during the reconnaissance phase. During exploitation, the tester(s) should notify the White Cell of when the exploits are taking place and if any problems occur. Likewise, the White Cell should be aware of when the controlled attack is occurring and should notify the tester(s) of any problems or disruptions. For emergency preparedness drills, this concept is known as *incident command*.

At the very least, the White Cell should include a manager who can intercept a detected intrusion attempt and prevent the test from escalating within the organization, especially to prevent reporting to law enforcement. If administrators are not included in the White Cell, someone with the needed authority and contacts to intercept a potential incident escalation must be included to interface with the site's administrators once the testing is in play. The White Cell team members from the facility or site should also be prepared to intercept any incident escalations that are detected by intrusion detection equipment or by system and network administrators who are not members of the White Cell.

2.1.3 Step 3: Designing and Conducting the Test

During the ROE establishment phase, the testers define the tools and techniques that will be used on the eligible set of systems. The first consideration in the design of the penetration test is role simulation. Role simulation involves deciding which role the tester(s) will assume during the test. Possible attacker simulation roles are:

1. An Outsider: The simulated attacker does not work for the target or any other third party (e.g., business partner, contractor, vendor, or regulatory agency). The goal of the penetration test is to determine if an outsider with no authorized access can subvert protections that are in place.
2. An Insider: The simulated attacker is someone who works for the facility or site.
3. An Associated Third Party: By simulating the role of an employee at one of the trusted partner sites, the tester(s) determines if penetration via one of the connections to trusted business partners is possible. If the third party will not allow a penetration test to originate from its site, the tester(s) could do a "paper/table top" or analytical assessment of the potential risks involved.

Once the various attacker simulation roles are chosen, a testing format is devised. An effective testing methodology can be divided into three distinct phases: reconnaissance, scenario development, and exploitation.

1. Reconnaissance: Involves gathering information about the target environment to plan an attack. During the establishment of the ROE, the facility or site representatives and the tester(s) determine what types of methods are acceptable.
2. Scenario Development: This phase uses the target information gathered during the reconnaissance phase to develop possible attacks and exploits. After developing the attack scenarios, the tester(s) should present them to the facility or site representatives. At this point, the decision is made as to whether some or all of the potential exploits should be tested and what limitation should be applied.

   Alternately, the facility or site may opt to accept the scenarios without having the tester(s) actively demonstrate potential exploits or test the viability of these scenarios. Analysis of the information presented and recommendations may be based on this passive (table top) analysis rather than an active penetration test. The extent to which this active participation also includes foreknowledge of the drill's intent must be decided by the facility representatives.
3. Exploitation: In this portion of the penetration test, the scenarios are carried out. The penetration test consists of reconnaissance and the test or target exploitation.

2.1.3.1   Reconnaissance

External reconnaissance (*recon*) is conducted by scouting, probing, scanning, and potentially mapping exposed perimeters. For cyber reconnaissance, this recon includes access to the public; mining information from Internet Web sites, newsgroups, e-mails, chat rooms, forums, or other open sources; and ultimately assembling a profile of the target's strengths and weaknesses. The intent is to discover access points and vulnerabilities that could potentially compromise the target.

Internal recon is very similar to external recon, except that all activity is based on a compromised situation that was exploited due to internal access. This can be in-depth or limited depending on the nature of the compromise, tools available, and the negotiated rules of engagement.

2.1.3.2   Test or Target Exploitation

This activity's purpose is to remotely gain control of or retrieve internal information from vulnerable systems. In general, target exploitation includes the following activities:

- Determine potential targets. The results from recon are analyzed to determine the most attractive targets.
- Acquire or develop exploits. Given the suspected vulnerabilities, exploits or exploit information are acquired and analyzed. Exploits are developed or modified as needed. Both normal operations and emergency operations should have exploits developed to test response.

- Execute the exploits. Exploits against the vulnerable areas are attempted. Exploits can involve such techniques as pushing systems or applications beyond capacity, and subverting intended function or use.
- Develop scenarios. Plausible scenarios are developed that maximize the utility of successful exploits and that can be used for follow-on testing.

The following table provides a sample list of elements to be analyzed in this phase.

Test or Target Exploitation

The purpose of this activity is to remotely gain control of or retrieve internal information from vulnerable systems. In general, target exploitation consists of the following:

Determine potential targets: The results from recon are analyzed to determine the most attractive targets.

Acquire or develop exploits: Given the suspected vulnerabilities, exploits or exploit information are acquired and analyzed from open sources, such as www.securityfocus.com, www.packetstormsecurity.org, www.insecure.org, and the bugtraq newsgroup. Exploits are developed or modified as needed.

Execute the exploits: Exploits against the vulnerable hosts are attempted. Exploits can involve such techniques as automated brute-force password guessing, parameter or command manipulation or injection, taking advantage of misconfigurations or default settings, pushing systems or applications beyond capacity, and subverting intended function or use.

Develop Scenarios: Plausible scenarios are developed that maximize the utility of successful exploits. This step typically involves escalation of access or user privileges, installation of root kits or remote access tools that provide the attacker with more direct control of a system, implementation of backdoors, obfuscation of presence, covering of tracks, and other activities that ensure the complete (and ideally undetected) compromise of a system.

Install additional tools

Investigate data and files

Investigate network and system configuration settings

Locate, obtain, and attempt to crack password files

Capture and analyze packets generated by or destined to the compromised host

Capture keystrokes

Manipulate system or application configurations/settings

Tests

Keyword searches: Keywords, such as *confidential, diagram, firewall, install, administrator, pin, password, modem, scada, download, vpn, ids, intrusion, vulnerability, https,* and others, can be entered into the company Web site search engine as well as Internet search engines, such as Google, AltaVista, Lycos, and WiseNut.

"Who is" searches: Internet registry information for the facility can be retrieved from Web-based who is providers, such as the American Registry for Internet Numbers (ARIN) and InterNIC.

Company financial research: The facility's financial profile or synopsis information can be obtained from financial Web sites, such as http://finance.yahoo.com.

Connect to common ports: Some connections to common ports (e.g., Web server port 80, mail server port 25, ftp server port 21) can be made via telnet, Web browser, or other tools to discover accessibility and version information from banners.

Trace route to hosts owned by the facility: Some traceroutes can be performed for hosts and/or IP addresses associated with the facility. Traceroute information can be used to help determine firewall and general perimeter topology.

Resolve IP addresses owned by the facility: Common tools, such as nslookup, host, and dig, are used to resolve IP addresses associated with the facility.

(*continued*)

Test or Target Exploitation (Continued)

Acquire e-mail message from the facility: The header of an email message can provide useful information about e-mail routing, internal hosts, and filtering applications such as virus programs.

Backwards navigate facility Web sites: A Web server's directory structure can be traversable by backwards navigating the Web site. For instance, given the URL, http://www.company.com/projects/docs.html, a user can probe higher-level directories by iteratively deleting the last subdirectory or file from the URL. For example, by deleting docs.html from the URL and inputting http://www.company.com/projects/, the user can see the entire contents of the /projects/ directory. Similarly, by deleting projects/, the user can see the entire contents of the /company/ directory. Note that Web servers can be configured to return the default Web page (e.g., index.html) for that directory or an error message. When the response is a directory listing, backwards navigation is useful for discovering documents, scripts, and other files not necessarily linked to the target Web site.

Acquire URLs of scripts or applications from facility Web sites: Scripts that provide input fields or perform actions on the server can be useful points of entry for a skilled adversary. Scripts often contain bugs that can allow an adversary to manipulate data or compromise a host.

Scan IP addresses registered to the facility: Various scans can be performed using assorted scanners, such as nmap, nessus, and whisker. Scans are used to determine live/responding hosts, open ports, and potential vulnerabilities.

2.1.4   Step 4: Report

The final report prepared by the tester(s) should contain detailed descriptions of the findings from each test phase. For the reconnaissance phase, the report should describe information obtained, and how or from whom the information was obtained. For the test or target exploitation phase, descriptions of the scenarios and outcomes should be discussed. Another important element of the report is recommendations. The tester(s) should describe how the vulnerabilities discovered can be fixed or mitigated.

2.2   PHYSICAL PROTECTIVE SYSTEMS AND PHYSICAL SECURITY

Several basic protective physical systems (PPSs) and security concepts (as well as some advanced concepts) should be used to provide an effective security approach for a given facility. Information on the facility structure includes the materials used in construction and the location and types of doors, gates, entryways, utilities, windows, and emergency exits.

One of the most basic physical security concepts is to identify the most important function or asset to protect. A consequence analysis, asset value ranking, or criticality review may be required. An early step in security system analysis is to describe thoroughly the facility, including the site boundary, building locations, floor plans, access points, and physical protection features, as well as the processes that take place within the facility. Weather conditions for the region and time of the year, and a description of adjacent residential or commercial areas are also key components.

This information can be obtained from several sources, including design blueprints, process descriptions, the process hazard analysis (PHA) reports, the Environmental Protection Agency (EPA) Risk Management Plan (RMP) documents, piping and instrumentation drawings/diagrams (P&ID), and site surveys (U.S. Department of Defense [DoJ], 2002).

Physical security assessments evaluate the systems in place (or being planned) and identify potential improvements. Physical security systems are reviewed for design, installation, operation, maintenance, and testing. The physical security assessment should focus on aspects directly related to critical facilities, including information systems, interstitial utilities (ventilation systems, plumbing), and all assets required for operation.

Physical security programs should be uniform across an organization. Physical security resources should be optimized around critical assets. A protection strategy should be developed to:

- Project an image of a hardened, secure facility to potential intruders
- Detect and delay intrusion
- Provide alarms and programmed responses to those alarms

2.2.1 Detect and Delay

An effective security system must be able to detect the adversary and delay the adversary long enough for a response force to arrive and neutralize the threat force before their mission is accomplished. The appropriate level of physical security is contingent upon the value of facility or site assets, the potential threats to these assets, and the cost associated with protecting the assets. These relationships must be considered when conducting an assessment. For example, low-cost physical security elements (e.g., locking doors, identification badges, escorting visitors) should be in place at critical assets. These types of security elements improve the security posture without significant cost and also develop a "security state of mind" for employees. Consideration of more stringent security elements (barriers, blast resistance, access control points, cameras/alarms, guard force) requires a benefit-cost approach.

Delay can be accomplished by fixed or active barriers (i.e., doors, vaults, and locks) or by sensor-activated barriers (i.e., dispensed liquids and foams). Entry control, to the extent control includes locks, may also be a delaying factor. Security personnel can be considered an element of delay if fixed and in well-protected positions. The measure of delay effectiveness is the time required by the adversary (after detection) to bypass each delay element (DoJ, 2002).

Criteria worksheets are used to determine physical security programs that protect facility or site assets. Equipped with the prioritized list of facility or site assets and information from interviews, the tester(s) use the worksheets to review security elements at a facility and make initial judgments as to the appropriate level of physical security.

The following worksheets are included in Appendix A to provide an example of worksheets used to collect physical security information:

- A.1 Physical Security Program (General)
- A.2 Physical Security Barriers
- A.3 Physical Security Access Controls/Badges
- A.4 Physical Security Locks/Keys
- A.5 Physical Security Intrusion Detection Systems (IDSs)
- A.6 Physical Security Communications Equipment
- A.7 Protective Force/Local Law Enforcement Agency
- A.8 Entrances into Critical Asset Areas
- A.9 Vehicle Gates through Critical Asset Area Fences
- A.10 Fences Surrounding Critical Assets

2.2.2 Assessment

The steps for conducting the physical security assessment are:

1. List critical facility or site assets as identified in the Critical Asset Identification step. The list should be prioritized.
2. Discuss with facility or site personnel the strengths and weaknesses of security programs protecting the critical assets.
3. Request through the RFI and then review documentation associated with the physical security programs for the critical assets. Complete the criteria worksheets. Initially complete the worksheets during documentation reviews. Confirm the information contained in the worksheets during interviews with personnel responsible for the physical security programs.
   - If the facility or site appears to have a functioning security organization, a check of facility or sitewide plans and procedures occurs to verify that plans and procedures are being implemented.
   - If the facility or site does not have a functioning security organization, most of the assessment is spent identifying the appropriate staffing and funding necessary to implement programs. The worksheets found in Appendix A.1 through A.7 can be used for determining which security programs are appropriate.
4. Conduct interviews with personnel responsible for the physical security programs present. Verify the information recorded in the assessment document, "Criteria Worksheets to Evaluate Physical Security Programs."
5. Review implementation documentation associated with the specific physical security elements. Complete the assessment worksheet found in Appendix A.8.
   - For companies with an insufficient security infrastructure, research into specific security element deficiencies is limited to finding just enough examples to support any staffing/funding recommendations.
   - Comprehensive assessment activities to review physical security elements protecting critical assets is conducted only for companies with a solid security infrastructure (staffing, plans/procedures, funding). The security staff can take actions to correct deficiencies found at the facilities, once reported.

6. Conduct tours of the critical assets. Verify the information recorded in the assessment worksheets, "Physical Security Elements Protecting Critical Assets." The worksheets found in Appendix A.1 through A.8 contain lists of security elements that are used to implement different levels of physical security.

Initial judgments are discussed with facility or site security personnel and other assessment team members to develop final recommendations.

2.2.3 Physical Asset Analysis

Physical asset analyses examine the systems and physical operational assets. Asset use, system redundancies, and emergency operating procedures are reviewed. Topology and operating practices for information systems, utility supply nodes, HVAC (heating, ventilation, and air conditioning), and major infrastructure elements, including interstitial spaces, are evaluated. Interstitial spaces are those areas within a facility that are not normally accessed and yet are conduits for air contaminant dispersal. Examples are plenums, vertical corridors for utility and HVAC features that are hidden behind walls, horizontal transmission corridors that intercommunicate between building areas, and plumbing system conduits.

The proposed methodology for physical assets is based on a macro-level approach. The analysis can be performed with historical and current facility or site data, public data, or both. The historic data analysis should be supplemented with on-site interviews and visits. Items to focus on during a site visit include the following:

- Trends in field staffing and training afforded to operations and maintenance staff
- Trends in maintenance expenditures
- Trends in infrastructure investments including retrofits of existing systems
- Historic infrastructure incidents, including utility outages, loss of HVAC system adequacy, breaches in water supply systems, infrastructure failure of information system hardware, including that required to automate building systems and emergency response
- Critical system components and potential system bottlenecks
- Overall system operation controls and protection features
  - Access to the process control system
  - List of authorized users
- Linkages of operation staff with physical and IT (information technology) security

The types of documentation include the following policy and procedure documents:

- Unusual occurrence reports
- Existing threat assessment information
- Results from past security surveys and audits

- Building blueprints and plans for future structures
- Site plans for detection, delay, and assessment systems

Site plans can help identify:

- Property borders
- Entrance and exit routes to and from the facility
- Specific vulnerable areas in and around the facility (i.e., adjacent buildings that a sniper could use to target the building)
- Adjacent parking lots and related security countermeasures
- Building locations and characteristics (i.e., the purpose of the building, who is allowed access, and operational conditions or states)
- Existing physical protection features
- Means and routes of access (DoJ, 2002)

2.2.4 Consolidation of Findings

The first step in developing findings is to define, "What is a finding?" Findings that infer increases in security cost must have a strong benefit-cost rationale to support the finding. The development of findings should occur through two steps.

1. Initially, findings should be identified through the completion of the assessment worksheets. These should be discussed with facility or site personnel to confirm the accuracy of the information.
2. The second step in formalizing a finding is an open discussion with the assessment team. Many security findings are interrelated across security functional areas and a specific finding may be a symptom of a larger issue; thus the need for discussion. The assessment team validates findings to the point where the findings can be documented in preparation for inclusion into a final report.

2.2.5 Recommendations

The process used to develop recommendations is similar to the development of findings and may happen concurrently with the development of findings. Detailed knowledge of facility or site operations is critical to sound recommendation development and is, thus, crucial to remediation success.

Initial and annual costs to implement recommendations must be developed. Analysis must show that the security cost is an attractive value to the facility or site because of the protection provided to the critical assets.

A recommendation is formalized during the open discussion with the assessment team. Several recommendations may be combined, or modified, when all assessment topical areas are viewed together. Implementation of one recommendation may eliminate the need for others. The assessment team discussion validates recommendations for the final report.

2.3   COMPUTER AND DETECTION SYSTEMS

Routine system-level security reviews or vulnerability assessments of internal or trusted systems, including computer systems, should be conducted. The intrinsic value of routine and frequent network architecture analysis is the ongoing resultant limitation or reduction in vulnerability of information systems.

Information examined should include network topology and connectivity (including subnets), principal information assets, interface and communication protocols, function and linkage of major software and hardware components (especially those associated with information security such as intrusion detectors), and policies and procedures that govern security features of the network.

Procedures for information assurance in the system, including authentication of access and management of access authorization, should be reviewed. The assessment should identify any obvious concerns related to architectural vulnerabilities, as well as operating procedures. Existing security plans should be evaluated, and the results of any prior testing should be analyzed. Results from the network architecture assessment should include potential recommendations for changes in the information architecture, functional areas and categories where testing is needed, and suggestions regarding system design that would enable more effective information and information system protection.

Documentation such as network diagrams, sample system reports, results from previous assessments, and other RFI questionnaires (see Appendix A in Chapter 1) provided by the facility serve as an introduction to and preliminary background information for the facility's current network architecture. The primary site visit consists of interviews and interaction with facility employees, including technical staff, managers, and the information technology department representatives.

2.3.1   Tours and Physical Inspections

Tours and physical inspections are made at key facilities. During these tours, the network architecture assessment focuses primarily on:

- Networking equipment: hubs, switches, routers, firewalls
- Production equipment: key servers, workstations
- Visible connection points: attached terminals, external modems
- Physical location of equipment: wiring closets, raised floors, control rooms, Halon or other fire-protected zones

These inspections provide additional information as well as verification of information previously obtained. Verification steps are used to evaluate that appropriate external and internal intrusion detection are in place. Consistent implementation or screening security measures for remote access, monitoring, and maintenance are also evaluated.

2.3.2 Detection

The discovery of adversary action, which includes sensing covert or overt actions, must be preceded by the following events:

- A sensor (equipment or personnel) reacts to an abnormal occurrence and initiates an alarm
- Information from the sensor and assessment subsystems is reported and displayed
- Someone assesses the information and determines whether the alarm is valid or invalid

Methods of detection include a wide range of technologies and personnel:

- Physical Protection: Entry control—means of allowing entry of authorized personnel and detecting the attempted entry of unauthorized personnel and contraband is part of the detection function of physical protection. Entry control works best when entry is permitted only through several layers of protection that surround targets of malevolent attacks. Entry to each layer should be controlled to filter and progressively reduce the population that has access. Only individuals who need direct access to the target should be allowed through the final entry control point.
- Search: Searching for metal (possible weapons or tools) and explosives (possible bombs or breaching charges) is required for high-security areas. This may be accomplished using metal detectors, x-ray screeners (for packages), and explosive detectors.
- Staffing: Security personnel at fixed posts or on patrol may serve a vital role in detecting an intrusion. Other personnel can contribute to detection if trained in security concerns and if in possession of means to alert the security force in the event of a problem (i.e., radio or cell phone).
- Alarms: An effective detection alarm assessment system provides information about whether the alarm is valid or a nuisance and details about the cause of the alarm.

The effectiveness of the detection function is measured both by the probability of sensing adversary action and by the time required for reporting and assessing the alarm.

2.4 OPERATIONS CONDITIONS AND OPERATIONAL SECURITY

Operations Conditions and Operational Security (OPSEC) assessments include human factors and issues associated with outward presentation of information. For example, a policy should be developed and procedures reviewed for placement and removal of various types of information on Web sites. The facility should develop procedures and classifications for marking and destroying sensitive materials. These features, or lack thereof, and others must be evaluated during the OPSEC assessment.

To know how operations at the facility can be interrupted, control procedures and features required for the site to operate effectively must be assessed. The operation and location of equipment and safety features must be documented.

Operational conditions are described by:

- The length and number of day and night shifts
- Activities typical to each shift and the associated security implications
- The number of employees, contractors, and visitors in the area during each shift and the level of access to the facility during weekdays, weekends, and holidays
- The use of batch versus continuous processes

Procedural information to be obtained includes:

- Entry control procedures to the facility for visitors, delivery persons, contractors, and vendors.
- Evacuation procedures.
- Emergency operations procedures in case of evacuation.
- Policies related to alarm assessment and communication to responding security personnel or local law enforcement. The availability of security and safety personnel, including local law enforcement (DoE, 2002).

The following worksheets are included in Appendix B to provide an example of worksheets used to collect operations security information:

- B.1 Human Resources Security Procedures
- B.2 Facility Engineering
- B.3 Facility Operations
- B.4 Administrative Support Organizations
- B.5 Telecommunications and Information Technologies
- B.6 Publicly Released Information
- B.7 Trash and Waste Handling

2.4.1 Assessment

The first stage of an OPSEC assessment review is to learn what critical assets require protection given operations at the facility or site. This review includes interviews with select staff members. The interviews provide details concerning facility or site security, everyday business practices, existing security measures, threat awareness, and potential countermeasures. If the information is not forthcoming, then summaries should be requested through the RFI. Information of interest may include:

1. Criteria for critical assets and information. Compare that information to the list of critical assets and determine if any critical assets need to be added.
2. Critical assets that are protected by law and/or protected by contractual agreements.
3. IT security information assets including operational data, telecommunications information, databases, and security software in use.

4. Legal documents, economic or financial data, and facility or site confidential documents associated with privacy laws, including payroll.
5. Threat against critical assets from insiders (disgruntled employees or in collusion with outside threat), criminals, psychotics, hackers, terrorists, and extremists.
6. Adversary access to assets via:
   • Publications: Web site, mail, press interviews, telephone and radio communications, modems (remote maintenance, computers, copier), trash
   • Contact with employees (direct or phone)
   • Requests for proposals
   • Job vacancy announcements
7. Training. Ensure employees are trained on what is sensitive, who the threat is against the assets, and what their responsibilities are to protect the asset.

2.4.2 Output

Typical OPSEC assessment output includes lists of critical:

• Assets
• Staff
• Business systems and sensitive business information (e.g., system diagrams, personal information, medical information, salary information, competitive bids, customer information, disaster recovery plans, and security procedures designed to protect critical assets)

2.5 POLICIES AND PROCEDURES ASSESSMENT

Policies and procedures assessments develop a comprehensive understanding of a facility's planned protection for critical assets, as described in their policies and procedures. Policies and procedures should:

• Define and communicate roles, responsibilities, authorities, and accountabilities for all individuals and organizations that interface with critical systems
• Address the key factors affecting security
• Provide a basis for identifying and resolving issues
• Enable effective compliance, implementation, and enforcement
• Reference or conform to established standards
• Provide clear and comprehensive guidance

The facility should develop a standardized process for developing, approving, and issuing policies and procedures. Employee education and awareness programs should be conducted on an ongoing basis to ensure policies and procedures are understood and followed. Understanding and assessing policies and procedures identifies strengths and areas for improvements. Improvement may include creation, modification, or further implementation of policies and procedures. Cancellation of policies and procedures that are no longer relevant or are inappropriate may constitute an improvement strategy (DoE, 2002).

2.5.1 Assessment

The assessment focuses on determining the infrastructures and infrastructure connections important to the facility. Specifically, the assessment focus includes:

1. Determining infrastructure linkages to the facility's functions
2. Determining linkages between these functions and specific facilities or types of facilities
3. Determining infrastructure connections
4. Assessing infrastructure vulnerabilities
5. Identifying possible infrastructure mitigation measures

The assessment is based on interviews with key personnel managing the facility's critical functions, facilities, and infrastructures, along with observing infrastructure components and tracing connections.

The methodology used for conducting the policy and procedures portion of the vulnerability assessment includes:

1. Review the facility's or site's "document tree" to determine policies and procedures that are implemented or in the development and review cycle, and where gaps exist. Policies and procedures that address numerous topics should be reviewed. Each assessment will be driven by the unique characteristics of the facility or site participating in the survey.
2. Review of the public Web site and other public sources to gain additional information about the facility or site and a better understanding of its organizational structure and affiliations.
3. Visit the facility or site to observe the impact of the policies and procedures, to acquire operational and functional information, and to identify where the facility or site's security posture would benefit from the implementation of new policies and procedures or the elimination or modification of policies and procedures.
   - Brief interviews should be scheduled with several employees representing the general employee population. The interviewer should draft a protocol for the interviews, and prior to the interview, provide that protocol to all interviewees. Each interviewee should be asked whether the interviewer may make follow-up telephone calls to clarify information noted during the interview.
   - Experts in Vulnerability Assessment Program (VAP) implementation policies and procedures should be interviewed. Potential experts include: security officers, emergency management officer, corporate communications officers, public relations officers, human resource managers, network/system administrators, policies and procedures developers and administrators, information security managers, training managers, contracts officers, and corporate attorney(s). Online supervisors should also be considered experts if these individuals directly contribute to the development of policies and procedures. Otherwise, an online supervisor input is captured in the assessment of implementation items rather than program development (DoE, 2002).
4. Calibrate information gathered during the interviews and the document reviews with other assessment team members.
5. Prepare the assessment report.

A set of checklists should be completed for the facility and for critical assets. All checklists presented for this section should be made site specific prior to the assessment. The worksheets and checklists included in Appendix C are:

- C.1 Infrastructure Oversight and Procedures
- C.2 Electric Power Supply and Distribution
- C.3 Petroleum Fuels Supply and Storage
- C.4 Natural Gas Supply
- C.5 Telecommunications
- C.6 Transportation
- C.7 Water and Wastewater
- C.8 Emergency Services
- C.9 Computers and Servers
- C.10 HVAC System
- C.11 Fire Suppression and Fire-Fighting System
- C.12 Supervisory Control and Data Acquisition (SCADA) System. Note: SCADA may describe systems of any size or geographical distribution. SCADA systems are typically used to perform data collection and control at the supervisory level. Some systems are called SCADA despite only performing data acquisition and not control. The process can be industrial, infrastructure, or facility based. SCADA usually refers to centralized systems that monitor and control entire sites, or complexes of systems spread out over large areas (on the scale of kilometers or miles). Most site control is performed automatically by remote terminal units (RTUs) or by programmable logic controllers (PLCs). Host control functions are usually restricted to basic site overriding or supervisory level intervention. The system is not critical to control the process in real time. For real-time needs, a separate or integrated real-time automated control system that can respond quickly to compensate for process changes within the time constants of the process is maintained.
- C.13 Physical Security System
- C.14 Financial System

2.6 IMPACT ANALYSIS

Impact analysis determines when failure of critical facilities or information systems will affect an organization's operations. In general, impact analysis requires understanding:

- Critical node applications
- Information processing and decisions influenced by this information
- Independent checks and balances
- Factors that might mitigate security breaches
- Secondary impact

For power systems, the physical chain of events following disruption, including the primary, secondary, and tertiary impacts of disruption, should be examined.

2.7 INFRASTRUCTURE INTERDEPENDENCIES

Infrastructure interdependencies refers to the physical and electronic (cyber) linkages within and among critical infrastructures. Critical infrastructures may include: energy (electric power, oil, natural gas), telecommunications, transportation, water supply systems, banking and finance, emergency services, and government services.

Direct infrastructure linkages between and among the infrastructures that support critical facilities must be identified. A detailed understanding of an organization's functions, internal infrastructures, and relationships to external infrastructures is required. Infrastructure interdependencies assessments examine and evaluate the infrastructures (internal and external) that support critical facility functions, along with their associated interdependencies and vulnerabilities.

2.7.1 Computer Systems

The impacts that exploitation of unauthorized access to critical facilities or information systems could have depend on, among other factors, the orientation of the observer. With the increasing use of computers and communication systems in all economic activities, a blackout affects all sectors. Critical systems usually have backup power sources, although most are not designed for an extended blackout, when the operating environment becomes more of a concern.

2.7.2 Utilities

The impacts of electric and gas outages are both direct (the interruption of an activity, function, or service that requires electricity or natural gas) and indirect (the interruption of activities or services). Examples of direct impacts include business shutdowns, food spoilage, damage to electronic data and equipment, and the inability to operate life-support systems in hospitals and homes. Indirect impacts include property losses resulting from arson and looting, overtime payments to police and fire personnel, and potential increases in insurance rates. Direct and indirect impacts are quantifiable in monetary terms (economic impacts); relate to interruption of leisure or occupational impacts (social impacts); or result in organizational, procedural, and other changes in response to outage conditions.

The risk and costs due to an interruption in electric or natural gas service vary depending on many factors. In accordance with the tolerance for those costs, reliable service may be a critical element. For example, in one industry, an outage may only imply postponed production, whereas in another industry, loss of power can ruin production or equipment (e.g., computer chip manufacturing) (DoE, 2002).

2.7.3 Costs

Costs of interruptions vary significantly by sector. Industrial sector costs are more directly measurable in terms of equipment damage, loss of materials, cost of idle resources, and human health and safety effects. For many commercial utility customers, any electric outage of more than a few seconds has a significant cost due to computer problems, equipment jamming, or ruined product. For example, automatic teller machines at banks will shut down during a power outage.

The difficulty in quantifying the external, nonmonetary component of value has led analysts to quantify only the direct, internal, or monetary component of power interruption costs, while continuing to recognize the existence and importance of external costs in defining an appropriate level of system reliability. Numerous estimates have been made of the direct internal costs of electric service interruptions to various customer classes. The major approaches used to establish these costs include:

- Production factor analysis
- Economic welfare analysis
- Empirical analysis of customer surveys

2.8 RISK CHARACTERIZATION

Risk characterization provides a framework for assessing vulnerabilities, threats, and potential impacts (determined in the other tasks). The recommendations for each task area are judged against a set of criteria to prioritize recommendations and determine actions.

Estimates of potential consequences and their economic implications are critically important in applying risk management principles to options faced by the facility or site. The task leaders are encouraged to list not only those recommendations for consideration that make immediate sense from their perspective, but also recommendations that may further reduce risk but seem too expensive given current circumstances. In such cases, the entire list of recommendations should be examined now and again in future evaluations when conditions may have changed enough (e.g., increased threat) so that some recommendations might then be cost-effective (DoE, 2002).

After the severity (S) of each undesired event and the likelihood of attack (L_A) for each adversary group have been determined, these values are ranked in a matrix to derive the LS values. If, for example, an adversary group has a level 2 L_A for a specific undesired event and the undesired event has a severity level of 3, the likelihood and severity level (LS) would be 3. Priority cases would be those undesired event/adversary group pairs with a likelihood and severity (L_S) value closer to 1 than the value chosen by the chemical facility (CF). These priority cases should be analyzed further for protection system effectiveness (DoJ, 2002).

2.8.1 Benefit-Cost Analysis

Develop a list of important characteristics and then identify various risk-reduction recommendations given these characteristics. Apply benefit-cost analysis techniques to determine the feasibility of implementing the recommendations. Example characteristics are:

1. Implementation cost (includes one-time costs, such as equipment cost, labor cost to install, and changes to existing structures) and implementation time (in weeks). Obviously, cost is an important characteristic, and the time to implement may be an important consideration for some recommendations.
2. Change in operating cost associated with implementing each recommendation (includes maintenance cost, consumables, and staff time to monitor or supervise). If the recommendation resulted in substituting for some procedure or activity, the cost savings associated with the substitution is considered as well as the operating cost of the recommendation. This characteristic is another component of the total cost for a recommendation.
3. Attractiveness of asset. The attractiveness of the asset refers to different levels of desirability related to levels of potential impacts that might be achieved with successful exploitation. Asset attractiveness is characterized as shown in Appendix D.
4. Level of consequence. This characteristic refers specifically to different levels of potential economic consequences.
5. Likelihood of preventing an aggressor attempt before (Pb) and after (Pa) the recommendation is implemented. Given an attempt, the likelihood of preventing an aggressor success before (Sb) and after (Sa) the recommendation is implemented. The definitions of these probabilities are complex. The wording asks for probabilities of preventing (an attempt or success) before and after implementing a recommendation. So, an effective cost benefit analysis is more likely to prevent an aggressor's attempt and/or success after implementing the recommendation.

   In many cases, Pb should be low—maybe even 0. For example, a transformer out in the country is an easy target, and with no significant obstacles that would discourage an aggressor from trying to do harm. In this case, interfering with line-of-sight to the transformer (e.g., by installing a screen, a barrier, or a berm) could have two benefits: (1) discourage an attempt and (2) delimit successful aggression against the asset (in such a case, Pa > Pb and Sa > Sb).
6. The technical and cultural difficulty associated with implementation of the recommendation. This characteristic refers to the possibility that the recommendation may make good sense but requires some cultural adjustments to be effective. Difficulty ranges include:
   - 5 Major system modifications and operating practice revisions
   - 4 Significant training and system modifications required
   - 3 Modest attention and study required for proper implementation and operation
   - 2 Minimal problems
   - 1 No problem
7. Dependency on other infrastructures. This characteristic refers to the extent to which the recommendation changes dependencies on other infrastructures, including telecommunications, water, road, rail, emergency services, banking and finance, and

government services. In addition, this measure addresses the change in dependence between the energy infrastructures (e.g., natural gas or electricity and vice versa).

- 5   Large increase in dependency
- 4   Small increase in dependency
- 3   No change in dependency
- 2   Reduces dependency
- 1   Greatly reduces or eliminates dependency

2.8.2 Presentation

Just describing the recommendations with respect to the characteristics is a major step toward structuring the findings. In addition, some groupings and observations have been made with respect to recommendations that have high consequence potential or are likely to have low costs to implement. A large number of recommendations are presented, some of which may become more desirable in the future.

Desirable cost benefit outcomes include:

- Low cost (relatively low combined implementation and annual operating cost)
- Large increase in probability of preventing an aggressor attempt
- Large increase in probability of preventing an aggressor success
- Most attractive assets (recommendations that address assets having high levels of potential impact if successfully exploited)
- Large consequences (recommendations that address potentially large impacts)
- High benefit-cost ratios

Just having a low cost recommendation does not necessarily mean the recommendation is worthwhile to implement. Figure 2.1 illustrates how to calculate expected damages before and after a recommendation is implemented.

The top half, which addresses the situation before implementing a recommendation, indicates that unknown probability "A" exists that an aggressor intends to attack an asset. If no intention is present, no damage is anticipated.

- Given intention, conditions surrounding the asset can either prevent an attack (with probability Pb), which does not result in damage, or fail to prevent an attack (with probability 1 − Pb).
- Given that circumstances do not prevent an aggressor's attempt, circumstances can either prevent success (with probability Sb), which does not result in damage, or fail to prevent success (with probability 1 − Sb). Therefore, the expected damage level before implementing the recommendation is:

$$Db = A * (1 - Pb) * (1 - Sb) * (Asset\ Value)$$

The bottom half of Figure 2.1 addresses the situation after implementing a recommendation. The unknown probability A that an aggressor intends to attack an asset is the same as that before implementing a recommendation and the logic is also the same. Therefore, the expected damage level after implementing a recommendation is:

$$Da = A * (1 - Pa) * (1 - Sa) * (Asset\ Value)$$

Before: Expected Damage = $D_b = A * (1 - P_b)* (1 - S_b)*$ (Asset Value)

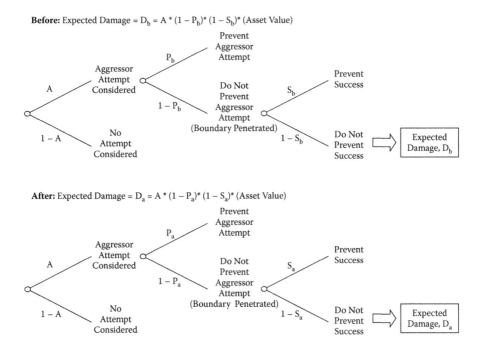

After: Expected Damage = $D_a = A * (1 - P_a)* (1 - S_a)*$ (Asset Value)

Figure 2.1 Risk consequence.

The benefit, B, of implementing a recommendation is the difference between Da and Db:

$$B = Db - Da$$

Finally, to get a variable-free (i.e., numeric) value for B, the value of A is momentarily set at 1 and the asset value is set at the consequence level addressed by a recommendation. This results in an upper bound on the benefit-cost ratio.

A recommendation having B < 0 cannot be justified on the basis of this type of an analysis. Other factors could override the effects of B < 0.

The cost, C, to implement a recommendation is the sum of the implementation cost, IP, and the net present value (NPV) of recurring operating costs (OC):

$$C = IP + NPV(OC).$$

To perform NPV calculations, the cost of capital and the time horizon is set. Two numbers are listed for each recommendation—the benefit-cost ratio and Acrit. Acrit is the value of A, the unknown probability of an aggressor attack, which is set to 1; this makes benefits equal to costs. Acrit is the reciprocal of the benefit-cost ratio. If the actual value of A is judged to be > Acrit, the recommendation is worth implementing. Determine which recommendation has the highest benefit-cost ratio and the smallest Acrit.

2.9   FINAL REPORT

Items suggested for inclusion in the final report are:

1. Screening process results
2. Facility characterization matrix and critical activities analyzed
3. Severity Level Definition Table
    - Severity level for each undesired event
4. Threat Definition Table
5. Likelihood of Attack Level Definition Table
    - L_A levels for each undesired event and/or adversary group
6. LS (likelihood and severity) Priority Ranking Matrix
    - LS levels for each undesired event and/or adversary group
7. Priority analysis of undesired event and/or adversary group
8. Most vulnerable adversary scenarios for:
    - Both physical and process control paths
    - Each prioritized undesired event and/or adversary group
9. L_{AS} Definition Table, L_{AS} levels for:
    - Both physical and process control paths
    - Each prioritized undesired event and/or adversary group
10. Risk priority ranking matrix and risk levels for:
    - Both physical and process control paths
    - Each prioritized undesired event and/or adversary group
11. Recommendations to reduce risk levels

2.10   POST-ASSESSMENT

The post-assessment phase involves prioritizing assessment recommendations, developing an action plan, capturing lessons learned and best practices, and conducting training. The risk characterization element results provide the basis for the post-assessment by providing prioritized lists of recommendations that are ranked by key criteria.

The next step is to develop an action plan that includes timelines, staffing assignments, and budgets to implement the proposed recommendations. Lessons learned should be captured along the way to improve the overall process in the future. Training and other technical support activities, such as workshops, are also appropriate throughout the process.

REFERENCES

U.S. Department of Energy, Office of Energy Assurance, 2002, *Vulnerability Assessment Methodology*, Electric Power Infrastructure.
U.S. Department of Justice, Office of Justice Programs, 2002, *A Method to Assess the Vulnerability of U.S. Chemical Facilities.*

APPENDIX A

Table A.1 Physical Security Program

General
1	A mission statement establishing the physical security program exists.

General

1 A mission statement establishing the physical security program exists.
2 An organizational chart, with areas of responsibility, is developed.
3 Adequate resources (budgetary and staffing) are present for the physical security program.
4 An executive (senior management) protection plan has been developed.
5   Bomb Threat Worksheets are available to personnel.
6 Security staff personnel are provided with adequate training to accomplish their functions.
7 A threat analysis is included in the security program designing process.
8 Company assets requiring security are defined.
9 The level of acceptable risk is defined.
10 Protection strategies are based on results of a documented vulnerability analysis.
11 Protection strategies are adequately defined in security planning documents.

Alarms

12 IDS (alarms/cameras) and physical security elements (barriers/guards) are adequate.
13 IDS (alarms/cameras) and physical security elements (barriers/guards) are routinely tested.

Note: IDS = intrusion detection systems.

Table A.2 Physical Security

Barriers

1 Physical barriers (fences, walls, or doors) define physical boundaries.
2 Physical barriers delay unauthorized access, and direct the flow of personnel.
3 Physical barriers direct pedestrians through designated portals into critical asset areas.
4 Physical barriers direct pedestrian traffic through designated portals into critical asset areas.
5 Vehicle barriers are used to preclude, deter, and (as needed) prevent penetration into critical asset areas.
6 Physical protection elements are implemented at storm sewers.
7 Physical protection elements are implemented at drainage swells.
8 Physical protection elements are implemented where site utilities intersect the fence perimeter.
9 Fence lines are kept clear of vegetation and trash.
10 Fence lines are kept clear of equipment that could impede observation.
11 Fence lines are kept clear of other objects that could impede observation.
12 Fences are within 2 inches of firm, hard ground.
13 Fence lines are free of objects that would aid in traversing the fence.
14 Fence line is free of holes and gullies.
15 Fence line is free of areas that would aid in traversing the fence.
16 Notification signs are posted as required.
17 Walls that constitute exterior barriers of critical asset areas are without margin gaps (floor to ceiling gaps).
18 Walls that constitute exterior barriers of critical asset areas and have gaps have other security.

Table A.3   Physical Security Access

Control/Badges

1   Control measures are in place at automated asset control points.
2   Control measures are in place at attended asset control points.
3   Badges are worn at all times, above the waist, on the exterior garment.
4   Standard security badges are issued to employees.
5   Security badges indicate an employee's authorized access area.
6   Badges for visitors are visibly different than employees' badges.
7   Badges are destroyed in a manner that precludes reconstruction.
8   Used badges are stored in a secure manner until destroyed.
9   Individuals are provided security badge requirement's information.
10 Temporary badges are utilized for employees who have misplaced their permanent badges.
11 Company policy includes a process for requesting security badges.
12 Company policy includes a process for approving security badges issuance.
13 Company policy includes control of security badge stock.
14 Company policy includes retrieval of security badges from terminated employees.
15 Company policy includes lost or stolen security badges.

Table A.4   Physical Security

Locks/Keys

1   Security keys are protected at the same level as the asset under protection.
2   An inventory and accountability system for keys is implemented.
3   Panic hardware or emergency exit mechanisms used on critical asset's emergency doors are operable from inside.
4   Panic hardware or emergency exit mechanisms used on emergency doors meet applicable life safety codes.
5   Company procedure for keys provides a process for requesting keys.
6   Company procedure for keys provides a process for approving the issuance of keys.
7   Company procedure for keys provides control of key stock.
8   Company procedure for keys provides for retrieval of keys from terminated employees.
9   Company procedure for keys provides for lost or stolen keys.

Table A.5   Physical Security

Intrusion Detection Systems (IDS)

General

1   Systems are continuously monitored to assess alarms.
2   Systems are operated and maintained to ensure few false alarms.
3   An audible and optional visual alarm signal directs the protective force (when alarm stations are not manned).
4   Compensatory measures are employed when systems are not operational.
5   Records are kept on actual and/or false nuisance alarms.
6   Nuisance alarm records are reviewed and analyzed.
7   System malfunctions are corrected.
8   IDSs in adjacent detection zones overlap, preventing gaps in detection zones.
9   Dips, obstructions, and equipment do not provide a pathway to avoid alarm system detection.

Table A.5   Physical Security (Continued)

10	The detection zone of exterior alarm system is free of any other item that degrades IDS effectiveness.
11	If alarm's detection systems are obstructed, compensatory measures are taken to provide timely detection.
12	Operator acknowledgment of alarms is straightforward and easily performed.
13	The alarm control system calls the central alarm station (CAS) operator's attention to alarm-associated monitors.

Intrusion Detection Systems (IDSs)

1	The quality of pictures from closed-circuit television (CCTV) cameras allows discrimination between human and animal intrusion.
2	Video recorders are actuated by alarm signals and operate automatically.
3	Video recorders actuated by alarm signals are sufficiently rapid enough to record an actual intrusion.
4	CCTV cameras have tamper protection and loss-of-video alarm annunciation.
5	Remote assessment of a perimeter IDS when used has complete coverage (no gap zones).
6	Objects or shadows do not block CCTV camera field-of-view.
7	Intrusion detection systems are protected from tampering.
8	Tamper and system problem indicators are provided on the intrusion detection system.

Lighting

1	Adequate lighting is available for alarm/intrusion assessment by CCTV or other means.
2	Protective lighting is sufficient to permit detection and assessment.
3	Lighting does not produce glare in CCTV.

Power Sources—Maintenance Program

1	IDSs receive routine preventive maintenance.
2	IDSs are operability-tested routinely to demonstrate effectiveness.
3	IDSs are functionally tested in accordance with established procedures at a frequency that is documented.
4	Compensatory measures are implemented immediately when any part of the IDS is out of service.
5	Personnel who test and maintain the IDS in use have access authorization consistent with the critical asset being protected.

Table A.6   Physical Security Communications Equipment

1	Communications equipment provides for reliable information exchange between protective force personnel.
2	Alternate communications capabilities are available immediately upon failure of the primary communications system.
3	Systems remain operable in the event of loss of primary electric power.
4	Duress systems are available to protective force personnel.
5	Activation of the duress alarm is accomplished in as unobtrusive a manner as practicable.
6	Duress alarms do not annunciate at the post initiating the duress alarm.

Table A.7   Protective Force/Local Law Enforcement Agency

1   The number of protective force members on duty is sufficient to accomplish the mission.
2   The command structure is defined.
3   The protective force mission is defined.
4   Protective force policy/procedures exist.
5   Emergency response plans exist.
6   The protective force has the equipment necessary to complete the mission (vehicles, uniforms, vests, weapons, flashlights, communications equipment).
7   The protective force receives initial and continuing training required.
8   The protective force conducts drills and exercises with the local law enforcement agency (LLEA).
9   If an LLEA is the primary response agency, a Memorandum of Understanding or other form of agreement is in place.
10 The LLEA's protection responsibility is defined (What are the expectations of the company?).
11 The LEA participates in drills/exercises.

Table A.8   Entrances into Critical Asset Areas

	Security Level		
	Low	Medium	High

Door Construction
Wood
9-gauge wire mesh
Hollow core metal, no lock/hinge protection
or
Hollow core metal
Tempered glass panel
Security glass panel
Half height turnstile
or
½-inch steel plate
Aluminum turnstile
Steel turnstile
Class V or VI vault
Dispensable barrier

Locks
No lock, or lock not used
or
Door unlocked, attended by personnel when unlocked
ID-actuated lock
Padlock
High-security padlock
Keyed cylinder
Combination
Mechanically coded
or
Electronically coded
Two-person-rule lock systems

Table A.8 Entrances into Critical Asset Areas (Continued)

Lock inaccessible from door exterior
Personnel controlling asset
or
Personnel at post
No duress, unprotected
or
No duress, small arms protected
Duress, unprotected
or
Duress, small arms protected
Protective force on patrol

Identification Check
Casual recognition
Credential
or
Credential and PIN
Picture badge
Picture badge and PIN

Entrances into Critical Asset Areas
Exchange picture badge
Exchange picture badge and PIN
Retinal scan and PIN

Hand geometry and PIN
Speech pattern and PIN
Signature dynamics and PIN
Fingerprint and PIN

Explosives Detectors
None
or
Detectors
Animal olfaction
Vapor collection
Handheld vapor collection
Thermal neutron

Metal Detectors (Handheld or Portal)
None
or
Ferrous materials only
Ferrous and solid lead materials
Ferrous materials and all forms of lead

Item Searches
None
or
Cursory
Rigorous

(Continued)

Table A.8   Entrances into Critical Asset Areas (Continued)

Alarms

None

or

Door penetration sensor

Vibration

Glass breakage

Conducting tape

Grid mesh

Multiple sensors

AND

Door position monitor

Position switch

Balanced magnetic switch

Alarms

Door penetration sensor

Vibration

Glass breakage

Conducting tape

Grid mesh

Multiple sensors

AND

Door position monitor

Position switch

Balanced magnetic switch

Alarm Assessment

No assessment

or

Delayed deployment

Timely deployment

CCTV without instant replay

or

CCTV with instant replay

Posted protective force with duress alarm

Automatic deployment of protective force

Surface Construction

Chain link mesh

16-gauge metal

Wood studs and sheet rock

Wood studs and plywood

or

Clay Block

8-inch solid block

8-inch filled block

Windows

9-gauge expanded mesh

  1/2-inch diameter ×1-1/4-inch quarry screen

  1/2-inch diameter bars with 6-inch spacing

  3/16-inch × 2-1/2-inch grating

Table A.8 Entrances into Critical Asset Areas (Continued)

Rebar Block

8-inch filled rebar block

12-inch filled rebar block

2-inch precast concrete tee

4-inch reinforced concrete

8-inch reinforced concrete

12-inch reinforced concrete

24-inch reinforced concrete

Roof/Ceiling

20-gauge metal with insulation

1/2-inch wood roof

Interior drop ceilings

or

20-gauge metal built-up roof

Concrete built up roof with T beam

Interior drop ceiling does not extend outside critical asset area

or

5 1/2-inch concrete roof

8-inch concrete roof

3-foot earth cover

3-foot soil cement earth cover

Interior drop ceiling does not extend outside critical asset area

Window Alarms

Window Alarms (Windows accessible by foot or ladder)

None

or

Window penetration sensor

Vibration

Glass breakage

Conducting tape

Grid mesh

Multiple sensors

or

Interior intrusion sensors (to cover windows that are not alarmed)

Sonic

Capacitance

Video motion

Infrared

Ultrasonic

Microwave

Multiple noncomplementary sensors

Multiple complementary sensors

Surface Penetration Alarms

(Alarms provide detection upon penetration of any critical asset
  area boundary surface)

(Continued)

Table A.8   Entrances into Critical Asset Areas (Continued)

Interior Intrusion Sensors

Sonic
Capacitance
or

Video Motion

Infrared
Ultrasonic
Microwave
Multiple noncomplementary sensors
Multiple complementary sensors
or

Exterior Intrusion Sensors

Seismic buried cable
or

Electric field

Infrared
or

Microwave

Video motion
Multiple noncomplementary sensors
Multiple complementary sensors
Alarm Assessment
None
or
Delayed deployment
Timely deployment
CCTV without instant replay
or
CCTV with instant replay
Posted protective force with duress alarm
Automatic deployment of protective force

Table A.9   Vehicle Gates through Critical Asset Area Fences

	Security Level		
	Low	Medium	High
Gate Construction			
No gate closure			
Vehicle bar			
or			
8-foot chain link			
8-foot chain link with outriggers			
8- to 12-foot chain link with outriggers			
Over 12-foot chain link with outriggers			

Table A.9   Vehicle Gates through Critical Asset Area Fences (Continued)

Vehicle Barriers

None
or
Aircraft cable
Blocked by vehicle (gate open)
Hydraulic wedge

Locks

No lock, or lock not used
or
Gate unlocked, attended by personnel when unlocked
ID-actuated lock
Padlock
High-security padlock

Personnel Controlling Access through Gate (if used)

Personnel at post
No duress, unprotected
or
No duress, small arms protected
Duress, unprotected
or
Duress, small arms protected
Protective force on patrol

Identification Check

Casual recognition
Credential
or
Credential and PIN
Picture badge
Picture badge and PIN
Exchange picture badge
Exchange picture badge and PIN
or
Retinal scan and PIN
Hand geometry and PIN
Speech pattern and PIN
Signature dynamics and PIN
Fingerprint and PIN

Explosives Detectors

None
or
Animal olfaction
Vapor collection

(Continued)

Table A.9   Vehicle Gates through Critical Asset Area Fences (Continued)

Handheld vapor collection
Thermal neutron

Metal Detectors (Handheld or Portal)

None
or
Ferrous materials only
Ferrous and solid lead materials
Ferrous materials and all forms of lead

Item and Vehicle Searches

None
or
Cursory
Rigorous

Alarms

None
or
Fence sensors
  Taut wire
  Vibration
  Strain
  Electric field
  Multiple sensors
OR
Intrusion sensors
  Seismic buried cable
Electric field
  Infrared
Microwave
  Video motion
or
Multiple noncomplementary sensors
Multiple complementary sensors

Alarm Assessment

None
or
Delayed deployment
Timely deployment
CCTV without instant replay
or
CCTV with instant replay
Posted protective force with duress alarm
Automatic deployment of protective force

Table A.10   Fences Surrounding Critical Assets

	Security Level		
	Low	Medium	High

Fence Construction

8-foot chain link
or
8-foot chain link with outriggers
8- to 12-foot chain link with outriggers
Over 12-foot chain link with outriggers

Signs

"No Trespassing" signs posted

Vehicle Barriers

None
or
Aircraft cable
  Concrete blocks
  Guard rails
  Steel posts
  Concrete median
  Concrete median and ditch
or
Crash I beam
Train barrier

Alarms

None
or
Fence sensors
Taut wire
Vibration
Strain
Electric field
Multiple sensors
or
Intrusion sensors
Seismic buried cable
Electric field
Infrared
Microwave
Video motion
or
Multiple noncomplementary sensors
Multiple complementary sensors

Alarm Assessment

None

(Continued)

Table A.10   Fences Surrounding Critical Assets (Continued)

or
Delayed deployment
Timely deployment
CCTV without instant replay
or
CCTV with instant replay
Posted protective force with duress alarm
Automatic deployment of protective force

APPENDIX B

Table B.1   Human Resources Security Procedures

(a) Responsibilities
What internal offices or departments deal with security issues?
(b) Background Checks
What background checks are conducted on prospective employees?
How extensive are the background checks?
Do background checks vary with the sensitivity of the position?
(c) Insider Threats
What current "insider" conditions (low morale, layoffs, disputes) might create a threat?
What are the security procedures?
What security procedures are used for handling disgruntled or at-risk employees?
What are the security procedures are used for handling employee terminations?
How many employees have been terminated in the last year?
(d) Disciplinary Procedures
What are the policies and procedures for handling security incidents?
What are the policies and procedures for other disciplinary actions?
(e) Security Training
Does the initial and periodic security awareness training include:
Security contacts
Critical assets
Threats
Sensitive information
Reporting suspicious activities
Employee responsibility

Table B.2   Facility Engineering

(a) Responsibilities
What internal offices or departments are responsible for facility engineering?
(b) Facility Engineering Information
What facility engineering information (e.g., engineering drawings, site maps, utility
service lines, floor plans, entry paths into the facility) is considered sensitive?
What offices or departments have control of this information?
What other offices or departments are allowed access to this information?

Table B.2   Facility Engineering (Continued)

What external organizations (e.g., fire departments, environmental agencies) have been
  given access to this information?
Is any of the facility engineering information publicly available?
How is sensitive facility engineering information protected?
What facility engineering information can be accessed via the computer system or network?
How is the information disposed of when it is no longer needed?
How secure are facility design, configuration, and layout; utility service systems; and building
  floor plans?
(c) Public Access to Facility
Where are tours allowed within the facility? Describe what portions of the facility are open
  and who is allowed to tour.
What portion of the facility is open to the public or special interest groups?
What periodic meetings are held within the facility where outsiders are allowed inside the
  facility?

Table B.3   Facility Operations

(a) Responsibilities
What internal offices or departments are responsible for facility operations?
(b) Facility Operations Control
Is the operation of the facility controlled from a central point (or several central points)?
  Describe.
Is there an automated process control system, energy management system, or SCADA
  system? Is it isolated or is remote access possible?
What facility operations control and information are on the computer systems? How is it
  protected? What other internal organizations have access to operations control
  capabilities and information?
Can sensitive operations information be gathered through the telecommunications
  system (e.g., microwave, cell phones, radio, pagers, voicemail, teleconferencing)?
Is access to the control point(s) limited to operations personnel? If not, who else has
  access (e.g., maintenance, janitors, vendors), and how is that access controlled?
(c) Facility Construction, Repair, and Maintenance
Are construction, repair, and maintenance at the facility done by employees, contractors, or
  both? If contractors are used, describe procedures for screening and monitoring
  contractor personnel.
Are cleaning and building maintenance (e.g., janitorial service) done by employees of your
  facility?
Are cleaning and building maintenance (e.g., janitorial service) done by contractors?
Are contractors subject to screening and monitoring?

Table B.4   Administrative Support Organizations

Purchasing and procurement activities include generating need (e.g., requisitions or RFPs),
  selecting suppliers, documenting purchases, providing delivery of items or services, and
  payments.
(a) Procurement
What internal offices/departments review procurement activities from a security perspective?
What is the security review process for RFPs, contracts, and procurement documents?
How is the procurement information protected before release?

(Continued)

Table B.4 Administrative Support Organizations (Continued)

How are documents, files, copiers, facsimiles, and computer files secured?
What security-sensitive information is uniquely marked?
Are paper marks used?
Are electronic marks used?
How is security-sensitive procurement information destroyed?

(b) Legal

What internal offices/departments review legal department activities from a security
   perspective?
How are legal documents reviewed for security implications?
   Patents
   Environmental impact statements
   Safety reports
   Securities and Exchange Commission filings
   Federal Energy Regulatory Commission filings
How are these documents protected?
How are these documents destroyed?

(c) Budget and Finance

What internal offices/departments review budget and finance activities from a security
   perspective?
How are budget and finance documents reviewed for security implications?
How are these documents protected?
How are these documents destroyed?

(d) Marketing

What internal offices/departments review marketing activities from a security perspective?
How are marketing materials reviewed for security implications?
How are these documents protected?
How are these documents destroyed?

(e) Internal Information

What are the policies and procedures for handling "Internal Documents"?
Memos
Notes
Newsletters
How are these documents protected?
How are these documents destroyed?

Table B.5 Telecommunications and Information Technologies

Note that this part of the operations security survey must coordinated with the portions of the
   interdependencies survey that address the telecommunications and computer equipment.

(a) Telecommunications

What are the policies and procedures for communications security?
What particular equipment carries sensitive traffic? Is this equipment restricted to selected
   users?
What training is provided concerning security issues while using telecommunications
   equipment?
What level of awareness is there concerning telecommunications equipment being
   operated in reverse as eavesdropping equipment?
Is voicemail protected by passwords? Have users changed the vendor-supplied
   passwords? Is there a master password?

Table B.5 Telecommunications and Information Technologies (Continued)

How are FAX machines protected (e.g., logging, stored information, computer connectivity)?
Is encryption used on any telecommunications circuits?
Describe all connections to external radio nets, including paging nets.

(b) Information Technology
What are the policies and procedures for computing and information technology security?
What computer architecture information is available to outsiders?
What encryption is used for internal files and/or information transmission?
Are system administrators trained to recognize "social engineering attacks" designed to
  obtain passwords and other security information?
Describe how e-mail is monitored?

Table B.6 Publicly Released Information

(a) Responsibilities
What internal offices or departments are responsible for reviewing information (from a
  security perspective) that is to be released to the public?

(b) General Procedures
What is the process used to review information before release?
How is the information protected before release? Include documents, files, copiers,
  facsimiles, and computer files.

(c) Report Release
Who is responsible for reviewing reports released by the organization?

(d) Press Contacts
Who is officially designated to interact with the press?
How are they trained (including training on security issues)? Who trains them?

(e) Briefings and Presentations
Describe how briefings and presentations to be given by employees of the organization
  are reviewed for security issues.
This checklist covers information that is released to the public via corporate
  communications, press releases, the Internet, and other means.

(f) Public Testimony
Describe how public testimony that is to be given by employees of the organization is
  reviewed for security issues.

(g) Internet Information
Describe the policy for the review of information posted on the organization's Internet site
  for security issues.
What is the required review process for information before it is posted on the Web site?

Table B.7 Trash and Waste Handling

(a) Responsibilities
What internal offices or departments are responsible for the security of trash and waste?
Describe established policies for trash and waste handling.

(b) Trash Handling
Where is trash accumulated?
Is the trash accessible to outsiders?
Who collects the trash?
Where is the trash taken?

(Continued)

Table B.7   Trash and Waste Handling (Continued)

(c) Paper Waste Handling
> Where is paper waste accumulated?
> Describe the availability and use of shredders throughout the facility.
> What paper waste is accessible to outsiders?
> Who collects the paper waste?
> Where is the paper waste taken? Is it sent for recycling?
> Describe any on-site destruction of paper waste. How it is protected until destroyed?

(d) Salvage Material Handling
> Does salvage material (e.g., serviceable equipment no longer needed, surplus
>   equipment) potentially contain sensitive information?
> Describe the procedures for inspecting salvage material before release.

(e) Dumpster Control
> Describe how dumpsters (for trash, paper waste, and salvage materials) that are
>   accessible to the public are monitored to prevent "dumpster diving."
> How are publicly accessible dumpsters sampled for sensitive information?

APPENDIX C

Table C.1   Infrastructure Oversight and Procedures

Does the facility have a central office or department responsible for overseeing the
  infrastructures?
Indicate the office/department.
List the infrastructures.
List who has responsibility for the infrastructures.
Describe the extent of responsibility.
What coordination or oversight role does the physical security office have for infrastructures?

Table C.2   Electric Power Supply and Distribution

Primary Source of Electric Power
Is the primary source of electric power a commercial source?
Are multiple independent feeds provided?
> If so, describe the feeds and their locations.
Is the primary source of electric power a system operated by the asset/facility?
> What type of system is operated?
If a facility-operated primary electric generation system is used, what are the fuel or fuels
used?
Is petroleum fuel used for the primary electric generation system?
> What quantity of fuel is stored on-site for the primary electric generation system?
> How long would the fuel stores last under different operating conditions?
If the fuel is stored on-site, are arrangements and contracts in place for fuel resupply and
management?

Electric Distribution System
Are the electric system components located outside of buildings protected from vandalism or
  accidental damage by fences or barriers?
> Generators

Table C.2   Electric Power Supply and Distribution (Continued)

Fuel storage facilities
Transformers
Transfer switches
If so, describe the protection type and security level provided.
Can the electric power sources and the internal electric distribution system's components be isolated for maintenance or replacement?
Can isolation be accomplished without affecting the asset/facility's critical functions?
If not, describe the limitations.
Have any single points of failure been identified for the electric power supply and distribution system?
If so, list and describe.

Backup Electric Power Systems
Are additional electrical supply emergency sources provided beyond the primary system?
Multiple independent commercial feeds
Backup generators
Uninterruptible power supply (UPS)
If so, describe them.
Is a central UPS provided to support all the asset/facility's critical functions in terms of capacity and connectivity?
Specify for how long the UPS can operate on battery power. List any potentially critical functions that are not supported.
Is a backup generator system on-site?
Does the backup generator support all the facility's critical functions in terms of capacity and connectivity?
List any potentially critical functions that are not supported.
Specify the backup generator's fuel.
Is petroleum fuel used for the backup generator system?
If yes, specify the quantity stored on-site and how long the fuel would last.
Is the fuel stored on-site?
Are arrangements and contracts in place for fuel resupply and management?

Commercial Electric Power Sources
Is the area supplied by multiple substations?
How many substations feed the asset/facility's area and the asset/facility itself?
If more than one substation used, which substations have sufficient individual capacities to supply the asset/facility's critical needs?
How may distinct independent transmission lines supply the substations?
Indicate if an individual substation is supplied by more than one transmission line, and which substations are supplied by independent transmission lines.

Commercial Electric Power Pathways
Are the power lines into the asset/facility's area and into the asset/facility aboveground (on utility poles), buried, or a combination of both?
If both, indicate locations of aboveground portions.
Do the power lines from these substations follow independent pathways to the asset/facility's area?
If not, specify how often and where the power lines intersect or follow the same corridor.
Are power lines' paths co-located with the other infrastructures' rights-of-way?
If yes, indicate how often and where the power lines follow the same rights-of-way.
Are the power lines' paths located in areas susceptible to natural or accidental damage?
Overhead lines near highways
Power lines across bridges
Dams
Landslide areas

(Continued)

Table C.2   Electric Power Supply and Distribution (Continued)

If yes, indicate the potential disruptions' locations and types.

Commercial Electric Power Contracts

What type of contract does the asset/facility have with the electric power distribution company or transmission companies?

Specify the companies involved and whether a direct physical link (distribution or transmission power line) exists to each company facility.

Has electrical service been interrupted in the past? If yes, describe the circumstances and any effect the outages had on the asset/facility's critical functions and activities.

Is an interruptible contract (even in part) present?

What are the general conditions placed on interruptions?

  Minimum quantity that is not interruptible

  Maximum number of disruptions per time period

  Maximum duration of disruptions

Historical Reliability

Historically, how reliable has commercial electric power been in the area? Quantify the annual number of disruptions and their durations.

  Typically, are these outages of significant duration (as opposed to a few seconds or minutes)?

  Quantify the duration of the outages.

Have electric power outages of sufficient frequency and duration occurred so as to affect the asset/facility's critical functions and activities?

Are remote components to the radio communications system (such as relay towers) used? If yes, describe the components and their uses.

Table C.3   Petroleum Fuels Supply and Storage

Uses of Petroleum Fuels

Are petroleum fuels used in normal operations at the asset/facility?

  If yes, specify the types and uses.

Are petroleum fuels used during contingency or emergency operations (i.e., backup equipment or repairs)? If yes, specify the fuel types and their uses.

Reception Facilities

How are the various petroleum fuels normally delivered to the asset/facility?

  Indicate the delivery mode and normal shipment frequency for each fuel type.

Under maximum use-rate conditions, are sufficient reception facilities provided to keep up with maximum contingency or emergency demand)?

  Truck racks

  Rail sidings

  Surge tank capacity

  Barge moorings

  If no, explain where the expected shortfalls would be and their impacts.

Are the petroleum fuel delivery pathways co-located with the other infrastructure's rights-of-way?

Are the petroleum fuel delivery pathways located in areas susceptible to natural or accidental damage?

  Across bridges

  Across dams

  In earthquake areas

  In landslide areas

If yes, indicate the locations and types of potential disruptions.

Table C.3 Petroleum Fuels Supply and Storage (Continued)

Are contingency procedures in place to allow for alternative delivery modes or routes?
If yes, describe these alternatives and indicate whether routes have sufficient capacity to fully
  support the asset/facility's critical functions and activities.

Supply Contracts
Are contracts in place for petroleum fuels supply?
  Pipeline
  Rail car
  Tank truck
  Specify the contractors, the types of contracts, the transport mode, and the normal
    shipment frequency.
Are arrangements for emergency petroleum fuel deliveries in place?
  Indicate the basic contract terms including maximum time to delivery, and the minimum
    and maximum quantity per delivery.
Also, indicate if these terms may have effects on the asset/facility's critical functions and
  activities.

Table C.4 Natural Gas Supply

Sources of Natural Gas
How many city gate stations supply the natural gas distribution system in the asset/facility's
  area and to the asset/facility?
  If more than one, which ones are critical to maintaining the distribution system?
How many distinct independent transmission pipelines supply the city gate stations?
Indicate if an individual gate station is supplied by more than one transmission pipeline and
  which stations are supplied by independent transmission pipelines.

Pathways of Natural Gas
Do the distribution pipelines from the individual city gate stations follow independent
  pathways to the asset/facility's area?
  If not, specify how often and where intersections occur or follow the same corridor.
Are the pipelines' paths co-located with the other infrastructures' rights-of-way?
  If yes, indicate how often and where the same rights-of-ways occur.
Are the pipelines' paths located in areas susceptible to natural or accidental damage?
  Across bridges
  Across dams
  In earthquake areas
  In landslide areas
  If yes, indicate the potential disruptions' locations and types.
Is the local distribution system well integrated (i.e., can gas readily get from any part of the
  system to any other part of the system)?

Natural Gas Contracts
Does the asset/facility have a firm delivery contract, an interruptible contract, or a mixed
  contract with the natural gas distribution company or the transmission companies?
Specify the companies involved and whether a direct physical link (pipeline) to each company exists.
Is an interruptible contract (even in part) provided?
What are the general conditions placed on interruptions?
  Minimum quantity that is not interruptible
  Maximum number of disruptions per time period

(Continued)

Table C.4   Natural Gas Supply (Continued)

Maximum duration of disruptions

Has natural gas service been interrupted in the past?

If yes, describe the circumstances and any effect the outages had on the asset/facility's critical functions and activities.

Does the asset/facility have storage or some other special contracts with natural gas transmission or storage companies?

If yes, briefly describe the potential for sustaining a continuous supply of natural gas to the asset/facility.

In case of a prolonged natural gas supply disruption, are contingency procedures in place to allow for alternative fuels use?

On-site propane-air

Liquefied petroleum gas

Petroleum fuels

If yes, describe these alternatives and indicate whether sufficient capacity is provided to fully support the asset/facility's critical functions and activities.

Historical Reliability

Historically, how reliable has the natural gas supply been in the area?

Quantify by describing any unscheduled or unexpected disruptions.

Were effects on the asset/facility's critical functions and activities observed?

If operating under an interruptible service agreement, has natural gas service ever been curtailed?

If yes, how often, for how long, and where were the asset/facility's critical functions and activities affected?

Table C.5   Telecommunications

Internal Telephone System

What telephone system types are used within the asset/facility?

Are multiple independent telephone systems provided?

Specify the system types, their uses, and whether copper-wire or fiber-optic based systems are used.

If multiple independent telephone systems are provided within the asset/facility, is each one adequate to support the critical functions and activities?

Indicate any limitations.

Are multiple (from independent systems) or redundant (from built-in backups) switches and cables provided, and physically separated and isolated to avoid common failures?

Are the telephone switches located in limited-access or secured areas away from potential damage due to weather or water leaks?

Specify protection types provided.

Data Transfer

For large-volume and high-speed data transfer within the asset/facility, is a separate switch and cables system provided within the asset/facility?

Specify the system type, and whether the system is copper-wire or fiber-optic based.

Is a separate system provided for large-volume and high-speed data transfer, and are redundant switches and cables included?

If yes, describe the situation.

Are redundant switches and cables used?

Are these systems physically separated and isolated to avoid common failures?

Are the data-transfer switches located in limited-access or secured areas away from potential damage due to weather or water leaks? Specify the protection types provided.

Table C.5   Telecommunications (Continued)

Cellular/Wireless/Satellite Systems

Are cellular/wireless telephones and pagers in widespread use within the asset/facility? If yes, briefly describe their uses.

If cellular/wireless telephones and pagers are in widespread use, are these systems adequate to support the critical functions and activities? Specify any limitations.

Are satellite telephones or data links in widespread use within the asset/facility?
  If yes, briefly describe the satellite phone uses.

If satellite telephones or data links are in widespread use, are these systems adequate to support the critical functions and activities?

Specify any limitations.

Intranet and E-mail System

Is the asset's/facility's Intranet and e-mail system dependent on the asset's/facility's computers and servers?
  If yes, describe the dependence.

Is the asset's/facility's Intranet and e-mail system dependent on the asset's/facility's telephone system? If yes, describe the dependence?
  If yes, describe the dependence.

Are the asset's/facility's Intranet and e-mail system separate from other systems?

If the asset's/facility's Intranet and e-mail system is a separate system, does this system have dedicated backup electric power supply, such as local UPSs?

If yes, specify under what conditions and for how long.

Does the asset's/facility's central HVAC system provide environmental control for important components or have an independent environmental control system?
  If so, specify the type.

If the asset's/facility's Intranet and e-mail system is a separate system, can this system operate with an environmental control loss?
  If yes, specify for how long under various conditions.

If the asset's/facility's Intranet and e-mail system is a separate system, are backup environmental controls available explicitly for the system?
  If yes, indicate the backup type and the expected maximum duration of operation.

If the asset's/facility's Intranet and e-mail system is a separate system, is special physical security provided for the important components?
  If yes, specify the security type and the protection level provided.

Is special fire suppression equipment provided for the Intranet and e-mail system? If yes, specify the system type:
  Halon
  Inergen
  Inert gases
  Carbon dioxide

Are special features or equipment in the important components' are provided to limit flooding or water intrusion during activation of five suppression system?
  If yes, indicate the precautions taken.

Are alarms provided for the important components for _____?
  Unauthorized intrusion
  Loss of electric power
  Loss of environmental control
  Fire
  Flooding
  Water intrusion

(Continued)

Table C.5   Telecommunications (Continued)

If yes, specify the alarm types, how the alarms are monitored, and the response procedure.

Redundant Access to Intranet and E-mail System
Does the asset/facility have a backup or redundant Intranet and e-mail system? If yes, describe the system and the amount of backup provided.
Do areas where critical functions and activities take place have multiple or redundant access to the Intranet and e-mail system?

On-site Fixed Components of the Microwave/Radio System
Are multiple or redundant radio communications systems in place within the asset/facility?
  If yes, specify the system types and their uses.
If multiple radio communications systems are in use, is more than one system adequate to support all the asset/facility's critical functions and activities?
  Specify any limitations.
Are provisions within the asset's/facility's primary electric power supply and distribution system provided to supply power for the radio communications systems?
  If yes, indicate under what conditions and for how long.
Do the radio communications systems have dedicated backup electric power supply?
  If yes, specify the type and how long the system can operate.
Are the components of the system located outside buildings (e.g., antennae, on-site towers) protected from vandalism or accidental damage by fences or barriers?
  If protected, specify the protection types and security level provided.

Mobile and Remote Components of the Microwave/Radio System
Are mobile components to the radio communications system (on vehicles or vessels) provided? If yes, describe the mobile components.
Are the radio communications system's mobile components protected from vandalism or accidental damage by locked boxes or lockable vehicle cabs?
  Specify the protection types and security level provided.
Are remote components to the radio communications system (such as relay towers) used?
  If yes, describe the components and their uses.
Are backup electric power sources provided for these remote components?
  If yes, indicate the backup type, the fuels used, and the expected length of operations.
Are environmental controls required for the remote components (i.e., heating, cooling)?
  If yes, describe.
Are backup environmental controls provided for these remote components?
  If yes, indicate the backup type, the fuels used, and the expected length of operations.
Is physical security provided for the radio communication system's remote components?
  If yes, specify the security type and the protection level provided.
Are alarms at the radio communications system's remote components activated by _____?
  Intrusion
  Loss of electric power
  Loss of environmental control
  Fuel reserves
  If yes, specify the alarm types, how alarms are monitored, and the response procedure.

Commercial Telecommunications Carriers
Are multiple telecommunications carriers used by the asset/facility (possibly commercial, contracted, or organization-owned)?
List and specify the service provided or the information type carried _____
  Analog telephone voice and FAX
  Digital telephone voice
  Internet connections
  Dedicated data transfer

Table C.5   Telecommunications (Continued)

List the media type used
  Copper cable
  Fiber-optic cable
  Microwave
  Satellite

Pathways of Commercial Telecommunications Cables
Are the telecommunications cables into the asset/facility's area and into the asset/facility
  itself aboveground (on utility poles), buried, or a combination?
If both, indicate locations of the portions that are aboveground.
Do the telecommunications cables follow independent pathways into the asset/facility's area
  and into the asset/facility itself?
 If not, indicate how independent the pathways are (some common corridors, intersect at one
  or more points).
Are the paths of the telecommunications cables co-located with other infrastructures' rights-of-
  way?
If yes, describe the extent of the co-location and indicate the other infrastructures.
Are the telecommunications cables' paths located in areas susceptible to natural or
  accidental damage?
  Overhead cables near highways
  Cables across bridges
  Dams
  Landslide areas
If yes, indicate the locations and potential disruption types.
To reach the communications transmission backbones, do the telecommunications carriers
  and cable pathways use separate independent _____
  End offices (EO)
  Access tandems (AT)
  Points of presence (POP)
  Network access points (NAP)
Briefly describe the independence (extent).

Historical Reliability of Commercial Carriers
Historically, has the public switched network (PSN) telephone system in the area been
  reliable? Quantify both complete outages and dropped connections.
Typically, when telephone outages occur, are outages of significant duration (as opposed to
  just a few seconds or minutes)?
  Quantify potential effects on the critical functions and activities at the asset/facility.
Historically, have the Internet and dedicated data transfer systems in the area been reliable?
  Quantify both complete outages and dropped connections.
When Internet or data transfer connectivity outages or disruptions occur, are these outages
  of significant duration (as opposed to just a few seconds or minutes)?
  Quantify potential effects on the critical functions and activities at the asset/facility.

Backup Communications Systems
Are redundant or backup telephone systems in place if the primary system is disrupted?
  Specify the extent to which the secondary systems can support the critical functions and
    activities at the asset/facility.
Are redundant or backup Internet and dedicated data transfer systems in place if the primary
  systems are disrupted?
  Specify the extent to which the secondary systems can support the critical functions and
    activities at the asset/facility.

Table C.6   Transportation

Road Access

Are multiple roadways provided into the asset/facility area from the major highways and interstates?

Describe the routes and indicate any load or throughput limitations with respect the asset/facility's needs.

Are choke points or potential hazard areas located along these roadways?

   Tunnels

   Bridges

   Dams

   Low-lying fog areas

   Landslide areas

   Earthquake faults

Describe the constrictions or hazards and indicate if, historically, closures have occurred somewhat regularly.

Road Access Control

Could intruders or others determined to do damage to the asset/facility gain access to the asset/facility or nearby areas by road without being readily identified and controlled?

If yes, describe the access means.

Indicate limitations on the number of people, vehicle size and number, and material size/quantity that could approach the asset/facility via road.

Are uncontrolled parking lots or open areas for parking near the facility? Could vehicles park without drawing significant attention?

If yes, indicate the number of vehicles and the vehicle size or types that would begin to be noticed.

Rail Access

Are multiple rail routes present into the asset/facility's area from the nearby rail yards or switchyards?

Describe the route or routes and indicate any load or throughput limitations with respect the asset/facility's needs.

Are any choke points or potential hazard areas present along these rail right-of-ways?

   Tunnels

   Bridges

   Dams

   Landslide areas

   Earthquake faults

Is sufficient rail siding space provided to accommodate rail cars at or near the asset/facility:

   if the number of incoming cars exceeds normal expectations?

   if outgoing cars are not picked up as normally scheduled?

Indicate the excess capacity magnitude in terms of the time period before the asset/facility's critical functions or activities would be affected.

Rail Access Control

Could intruders or others determined to do damage to the asset/facility gain access to the asset/facility or nearby areas by rail without being readily identified and controlled?

If yes, describe the access means, and indicate any limitations on the number of people and rail cars that could approach the asset/facility by rail.

Are railroad tracks or sidings present near the asset/facility? Could rail cars be positioned without drawing significant attention?

If yes, indicate the number and the rail car types that would begin to be noticed.

Table C.6   Transportation (Continued)

Airports and Air Routes

Are multiple airports present in the area? Are these airports of sufficient size and with sufficient service to support the asset/facility's critical functions and activities? Enumerate the airports and indicate any limitations.

Are any regular air routes used that pass over or near the asset/facility? Could these routes' usage present a danger to the asset/facility (if an air disaster occurs)? Record any concerns.

Waterway Access

Are multiple water routes present from the asset/facility's provided ports, harbors, or landings to the open ocean or major waterway?

Describe the routes and indicate any load, draft, beam, or throughput limitations with respect to the organization's needs.

Are choke points or potential hazard areas present along these waterways such as bridges, draw or lift bridges, locks and dams, low-lying fog areas, or landslide areas?

Bridges

Draw or lift bridges

Locks and dams

Low-lying fog areas

Landslide areas

Is sufficient mooring, wharf, or dock space provided at the ports, harbors, or landings to accommodate ships or barges:

if the number of incoming vessels exceeds normal expectations?

if outgoing barges are not picked up as normally scheduled?

Indicate the excess capacity magnitude in terms of the time period before the asset/facility's critical functions or activities would be affected.

Waterway Access Control

Could intruders or others determined to do damage to the asset/facility gain access to the asset/facility or nearby areas by water without being readily identified and controlled?

If yes, describe the access means.

Indicate any limitations on the number of people, the vessel size and number, and the material size/quantity that could approach the asset/facility via water.

Are uncontrolled docks/mooring areas located near the asset/facility or the ports/harbors/landings?

Are these uncontrolled docks/moorings used by the asset/facility, and could vessels moor without drawing significant attention?

If yes, indicate the number of vessels and the size or vessel types that would begin to be noticed.

Pipeline Access

What materials, feedstocks, or products are supplied to or from the asset/facility via pipeline transportation?

For the above, do include as products; crude oil, intermediate petroleum products, refined petroleum products, or liquefied petroleum gas.

For the above, do not include water, wastewater, or natural gas unless special circumstances exist related to these items.

Are multiple pipelines and pipeline routes provided into the asset/facility area from major interstate transportation pipelines?

If yes, indicate which pipelines or pipeline combinations have sufficient capacity to serve the asset/facility.

(Continued)

Table C.6   Transportation (Continued)

List the pipeline owners/operators, and indicate the service types (dedicated or scheduled shipments).

Describe the route or routes, and indicate any capacity limitations with respect the asset/facility's needs.

Are bottlenecks or potential hazard areas located along these pipeline or pipeline routes?
  For the above, do include interconnects, terminals, tunnels, bridges, dams, landslide areas, or earthquake faults.

Describe the constrictions or hazards and indicate if, historically, outages or delays have occurred somewhat regularly.

Pipeline Access Control

Could intruders or others determined to bring down the asset/facility gain access to the pipeline near the asset/facility or elsewhere along the pipeline route?

Describe the protective measures in place, and indicate any pipeline segments or facilities (pump stations, surge tanks) of concern.

Table C.7   Water and Wastewater

Primary Domestic Water System

Does the asset/facility have a domestic water system?
  If yes, specify the water uses (e.g., restrooms, locker rooms, kitchens, HVAC makeup water).

Does the water supply for the domestic water system come from an external source (community, city, or regional water mains) or from an internal system (wells, river, or reservoir)?
  If internal, describe the system.

Domestic Water Supply (External)

What external water supply system type provides the domestic water?
  Indicate if the domestic water system is public or private.
  Indicate the general size (community, city, or regional).

Are on-site pumps and/or storage tanks used to boost the pressure or provide for periods of peak usage?
  If yes, briefly describe the equipment and the equipment's purpose.

Are the on-site booster water pumps normally dependent on the asset's/facility's primary electric power supply and distribution system?

Are multiple electric supply sources provided explicitly for the on-site booster water pumps? If yes, specify.
  Multiple independent commercial feeds
  Backup generators
  UPSs

Is a special UPS provided?
  Can this UPS support the on-site booster pumps at required levels?
  Specify for how long the UPS can operate on battery power.

Is a special backup generator system provided?
  Can this backup generator support the on-site booster pumps at required levels?
  Also indicate the fuel type used.

Is petroleum fuel used for the dedicated backup generator (for the booster pumps)?
  Indicate the fuel quantity stored on-site and how long the fuel would last.
  Is the fuel for the booster pumps' dedicated backup generator stored on-site?

Are arrangements and contracts in place for fuel resupply and management?

Table C.7   Water and Wastewater (Continued)

Domestic Water Supply (Internal)

Indicate the water source (e.g., wells, river, or reservoir), the adequacy of the supply's
  capacity, and whether water is gravity fed or requires active pumps (generally electric).
Are the on-site domestic water system pumps independent of the asset's/facility's primary
  electric power supply and distribution system?
Are multiple electric supply sources provided explicitly for the on-site domestic water system
  pumps? If yes, specify.
   Multiple independent commercial feeds
   Backup generators
   UPSs
Is a special UPS provided? Can the UPS support the on-site domestic water system pumps
  at required levels? Specify for how long the UPS can operate on battery power.
Is a special backup generator system provided?
   Can the backup generator support the on-site domestic water system pumps at the
     required levels?
   Also, indicate the fuel or fuel type used.
Is petroleum fuel is used for the dedicated backup generator system (for the on-site domestic
  water system pumps)?
   Indicate the quantity stored on-site and how long the fuel would last.
Are arrangements and contracts in place for fuel resupply and management?

Backup Domestic Water System

Is an independent backup water source provided for the primary domestic supply system?
If yes, specify the backup system type.
   Wells
   River
   Reservoir
   Tank truck
Describe the specific water source. Indicate the adequacy of the backup supply's capacity.
Indicate if the water is gravity fed or requires active pumps (generally electric).
Are the independent backup water source system pumps independent of the asset's/facility's
  primary electric power supply and distribution system?
Are multiple electric supply sources provided explicitly for the backup water source system
  pumps? If yes, specify.
   Multiple independent commercial feeds
   Backup generators
   UPSs
Is a special UPS provided?
   Can this UPS support the backup domestic water source pumps at the required levels?
Specify for how long the backup UPS can operate on battery power.
Is a special backup generator system provided?
   Can this backup generator support the backup domestic water source system pumps at
     the required levels?
Also, indicate the fuel types used.
Is petroleum fuel used for the dedicated backup generator system (for the backup water
  source system pumps)?
   Indicate the quantity stored on-site and how long the fuel would last.
Are arrangements and contracts in place for fuel resupply and management?

Primary Industrial Water System

Does the asset/facility have an industrial water system?
   If yes, specify the water uses (e.g., wash water, process water, process steam generation, cooling).

(Continued)

Table C.7   Water and Wastewater (Continued)

Does the water supply for the industrial water system come from an external source
  (community, city, regional water mains) or from an internal system (wells, river, reservoir)?
    If internal, describe the system.

Industrial Water Supply (Internal)
What external water supply system type provides the industrial water?
  Indicate whether this system is public or private. Indicate general size (community, city,
    regional).
Are on-site pumps and/or storage tanks used to boost the pressure or provide for peak
usage periods?
  If yes, briefly describe the pumps and their purpose.
Are the on-site booster water pumps for the industrial water system independent of the
  asset's/facility's primary electric power supply and distribution system?
Are multiple electric supply sources provided explicitly for the on-site booster water pumps? If
yes, specify.
  Multiple independent commercial feeds
  Backup generators
  UPSs
Is a special UPS provided?
  Can the UPS support the on-site booster pumps at required levels?
Specify for how long the UPS can operate on battery power.
Is a special backup generator system provided?
  Can the backup generator support the on-site booster pumps at required levels?
  Also, indicate the fuel types.
Is petroleum fuel used for the dedicated backup generator system (for the booster pumps)?
  Indicate the quantity stored on-site and how long the fuel would last.
Are petroleum fuel arrangements and contracts in place for fuel resupply and management?

Industrial Water Supply (External)
Indicate the water sources (e.g., wells, river, reservoir), the adequacy of the supply's
  capacity, and whether the water is gravity fed or requires active pumps (generally electric).
Are the on-site industrial water system pumps independent of the asset's/facility's primary
  electric power supply and distribution system?
Are multiple electric supply sources provided explicitly for the on-site industrial water system
  pumps? If yes, specify.
  Multiple independent commercial feeds
  Backup generators
  UPSs
Is a special UPS provided?
Can the UPS support the on-site industrial water system pumps at required levels?
Specify for how long the UPS can operate on battery power.
  Is a special backup generator system provided?
Can the backup generator support the on-site industrial water system pumps at the required
  levels?
  Also, indicate the fuel types.
Is petroleum fuel used for the dedicated backup generator system (for the on-site industrial
  water system pumps)?
  Indicate the quantity stored on-site and how long the fuel would last.
Are arrangements and contracts in place for fuel resupply and management?

Backup Industrial Water System
Is an independent backup water source provided for the primary industrial water supply
  system?

Table C.7   Water and Wastewater (Continued)

If yes, specify the backup system type (e.g., wells, river, reservoir, tank truck).
Describe the specific water source.
Indicate the adequacy of the backup supply's capacity.
 Indicate if the water is gravity fed or requires active pumps (generally electric).
Are the independent backup water source system pumps independent of the asset's/facility's primary electric power supply and distribution system?
Are multiple electric supply sources provided explicitly for the backup water source system pumps? If yes, specify.
 Multiple independent commercial feeds
 Backup generators
 UPSs
Is a special UPS provided?
 Can the UPS support the backup industrial water source pumps at the required levels?
Specify for how long the UPS can operate on battery power.
Is a special backup generator system provided?
 Can the backup generator support the backup industrial water source system pumps at required levels?
 Also, indicate the fuel types.
Is petroleum fuel used for the dedicated backup generator system (for the backup water source system pumps)?
 Indicate the quantity stored on-site and how long the fuel would last.
Are arrangements and contracts in place for fuel resupply and management?

Primary Industrial Wastewater System
Does the asset/facility have an on-site industrial wastewater system?
 If yes, specify the wastewater types that are processed and the processes used.
Are the on-site industrial wastewater lift pumps independent of the asset's/facility's primary electric power supply and distribution system?
Are multiple electric supply sources provided explicitly for the on-site industrial wastewater lift pumps? If yes, specify.
 Multiple independent commercial feeds
 Backup generators
 UPSs
Is a special UPS provided?
Can the UPS support the on-site industrial wastewater lift pumps at required levels?
Specify for how long the UPS can operate on battery power.
Is a special backup generator system provided?
 Can the backup generator support the on-site industrial wastewater lift pumps at the required levels?
 Also, indicate the fuel types.
Is petroleum fuel used for the dedicated backup generator system (for the on-site industrial wastewater lift pumps)?
Indicate the quantity stored on-site and how long the fuel would last.
Are arrangements and contracts in place for fuel resupply and management?

Backup Wastewater System
Is an independent backup system provided to handle industrial wastewater?
If yes, specify the backup system type.
 Redundant system
 Holding ponds
 Temporary unprocessed wastewater discharge

(Continued)

Table C.7   Water and Wastewater (Continued)

Describe the specific process. Indicate the adequacy of the backup's capacity and any
  limitations on how long the backup generator can operate.
Indicate if the process is gravity fed or requires active lift pumps (generally electric).
Are backup lift pumps independent of the asset's/facility's primary electric power supply and
  distribution system?
Are multiple electric supply sources provided explicitly for the backup wastewater lift pumps?
  If yes, specify.
    Multiple independent commercial feeds
    Backup generators
    UPSs
Is a special UPS provided?
Can the UPS support the backup industrial wastewater system at the required levels?
Specify for how long the UPS can operate on battery power.
Is a special backup generator system provided?
    Can the backup generator support the backup industrial water source system pumps at
      required levels?
    Also, indicate the fuel types.
Is petroleum fuel used for the dedicated backup generator system (for the backup wastewater
  lift pumps)?
Indicate the quantity stored on-site and how long the fuel would last. Are arrangements and
  contracts in place for fuel resupply and management?

Commercial/Public Water Supply Reliability
Historically, has the city water supply in the area been reliable and adequate?
Quantify the reliability and specify any shortfall in the supply pressure or flow rate.
Typically, when disruptions in the city water supply occur, are these disruptions of significant
  duration (as opposed to just a few hours)?
Quantify potential effects on the critical functions and activities at the asset/facility.

Commercial/Public Wastewater System Reliability
Historically, has the public wastewater system in the area been reliable and adequate?
    Quantify the reliability and specify any shortfall in the system capacity.
Typically, when disruptions in the public wastewater system occur, are these disruptions of
  significant duration (as opposed to just a few hours)?
    Quantify potential effects on the critical functions and activities at the asset/facility.
Are contingency plans or procedures in place to handle domestic wastewater from the asset/
  facility if the public system is temporarily unable to accept the waste?
    If yes, describe the contingency plans.
Mention any limitations on wastewater quantity and outage duration that might affect the
  asset/facility's ability to maintain critical functions or activities.

Table C.8   Emergency Services

How are the local police involved in protecting the asset/facility?
What are typical response times and response capabilities?
Have local police provided services in the past?
    Has their response been helpful?

County/State Police
How are the county/state police involved in protecting the asset/facility?
What are typical response times and response capabilities?
Have county/state police provided services in the past?
    Has their response been helpful?

Table C.8   Emergency Services (Continued)

Federal Bureau of Investigation (FBI)
How is the FBI involved in protecting the asset/facility?
What are typical response times and response capabilities?
Has the FBI provided services in the past?
   Has their response been helpful?

Fire Department
How is the local fire department involved in protecting the asset/facility?
Do the local fire departments provide inspection and/or certification services?
What are their typical response times and response capabilities?
Have the fire departments provided services in the past?
   Has their response been helpful?

Emergency Medical Services
How is the local emergency medical or ambulance service involved in protecting/treating the
   personnel at the asset/facility?
Do the local medical services provide inspection and/or certification services?
What are their typical response times and response capabilities?
Have the local medical services provided services in the past?
   Has their response been helpful?

Table C.9   Computers and Servers

Are the asset's/facility's primary electric power supply and distribution systems provided to
   supply power for the computers and servers?
   If yes, indicate under what conditions and for how long.
Do the computers and servers have a dedicated backup electric power supply (local UPSs,
   generators)?
   If yes, specify the backup types and how long the backup power supplies can operate.

Environmental Control
Does the asset's/facility's central HVAC system provide environmental control to the
   computer and server areas?
Do the computer and server areas have a dedicated independent environmental control
   system?
   Specify the type.
Can the computers and servers operate with a loss of all environmental control?
   If yes, specify for how long and under what conditions.
Are any backup environmental controls provided explicitly for the computer and server areas?
If yes, indicate the backup type and the expected maximum operation duration.

Protection
Is special physical security provided for the computer and server areas?
   If yes, specify the security type and the protection level provided.
Is special fire suppression equipment (Halon, Inergen, inert gases, or carbon dioxide)
   provided in the computer and server areas?
   If yes, specify the type.
Are special features or equipment provided in the computer and server areas to limit flooding
   or water intrusion? If yes, describe.
Are alarms provided for the computer and server areas?
Are there alarms for unauthorized intrusion, electric power loss, environmental control loss,
   fire, and flooding or water intrusion?
   If yes, specify the alarm types, how the alarms are monitored, and the response procedure.

Table C.10   HVAC System

Primary HVAC System

Can critical functions and activities that depend on environmental conditions continue without the HVAC system?

  If yes, specify which functions and how long these functions can continue under various external weather conditions.

Is the HVAC system that supplies the areas of the asset/facility where critical functions that depend on environmental conditions are carried out separate?

  For the above, *separate* means separate from or separable from the general asset-/facility-wide HVAC system.

Supporting Infrastructures

Does the HVAC system (or critical portion thereof) depend on the primary electric power supply and distribution system to supply electric power?

  Specify under what conditions and for how long.

Besides (or in addition to) electric power, what fuel or fuels does the HVAC system (or critical portion thereof) depend on?

Does the HVAC system (or critical portion thereof) depend on natural gas?

  Are provisions in place to provide alternative fuels during a natural gas outage?

  Specify the fuel and how long the HVAC system can operate.

Does the HVAC system (or critical portion thereof) depend on petroleum fuels for adequate operation?

  Specify the type of fuel and how long the HVAC system can operate on the fuel provided on-site.

Are arrangements and contracts in place for resupply and management of the fuel?

Does the HVAC system (or critical portion thereof) depend on water?

  Specify if the water need is continuous or for makeup purposes only and the quantities/rates involved.

Is a backup supply in place (e.g., well and pump, storage tank, or tank trucks)?

  Specify how long the HVAC can operate on the backup water supply system.

Backup HVAC Systems

Is a separate backup provided for the HVAC system?

If yes, describe the system and the energy and water supply systems the backup system requires.

Are contingency procedures in place to continue with the critical functions and activities that take place at the asset/facility during an HVAC outage?

  If yes, briefly describe.

How long can the critical functions and activities at the asset/facility continue using the backup HVAC system or under the contingency procedures?

Table C.11   Fire Suppression and Fire-Fighting System

Alarms

Does the entire asset/facility have a fire and/or smoke detection and alarm system?

  If yes, specify the system type, how the system is monitored, and the response procedure.

Fire Suppression

Does the entire asset/facility have a fire suppression system (e.g., overhead sprinkler system)?

  If yes, specify the medium (usually water).

  Is the sprinkler system the flooded-pipe or pre-armed type?

Table C.11 Fire Suppression and Fire-Fighting System (Continued)

Does the water supply for the fire suppression system come from city water mains or an on-site system (wells, rivers, reservoir)?

Does the water supply for the fire suppression system come from city water mains?

Specify whether separate city fire mains are provided, and if the pipe from the main to the asset/facility is separate from the domestic water supply.

Does the water supply for the fire suppression system come from an on-site system?

Specify the source, indicate the adequacy of the supply's capacity, and indicate if water is gravity fed or requires active pumps (generally electric).

Fire Fighting

Does the asset/facility have a fire-fighting department?

If yes, describe this department in terms of its adequacy to protect the asset/facility.

Are city or community fire-fighting services provided to the facility?

If yes, indicate the service type and the estimated response time.

Does the water supply for the fire-fighting hydrants come from city water mains?

If yes, specify the hydrant numbers, coverage, and accessibility.

Does the water supply for the fire hydrants come from an on-site system (wells, rivers, reservoir)?

Specify the source, indicate the adequacy of the supply's capacity, and indicate if water is gravity fed or requires active pumps (generally electric).

Also, specify the hydrants numbers, coverage, and accessibility.

Other Systems

Is special fire suppression equipment provided in certain areas such as computer or telecommunications areas?

Halon

Inergen

Inert gases

Carbon dioxide

If yes, indicate the types and adequacies of these special systems.

Table C.12 SCADA System

Type of System

Does the asset/facility make use of a substantial SCADA system (i.e., one that covers a large area or a large number of components and functions)?

If yes, indicate what functions are monitored and/or controlled, the system type, and the system extent.

Is the SCADA system independent of the asset's/facility's primary electric power supply and distribution system?

Is the SCADA system independent of the asset's/facility's telephone system?

Is the SCADA system independent of the asset's/facility's microwave or radio communications system?

Is the SCADA system independent of the asset's/facility's computers and servers?

Control Centers

Where is the primary control center for the SCADA system located?

Is a backup control center provided?

If yes, where is the backup control center located?

(Continued)

Table C.12   SCADA System (Continued)

Is the backup control center sufficiently remote from the primary control center to avoid
common failures (fires, explosions, other large threats)?

Are backups to the SCADA computers and servers provided at the backup control center or
at some other location?

   If yes, indicate the location(s) of the backup computers and servers.

   If yes, indicate whether they are completely redundant or cover only the most critical
   functions.

   If yes, indicate whether they are active "hot" standbys or have to be activated and initialized
   when needed.

Note: The following questions on electric power sources and communications pathways
apply to the control centers as well as the other SCADA system components.

Electric Power Sources

Are multiple electric supply sources provided explicitly for the SCADA system? If yes, indicate
the types.

   Multiple independent commercial feeds

   Backup generators

Communications Pathways

Is a special UPS provided?

Does the UPS support all the SCADA system functions in terms of capacity?

   Specify for how long the UPS can operate on battery power,

Is a special backup generator system provided?

   Does this backup generator system support all the SCADA system functions in terms
   of capacity?

What is the fuel or fuels used by the special SCADA backup generator system?

   If stored on-site, specify the quantity stored and how long it would last.

If the SCADA backup generator fuel is stored on-site, are arrangements and contracts in
place for fuel resupply and management?

Communications Pathways

Are dedicated multiple independent telephone systems or dedicated switches and cables
supporting the SCADA system provided?

   If yes, specify whether copper-wire or fiber-optic based.

If dedicated multiple independent telephone systems or dedicated switches and cables
supporting the SCADA system are provided, is each one individually adequate to support
the entire system?

   Specify any limitations.

Are the redundant telephone systems or switches and cables physically separated and
isolated to avoid common failures?

If not, indicate any potential common failure points.

Are the dedicated SCADA telephone switches and data-transfer switches located in a limited
access or secured area away from potential damage due to weather or water leaks?

   If so, specify the protection type.

Are dedicated multiple or redundant radio communications systems in place to support the
SCADA system? If yes, indicate the types.

If multiple radio communications systems are provided, is each one individually adequate to
support the entire SCADA system?

   If not, specify any limitations.

Are provisions in place within the asset's/facility's primary electric power supply and
distribution system to supply power for the special SCADA radio communications systems?

   If yes, specify under what conditions and for how long.

Do the special SCADA radio communications systems have dedicated backup electric
power supply?

Table C.12   SCADA System (Continued)

If yes, specify the type and how long the SCADA can operate.

Are the special SCADA radio communications system components (antennae, on-site towers) located outside of buildings protected (from vandalism/accidental damage) by fences or barriers?

If protected, specify the protection types and security level provided.

Remote Components

Are remote components to the special SCADA radio communications system (e.g., relay towers) provided?

If yes, identify the components and their locations.

Are backup electric power sources provided for these remote components?

If yes, indicate the backup type, the fuels used, and the expected length of operations.

Are environmental controls required for the special SCADA radio communications system's remote components (heating, cooling)?

If yes, describe the environmental controls.

Are backup environmental controls for the remote environmental components provided?

If yes, indicate the backup type, the fuels used, and the expected length of operations.

Is physical security provided for the special SCADA radio communications system's remote components?

If yes, specify the security types and the protection level provided.

Are alarms provided at the remote components of the special SCADA radio communications system for _____?

Intrusion

Loss of electric power

Loss of environmental control

Loss of fuel reserves

If yes, specify the alarm types, how alarms are monitored, and alarm response procedure.

Dedicated SCADA Computers and Servers

Are provisions within the asset's/facility's primary electric power supply and distribution system provided to supply power for the special dedicated SCADA computers and servers?

If yes, specify under what conditions and for how long.

Do the special dedicated SCADA computers and servers have a dedicated backup electric power supply, such as local UPSs?

If yes, specify the types and how long the SCADA can operate.

Does the asset's/facility's central HVAC system provide environment control for the separate special SCADA computer and server areas?

How long can the separate dedicated SCADA computers and servers operate with a loss of all environmental control?

Indicate the conditions that could affect the length of time.

Do the separate dedicated SCADA computer and server areas have dedicated independent environmental control system?

If yes, specify the type.

Are any backup environmental controls provided explicitly for the dedicated SCADA computer and server areas?

If yes, indicate the backup type and the expected maximum duration of operation.

Is special physical security provided for the separate SCADA computer and server areas?

If yes, specify the security type and the protection level provided.

Is special fire suppression equipment provided in the separate dedicated SCADA computer and server areas?

(Continued)

Table C.12   SCADA System (Continued)

Halon
Inergen
Inert gases
Carbon dioxide
If yes, specify the type of system.
Are special features or equipment provided in the separate SCADA computer and server
  areas to limit flooding or water intrusion?
  If yes, indicate the precautions taken.
Are alarms provided for the separate SCADA computer and server areas for _____?
  Unauthorized intrusion
  Loss of electric power
  Loss of environmental control
  Fire
  Flooding
  Water intrusion
If yes, specify the alarm types, how alarms are monitored, and the alarm response
  procedure.

Table C.13   Physical Security System

Electric Power Sources
Are the asset's/facility's monitoring and alarm systems normally dependent on the asset's/
  facility's primary electric power supply and distribution system?
  Is the asset's/facility's primary electric power supply and distribution system the primary
    electric power source?
Are multiple electric power sources provided for the monitoring and alarm systems?
  Asset's/facility's primary electric power supply and distribution system
  Backup or redundant systems or multiple independent commercial electric feed combinations
  Backup generators
  UPSs dedicated to support the monitoring and alarm systems
  Batteries dedicated to support the monitoring and alarm systems. Specify what electric
    power sources are in place.
Specify what electric power sources are in place.
Is a special UPS provided? Can the UPS support all the monitoring and alarm systems'
  functions in terms of capacity?
  Specify for how long the UPS can operate on battery power.
Is a special generator system provided?
  Can this generator system support all the monitoring and alarm systems' functions in terms
    of capacity?
  Also, indicate the fuel or fuels types used.
Is petroleum fuel used for the special security generator system?
  Indicate the quantity stored on-site and how long the fuel would last.
Are arrangements and contracts in place for fuel resupply and management?

Communications Pathways
Are the asset's/facility's monitoring and alarm systems normally dependent on the asset's/
  facility's telephone system?
Are multiple independent telephone systems or dedicated switches and cables provided for
  supporting the monitoring and alarm systems?
  Asset's/facility's telephone system
  Backup or redundant systems

Table C.13 Physical Security System (Continued)

Are the redundant telephone systems or switches and cables physically separated and isolated to avoid common causes of failure?

If not, indicate any potential points of common failure.

Specify the system types used and whether they are copper-wire or fiber-optic cable based.

Are the redundant telephone systems or switches and cables physically separated and isolated to avoid common failures? If not, indicate any potential common failure points.

Are the dedicated monitoring and alarm systems' telephone switches and data-transfer switches located in a limited access or secured area?

Is this area away from potential damage due to weather or water leaks?

If so, specify the protection type.

Are the asset's/facility's monitoring and alarm systems normally dependent on the asset's/facility's microwave or radio communications system?

Are multiple independent microwave or radio communications systems provided for supporting the monitoring and alarm systems?

Asset's/facility's primary microwave or radio communications system

Backup or redundant systems or multiple independent radios, antennae, and relay towers combinations

Specify the radio systems used.

Are multiple electric power sources provided for the microwave or radio communications systems dedicated to support the monitoring and alarm systems?

Electric power supply and distribution system

Backup or redundant systems or multiple independent commercial electric feed combinations

Backup generators

UPSs

Batteries dedicated to support the special microwave or radio communications systems

If yes, specify the types and how long they can operate.

Are the special radio communications system components dedicated to the monitoring and alarm systems located outside of buildings?

Note: For the above, *outside* is defined as antennae, on-site towers.

Are these protected from vandalism or accidental damage by fences or barriers?

If protected, specify the protection types and security level provided.

Are the special radio communications system remote components dedicated to the monitoring and alarm systems (e.g., relay towers)?

If yes, identify the components and their locations.

Are backup electric power sources provided for the remote components?

If used, indicate the backup type, the fuels used, and the expected length of operations.

Are environmental controls required for the special monitoring and alarm radio communications system (heating, cooling) remote components? If yes, describe.

Are backup environmental controls provided for the remote components?

If yes, indicate the backup type, the fuels used, and the expected length of operations.

Computer Support

Are the asset's/facility's monitoring and alarm systems normally dependent on the facility's main computers and servers?

Are multiple independent computers provided supporting the monitoring and alarm systems?

Asset's/facility's main computers and servers

Their backup or redundant systems or multiple independent computer combinations

Asset's/facility's main computers and servers and their backup or redundant systems or multiple independent computer combinations

Specify the computer types used.

(Continued)

Table C.13   Physical Security System (Continued)

Are multiple electric power sources provided for any computers dedicated to support the monitoring and alarm systems?
  Asset's/facility's primary electric power supply and distribution system
  Backup or redundant systems or multiple independent commercial electric feed combinations
  Backup generators, or UPSs dedicated to support the monitoring and alarm systems
  UPSs dedicated to support the monitoring and alarm systems
  If yes, specify the type and how long the UPS can operate.
Does the asset's/facility's central HVAC system provide environment control (for the separate dedicated computers) for the monitoring and alarm systems?
How long can the separate dedicated computers of the monitoring and alarm systems operate with a loss of all environmental control?
  Indicate the conditions that could affect the length of time.
Do the separate dedicated computers for the monitoring and alarm systems have a dedicated independent environmental control system? If yes, specify the type.
Are backup environmental controls provided explicitly for any monitoring and alarm systems' dedicated computers?
  If yes, indicate the backup type and the expected maximum duration of operation.

Table C.14   Financial System

Electric Power Sources

Are the asset's/facility's financial systems and functions normally dependent on the asset's/facility's primary electric power supply and distribution system?
Is the facility's electric power supply and distribution system the primary electric power source?
Are multiple electric power sources provided for the financial systems and functions?
  Facility's electric power supply and distribution system
  Backup or redundant systems or multiple independent commercial electric feed combinations
  Backup generators
  UPSs
Specify what electric power sources are in place.
Is a special UPS provided?
  Can the UPS support all the financial systems and functions?
  Specify how long the UPS can operate on battery power.
Is a special generator system provided?
  Can this generator system support all the financial systems and functions?
  Also, indicate the fuel types used.
Is petroleum fuel used for the special security generator system?
  Specify the quantity stored and how long the fuel would last.
Are arrangements and contracts in place for fuel resupply and management?

Communications Pathways

Are the asset's/facility's financial systems and functions normally dependent on the asset's/facility's telephone system?
Are multiple independent telephone systems or dedicated switches and cables provided supporting the financial systems and functions?
  Facility's telephone system and the telephone system's backup or redundant systems or multiple independent telephone system combinations
  Dedicated communications lines
Specify the system types used and whether the systems are copper-wire or fiber-optic cable based.

Table C.14   Financial System (Continued)

Are the redundant telephone systems or switches and cables physically separated and
isolated to avoid common failures?
   If not, indicate any potential failures.
Are the telephone switches and data-transfer switches that support the financial systems and
functions dedicated?
Are these located in a limited access or secured area away from potential damage due to
weather or water leaks?
   If so, specify the protection type.

Computer Support
Are the asset's/facility's financial systems and functions normally dependent on the facility's
main computers and servers?
Are multiple independent computers supporting the financial systems and functions
provided?
   Facility's main computers and servers
   Backup or redundant systems or multiple independent computer combinations
Specify the computer types used.
Are multiple electrical supply sources provided for any computers dedicated to support the
financial systems and functions?
   Asset's/facility's primary electric power supply and distribution system
   Backup or redundant systems or multiple independent commercial electric feeds
    combinations
   Backup generators
   UPSs dedicated to support the financial systems and functions
   If yes, specify the type and how long the systems can operate.
Does the asset's/facility's central HVAC system provide environmental control for any
separate, dedicated computers?
   Do these computers support the financial systems and functions?
How long can the separate, dedicated computers that support the financial systems and
functions operate with a loss of any environmental control?
Indicate the conditions that could affect the length of time.
Do the separate, dedicated computers that support the financial systems and functions have
a dedicated independent environmental control system?
If so, specify the type.
Are any backup environmental controls provided explicitly for the dedicated computers that
support the financial systems and functions?
If yes, indicate the backup type and the expected maximum duration of operation.

APPENDIX D

Table D.1   Asset Attractiveness Scale

Asset Attractiveness Scale Value	Example
Extremely attractive = 5	Potentially devastating economic and public confidence impacts
Very attractive = 4	Potentially substantial economic impacts and public confidence impacts
Attractive = 3	Potentially significant economic impacts and regional public confidence impacts
Less attractive = 2	Potentially nonnegligible economic impacts and some local bad publicity

CHAPTER 3

Regulations, Building Codes, Industry Standards
Compliance Documentation

**Martha Boss, Dennis Day, Charles Allen,
Robert Araujo, and Lawrence Fitzgerald**

CONTENTS

During vulnerability assessments, the various code requirements and current compliance should be assessed. Any building or site component (i.e., utility element, computer systems, chemical use) that makes a facility more vulnerable in a nonemergency situation, will likewise increase vulnerability during emergency situations. All utility elements, such as electrical transformers, emergency generators, backflow preventers and meters, are generally located with access convenient to the utility companies. Their location and the location of control elements should be evaluated given security considerations. These control elements must be readily accessible for some emergency situations and must also not be so hidden that untoward access is not noticed.

To address attendant chemical hazards, many of the chemical lists provided by compliance regulations can be used to assess vulnerability. Lists to examine are found in regulations promulgated by the U.S. Department of Homeland Security (DHS), the U.S. Environmental Protection Agency (EPA), the U.S. Occupational Safety and Health Administration (OSHA), and the U.S. Department of Transportation (DoT). If your facility has any of these listed items in any quantity, your facility may have some vulnerability—even if not specifically regulated under the specific regulations that included the lists of note.

3.1   DEPARTMENT OF HOMELAND SECURITY

The U.S. DHS is responsible for developing and implementing the National Infrastructure Protection Plan (NIPP) as well as coordinating the protection of critical infrastructure and key resources (CI/KR). Most of this responsibility stems from the Homeland Security Act of 2002 and a series of Homeland Security Presidential Directives (HSPDs) signed by President George Bush. For example, under HSPD-7, various federal entities such as the Departments of Defense, Agriculture, Interior, Treasury, Energy, Health and Human Services, and the EPA are assigned as coordinating agencies responsible for various sectors of the U.S. economy and society (for example, food sector or health care). However the secretary of DHS is the overall official responsible for coordinating the protection of all sectors and agencies. The result of these various regulations and HSPDs is the development of a series of sector-specific law, rules, regulations, and standards.

3.2 DHS AND CFATS

The DHS established the Chemical Facility Anti-Terrorism Standards (CFATS) in 6 CFR (Code of Federal Regulation) 27 on November 20, 2007. This standard applies to any facility that produces, consumes, stores, or otherwise possesses certain regulated chemicals in quantities above a regulatory threshold, and both the facility owner and the facility operator are responsible for compliance with CFATS. With regard to 6 CFR 27, a *chemical facility* is defined by those chemicals a facility controls, regardless of the overall use of the facility. The standard identifies a list of approximately 300 Chemicals of Interest (COIs) in Appendix A of CFATS. For each COI, DHS has designated the screening threshold quantities (STQ) for the following three different risk-based scenarios:

- Intentional release
- Theft for possible use in the development of chemical weapons
- Sabotage

If any of the chemicals listed in Appendix A of CFATS are identified on-site above the STQ for any of the above three scenarios, the facility may be determined to be "high-risk" by DHS. In addition, certain facilities with chemicals other than those listed as COI in Appendix A, could also be designated a high-risk facility by DHS based upon other criteria such as economic importance/impact, market share, and national defense considerations. The rule applies to solids, liquids, and gases, and applies if the chemicals are present in either concentrated or diluted form, as a component of a mixture, as a byproduct, or as a waste stream. Some chemicals are regulated in amounts less than 1 pound, while others require over 10,000 pounds to be regulated under CFATS.

Currently four main exemptions are provided under CFATS:

1. Facilities whose security is regulated by the U.S. Coast Guard under the Maritime Transportation Security Act (MTSA) under 33 CFR 105
2. Facilities whose security is regulated (on a facilitywide basis) by the Nuclear Regulatory Commission
3. Water treatment facilities whose security is regulated by the EPA
4. Wastewater treatment facilities whose security is regulated by the EPA

Discussion is extant about removing one or more of these exemptions to capture more facilities under CFATS, but currently the four above exemptions exist.

Provisions (6 CFR 27.203-4) for fuels mixtures have been added to the requirements to include fuels with (NFPA Hazard Codes 1, 2, 3, or 4) when stored in aboveground tank farms, including tanks farms that are part of pipeline systems," (6 CFR 27.203(b)(1)(v)). If these facilities possess fuels that contain 1% or more COI at or above the STQ, the facilities must submit a Top-Screen. For example, STQ for butane is 10,000 pounds. A facility that possesses 200,000 pounds (approximately 33,000 gallons) of gasoline containing 5% butane, including product within piping, would be required to complete a Top-Screen.

The final rule specifies three STQ categories, each carrying different applicability rules. Instructions on how to calculate applicable quantities of COI are included in the Federal Register notice. A higher STQ has been specified for propane.

Facilities should conduct an applicability review to determine if CFATS applies to each facility. If CFATS applies, the facility is required to file a *Top-Screen* submission to DHS within 60 days. The Top-Screen is a consequence assessment, meant to identify all facilities within the United States that use any of the chemicals of interest. The online form includes an analysis of the types of chemicals, the containers these chemicals are stored in, flammability ratings, toxicity ratings, and the types of threats that the chemicals pose.

The DHS reviews the Top-Screen forms and applies a risk analysis model. Facilities identified as high risk will be ranked into one of four tiers, with Tier 1 having the highest risk. DHS requires listed high-risk facilities to submit further documentation, including a Security Vulnerability Assessment (SVA) and a Site Security Plan. Note that the process of evaluation and compliance is ongoing, based on changes in the facility and policies. Failure to comply with CFATS can carry significant penalties including $20,000 per day for noncompliance and federal law enforcement officials ordering a facility closed.

Screening chemical facilities (CFs) has two purposes:

1. For individual CFs, the screening determines whether or not a vulnerability assessment (VA) should be conducted.
2. For organizations with more than one CF, the screening determines which CFs should undergo VAs and prioritizes them.

The screening process is based primarily on the possible consequences of potential terrorist incidents at CFs.

1. What is the desired event? For the information presented below, an off-site release was considered.
2. Will the loss of a facility have a significant impact on the nation (for example, is the facility a sole source for a chemical vital to national defense industries)? If the answer is *yes,* the VA information may need to be classified.
3. Does the facility have a total on-site inventory of threshold quantity (TQ) or greater of a chemical covered under Federal Regulation 40 CFR 68.130? If the answer is *no,* a VA probably is not needed, although a CF may decide to do a VA for other reasons. For companies with more than one CF, the screening process should proceed to the other facilities. If the answer is *yes,* further screening is done based on the estimated number of people that would be affected by the worst-case scenario for the Risk Management Plan (RMP).

Further screening is based on the number of people that would be affected by the worst-case scenario. Estimate how many people would be affected by the worst-case scenario from the RMP for toxic substances and assign levels.

Level 1: More than 100,000
Level 2: 10,000–100,000
Level 3: 1,000–9,999
Level 4: Less than 1,000

Other factors considered in the screening step include accessibility, recognizability, and importance to the company, the region, and the nation. The final screening step is to prioritize the CFs that need VAs from Level 1 (highest) to Level 4 (lowest) (DHS, 2007).

CFs that must submit RMPs most likely will be rated at severity Level 1.

3.3   MARITIME TRANSPORTATION SECURITY ACT

The Maritime Transportation Security Act (MTSA) of 2002 regulates security at certain maritime facilities, vessels, and ports under 33 CFR 101-105 and is enforced by the U.S. Coast Guard (USCG). Facilities that handle certain dangerous cargo (such as fuels) or large passenger vessels or foreign flagged vessels are required to conduct Facility Security Assessments (FSAs) and submit Facility Security Plans (FSPs) to USCG for review and approval.

3.4   SUPPORT ANTI-TERRORISM BY FOSTERING EFFECTIVE TECHNOLOGIES ACT (SAFETY ACT)

The Support Anti-Terrorism by Fostering Effective Technologies Act of 2002 (SAFETY ACT) is a mechanism that DHS utilizes to evaluate and potentially approve various security systems, approaches, and services. Firms and organizations that utilize SAFETY ACT–certified services may be eligible for legal relief in the event that certain terrorist actions cause harm to a facility, employees, or the general public.

3.5   DOT AND HAZARDOUS MATERIALS

The requirement for emergency response information is contained in Department of Transportation Regulation 49 CFR, Part 172, Subpart G (DOT, 2001). Written emergency response information must be appropriate for the hazardous material being transported. If the carrier's equipment has an emergency response guide on board, this information may suffice to provide a separate emergency response document.

The emergency information and phone numbers must be maintained at all times that a shipment is in transit. The use of beepers, answering machines, and switchboards is not authorized. The phone number must be to someone capable of providing information on the material being transported.

For transportation by highway, if a transport vehicle contains hazardous materials for which a shipping paper is required and the transport vehicle is separated from its motive power and parked at a location other than a facility operated by the consignee, consignor, or carrier, the carrier must:

- Mark the transport vehicle with the telephone number of the motor carrier on the front exterior near the brake hose or electrical connection; or
- Have the shipping paper and emergency response information readily available on the transport vehicle.

This requirement does not apply if the identification number for each hazardous material contained therein is marked on the outside of the vehicle on an orange panel or white square-on-point placard.

3.5.1 Security Plans

The security plan requirements in Part 172 Subpart I (DOT, 1994, 1996) of the Hazardous Materials Regulations (HMR) require each hazmat employer subject to the security plan requirements to establish and implement a security plan. The employer is also required to train their hazmat employees on the security plan. The purpose of these requirements is to enhance the security of hazardous materials transported in commerce.

The security plan requirements apply to each person who offers for transportation in commerce or transports in commerce one or more of the following hazardous materials:

1. A highway route–controlled quantity of a Class 7 (radioactive) material, as defined in 173.403 of this subchapter, in a motor vehicle, rail car, or freight container;
2. More than 25 kg (55 pounds) of a Division 1.1, 1.2, or 1.3 (explosive) material in a motor vehicle, rail car, or freight container;
3. More than one L (1.06 qt) per package of a material poisonous by inhalation, as defined in 171.8 of this subchapter, that meets the criteria for Hazard Zone A, as specified in 173.116(a) or 173.133(a) of this subchapter;
4. A shipment of a quantity of hazardous materials in a bulk packaging having a capacity equal to or greater than 13,248 L (3,500 gallons) for liquids or gases or more than 13.24 cubic meters (468 cubic feet) for solids;
5. A shipment in other than a bulk packaging of 2,268 kg (5,000 pounds) gross weight or more of one class of hazardous materials for which placarding of a vehicle, rail car, or freight container is required for that class under the provisions of subpart F of this part;
6. A quantity of hazardous material that requires placarding under the provisions of subpart F of this part; or
7. A select agent or toxin regulated by the Centers for Disease Control and Prevention under 42 CFR part 73.

The first six categories above are the same categories that require Federal Hazmat Registration. (Note: Number 7 is not listed for Federal Hazmat Registration.)

3.5.2   Hazardous Materials Registration

The U.S. DOT requires that certain individuals or entities who offer for transportation, or transport in foreign, interstate, or intrastate commerce: (a) any highway
route–controlled quantity of a Class 7 (radioactive) material; (b) more than 25 kg
(55 lbs.) of a Division 1.1, 1.2, or 1.3 (explosive) material in a motor vehicle, rail car, or
freight container; (c) more than 1 L per package of a material extremely poisonous by
inhalation; (d) a hazardous material in a bulk packaging having a capacity of 3,500 gal
for liquids or gases, or more than 468 cubic feet for solids; (e) a shipment in other than
bulk packaging of 5,000 lb. gross weight or more of one class of hazardous material for which the transport vehicle requires placarding; (f) any quantity of materials
requiring placarding. The following are excepted from the registration requirement:

1. An agency of the federal government
2. A state agency
3. An agency or political subdivision of a state
4. An employee of (1)–(3)
5. A hazmat employee (including an owner operator of a motor vehicle leased to a
    registered motor carrier for 30 days or more)
6. A person domiciled outside the United States who offers hazardous materials
    solely from outside the United States. (See 49 CFR 107.606(a)(6) for exceptions
    and reciprocity.)

Registration is required annually and includes a fee.

3.6   NERC

The North American Electric Reliability Council (NERC) is an affiliate of the
Federal Electric Regulatory Commission (FERC) and has enforcement capability.
NERC has established security standards for FERC-regulated electrical generation
and transmission and distribution facilities and utilities. While primarily focused
on cyber and information technology (IT) security issues, the NERC standards
also address some physical security elements such as access control and intrusion
detection.

3.7   EPA AND THE LIST OF LISTS

The Consolidated List of Chemicals Subject to the Emergency Planning and
Community Right-to-Know Act (EPCRA) and Section 112(r) of the Clean Air
Act (CAA), (also known as the List of Lists) was prepared to help facilities handling chemicals determine the need to submit reports under sections 302, 304,
or 313 of EPCRA and, for a specific chemical, what reports may need to be
submitted.

The List of Lists will also help facilities to determine compliance requirements if subject to accident prevention regulations under CAA section 112(r). These lists should be used as a reference tool, not as a definitive source of compliance information. Compliance information for EPCRA is published in the CFR, 40 CFR Parts 302, 355, and 372. Compliance information for CAA section 112(r) is published in 40 CFR Part 68.

3.8   EPA AND RMP

Under the authority of section 112(r) of the Clean Air Act, the Chemical Accident Prevention Provisions require facilities that produce, handle, process, distribute, or store certain chemicals to develop a Risk Management Program, prepare a Risk Management Plan, and submit the RMP to EPA. Covered facilities were initially required to comply with the rule in 1999, and the rule has been amended on several occasions since then, most recently in 2004.

The EPA using the RMProgam and RMPlan, regulates certain gaseous and vapor chemical hazards that may impact nearby communities in the event of an environmental release of these chemicals. The RMP, like the Spill Prevention Controls and Countermeasures (SPCC) and Facility Response Plan (FRP), requires certain security standards to provide a basic level of protection for these facilities. In addition, RMPs also include an off-site consequence analysis, and this analysis should be considered as confidential information and access to this analysis should be restricted.

Both RMP and the OSHA Process Safety Management (PSM) standards require process hazard analysis (PHA). The severity of consequences for each undesired event must be derived. For facilities that have conducted PHAs, the severity table created for the PHA should be considered first during vulnerability assessment. This table may need to be modified to account for the consequences of a malevolent (rather than an accidental) event.

Another source of data to help determine the severity of consequences is the RMP analysis of the off-site consequences of the worst-case and alternative-release scenarios. (The results of these analyses may also need to be modified.)

Regulated toxic and flammable substances under section 112(r) of the Clean Air Act are the substances listed in Tables 1, 2, 3, and 4, as described below. Threshold quantities for listed toxic and flammable substances are specified in the RMP regulation's tables. Table 1 *of Sec. 68.130, the List of Regulated Toxic Substances and Threshold Quantities for Accidental Release Prevention*, presents in alphabetical order 77 substances judged to be a potential off-site toxic risk at various threshold quantities. *Table 3 of Sec. 68.130, the List of Regulated Flammable Substances \1\ and Threshold Quantities for Accidental Release Prevention*, presents in alphabetical order 63 substances judged to be a potential off-site flammable risk at various threshold quantities. Note: Tables 2 and 4 mirror Tables 1 and 3 except the substances are listed by Chemical Abstracts Service (CAS) number.

3.9 OSHA AND PROCESS SAFETY MANAGEMENT

OSHA standard 29 CFR 1910.119 is required by the Clean Air Act amendments, as is the EPA's RMP. Employers who merge the two sets of requirements into their PSM program will better assure full compliance with each as well as enhancing their relationship with the local community.

The process safety management standard targets highly hazardous chemicals that have the potential to cause a catastrophic incident. This standard as a whole is to aid employers in their efforts to prevent or mitigate episodic chemical releases that could lead to a catastrophe in the workplace and possibly in the surrounding community. To control these types of hazards, employers need to develop the necessary expertise, experiences, judgment, and proactive initiative within their workforce to properly implement and maintain an effective process safety management program as envisioned in the OSHA standard.

An effective PSM program requires a systematic approach to evaluating the whole process. Using this approach, the process design, process technology, operational and maintenance activities and procedures, nonroutine activities and procedures, emergency preparedness plans and procedures, training programs, and other elements that impact the process are all considered in the evaluation. The various lines of defense that have been incorporated into the design and operation of the process to prevent or mitigate the release of hazardous chemicals are evaluated and strengthened to assure their effectiveness at each level. PSM is the proactive identification, evaluation, and mitigation or prevention of chemical releases that could occur as a result of failures in process, procedures, or equipment.

Appendix A to 29 CFR 1910.119 contains a listing of toxic and reactive highly hazardous chemicals that present a potential for a catastrophic event at or above the threshold quantity. However for flammable risks, the aggregate quantity of flammables is regulated. This approach is very different from that taken in the EPA RMP regulations, which list specific flammable chemicals and their threshold quantities that trigger the EPA regulations. OSHA PSM regulations provide that a process which involves a flammable liquid or gas (as defined in 1910.1200(c)) on-site in one location, in a quantity of 10,000 pounds (4535.9 kg) is regulated. Exceptions to this requirement include hydrocarbon fuels used as fuel, and flammable liquids maintained at atmospheric pressure in atmospheric tanks (without benefit of refrigeration). Certain retail facilities and oil and gas production wells and servicing operations, and remote facilities are also exempt.

The Hazard Communication standard 1910.1200(c) defines flammable as:

- "Aerosol, flammable" means an aerosol that, when tested by the method described in 16 CFR 1500.45, yields a flame projection exceeding 18 inches at full valve opening, or a flashback (a flame extending back to the valve) at any degree of valve opening;
- "Gas, flammable" means: (A) A gas that, at ambient temperature and pressure, forms a flammable mixture with air at a concentration of thirteen (13) percent by volume or less; or a gas that, at ambient temperature and pressure, forms a

range of flammable mixtures with air wider than twelve (12) percent by volume, regardless of the lower limit;

- "Liquid, flammable" means any liquid having a flashpoint below 100°F (37.8°C), except any mixture having components with flashpoints of 100°F (37.8°C) or higher, the total of which make up 99 percent or more of the total volume of the mixture.

3.10 EPA AND SPILL PREVENTION CONTROLS AND COUNTERMEASURES

The EPA enforces the Spill Prevention Controls and Countermeasures (SPCC) plan requirements and implementation of those plan elements. The SPCC regulations address security elements such as fencing, lighting, access control, and similar items. While the SPCC requirements are relatively straightforward to implement and achieve, these requirements must be understood in the context of other security requirements and objectives. For example, if a facility is in need of fencing, the ability to cite a regulatory requirement as part of the justification for a capital budget is important and may make the difference as to whether or not the fence is funded.

3.11 EPA AND FACILITY RESPONSE PLAN

The Facility Response Plan (FRP) requirements are enforced by the EPA and/or the USCG (if a maritime impact is expected) for those facilities that could release petroleum, oils, or chemicals to the environment and such a release could cause a "substantial harm." The FRP requirements are described in the Clean Water Act, and supporting regulations. Both the EPA FRP requirements and the USCG FRP requirements have security elements to address lighting, access control, detection of unauthorized access, and similar fundamental security elements.

3.12 EPA TITLE 40 AND U.S. OSHA TITLE 29

The United States Environmental Protection Agency (EPA) regulations under 40 CFR determine the legal requirements to ensure a healthful environment. The basic environment in which a facility is located may influence the overall vulnerability. Therefore, the past evidence of compliance on or near a facility's location should be investigated.

Through various regulations, the EPA provides tools for assessing chemicals stored, used, and handled on-site. Even if compliance is not required due to limited chemical usage, the tools provided by the EPA can greatly assist facilities in determining risk. Example regulations of note include: EPCRA, RMP, and recently promulgated VA regulations.

Building materials and materials stored in buildings should be evaluated to determine if the building interiors present inherent hazards should these materials be disturbed. Asbestos, formaldehyde, polychlorinated biphenyls (PCBs), chlorinated fluorocarbons, solder or flux containing more than 0.2% lead, water pipe or pipe fittings containing more than 85 lead, and paint containing more than 0.06% lead are examples of inherent hazards. When lead is found, the controls required by OSHA in 29 CFR 1926.62 must be implemented if the lead is being released to the environs. The guiding standards to control asbestos hazards are the OSHA and EPA regulations, in particular 29 CFR 1926.58, 40 CFR 61.140–157, and 49 CFR 171–180. In general, projects should be designed to avoid or minimize disturbance of hazardous materials, including building materials.

3.13   EPA AND CERCLA INSPECTIONS: DUE DILIGENCE AND ASTM

Various techniques have been developed to assess buildings. Often with some modification these inspection techniques can also be used to determine building vulnerabilities. Due diligence inspections were developed in response to the Comprehensive Environmental Response, Compensation and Liability Act (CERCLA), also known as Superfund. With the advent of CERCLA, potential property buyers realized that the purchase of contaminated commercial property could expose their businesses to liability. The American Society for Testing and Materials (ASTM) formulated the *Standard Practice for Environmental Site Assessments: Phase I Environmental Site Assessment Process (ASTM E 1527-05)* to address the need for these property inspections.

The ASTM realized that assessments included indoor areas and methods of transmittal within or from indoor areas, and the resultant assessment techniques are adaptable to vulnerability assessments. Phase I assessments conducted under ASTM include historical reviews and visual assessments of property. Prior to the on-site investigation, documentation reviewed includes the building configuration, size and use, and occupancy profile. The heating, ventilation, and air conditioning (HVAC) system type and components should also be reviewed for vulnerability assessments.

Due diligence investigations can also be conducted based solely on the needs of the buyer or seller of a property. In those instances ASTM and other investigative guidance can be used or not. To the extent practicable, the buyer and seller may communicate the investigation and vulnerability assessment dictates and requirements.

The *Phase I Environmental Assessment Process (ASTM E 1527-05)* was published in 2005 by the ASTM. E-1527 defines good commercial and customary practice for conducting an environmental site assessment of a parcel of commercial real estate with respect to the range of contaminants within the scope of the Comprehensive Environmental Response, Compensation and Liability Act (CERCLA), also known as Superfund, and petroleum products.

The innocent landowner defense (ILD) to CERCLA liability concept is provided in 42 United States Code (USC) § 9601(35) and § 9607(b)(3) and was included as part of the Superfund Amendments and Reauthorization Act (SARA) of 1986.

The E-1527 Standard codifies tasks that, when considered in concert with each other, may satisfy the ILD to CERCLA liability. This defense is that, *all appropriate inquiry into the previous ownership and uses of a property consistent with good commercial or customary practice in an effort to minimize liability* (§ 9601(35)(B)* has occurred.

The concept of a material threat is included within E-1527's definition of a recognized environmental condition (REC). E-1527 defines a material threat as a physically observable or obvious threat that is reasonably likely to lead to a release that, in the opinion of the environmental professional, is threatening, and might result in impact to public health or the environment.

In the nonmandatory information appendices to E-1527 exclusions to E-1527 are discussed for radon; asbestos; lead in drinking water; lead-based paint; and mold, fungi, and microbial growth in building structures. These exclusion discussions for radon and asbestos cite case law that establishes exclusions for naturally occurring radon and asbestos (as part of a structure not slated for demolition—ever). However, this case law was based solely on rulings as to remediation monies and liability after property transfer. These cases do not discuss release of these substances in terms of impact to public health or the environment, or the attendant liabilities associated with those releases.

Consequently, in the non-scope issues section of a Phase I report, any conditions that the environmental professional deems to potentially impact public health and that exceed *de minimus* levels of concern should be reported. These include concentration of radon within a building that exceeds natural levels; asbestos that in the future may be subject to demolition; and biological growth of such an extent that release of spores, vegetative structures, and mycotoxins into the environment constitutes a threat to health for a future occupant of the structure.

3.14 BUILDING CODES

Building code adoption varies from location to location. Research must always be conducted to determine which regulations and industry standards have been used to develop local building codes. In many cases, the development of local building codes may not reflect the most current industry standards, which may have been updated more frequently than the local building codes. Issues such as grandfathering to justify past practices and adoption of only portions of current (or near current) industry standards must be addressed in any vulnerability assessment. Despite the fact that buildings may be "to code," that code may not reflect current knowledge that could guarantee a safer building.

The requirements of the International Codes, as developed by the International Code Council (ICC), and the codes of the National Fire Prevention Association (NFPA), and the National Electrical Code (NEC), often form the basis for local building codes. These codes, whether state or local, must be investigated for the specific locality. Clauses associated with grandfathering (acceptance of prior codes for aged structures) are critically important, since extant and current codes may not be enforceable for older buildings. This lack of code enforcement may result in a latent and unknown vulnerability if the building user wrongly assumes that current code compliance has occurred. Note: In some ways the International Codes produced by the ICC and the NFPA codes are similar, and their adoption by local authorities and adoption by reference from OSHA may be similar.

3.14.1 Public Buildings Amendments of 1988, 40 U.S.C.

The Public Buildings Amendments of 1988, 40 U.S.C. 3312, requires that each building constructed or altered by a U.S. federal agency comply with nationally recognized model building codes and with other applicable nationally recognized codes. These requirements may also be useful as guidance criteria for other buildings and useful in assessing building vulnerability.

3.14.2 International Codes

The International Code Council (ICC) is a consolidated organization. The ICC was formerly the Building Officials and Code Administrators International, Inc. (BOCA), the International Conference of Building Officials (ICBO), and the Southern Building Code Congress International, Inc. (SBCCI). The ICC family of codes includes the International Building Code (IBC), International Fire Code (IFAN COIL), International Plumbing Code (IPC), International Mechanical Code (IMC), and the family of codes is available through www.intlcode.org.

The International Codes, or I-Codes, published by the International Code Council, provide minimum safeguards for people at home, at school, and in the workplace. The I-Codes are a complete set of comprehensive, coordinated building safety and fire prevention codes. Legislative bodies are not obligated to adopt model building safety or fire prevention codes, and may write their own code or portions of a code. A model code has no legal standing unless adopted as law by a legislative body (state legislature, county board, city council, etc.) Most U.S. cities, counties, and states choose I-Codes developed by the International Code Council.

3.14.3 NFPA and NEC

The National Fire Protection Association (NFPA) has established national model codes and standards, many of which have become the standard of practice or have been adopted by the local authority having jurisdiction (AHJ). NFPA codes should be used (unless the local AHJ has a more stringent set of standards or codes).

NFPA standards such as NFPA 72 address design, installation, and testing of fire alarm systems, including automatic smoke detection systems. NFPA 30, 16, and 13 address the design, installation, and testing of fire prevention systems such as sprinkler systems. NFPA guidance documents are available through www.nfpa.org/.

The U.S. 9/11 Commission, the Department of Homeland Security, and Congress have all endorsed NFPA 1600, the *American National Standard on Disaster/ Emergency Management and Business Continuity Programs.*

3.14.3.1 Life Safety Code (NFPA 101) and NFPA

In particular, the *Life Safety Code* (NFPA 101) is a standard requirement to ensure that appropriate emergency notification and facility evacuation are provided for in building design. Designs need to account for elements such as egress routes and stairway capacity, places of refuge, and similar concepts to ensure effective building evacuation by building occupants in the event of a fire or other emergency. NFPA 101 is not a fire protection and prevention standard. However, the concepts presented in NFPA 101 should be assessed to determine facility conformance to NFPA 101 guidelines during fire prevention program development and emergency response planning. Lack of effective Life Safety Code implementation should be viewed by both security and safety assessors as a vulnerability.

3.14.3.2 National Electrical Code (NEC)

The NFPA, National Electric Code (NFPA 70), presents standard requirements to ensure both fire and electrical safety. The National Electrical Code is available through www.nfpa.org.

3.14.4 UL Standards

Underwriters Laboratories (UL) is somewhat unique in their role in the security and fire detection industry. UL routinely tests various equipment and components to assess reliability. Equipment that meets UL's stringent standards are considered "UL Listed" and are authorized to display the UL label.

UL also has a certification service where UL inspectors verify the performance of security systems and fire detection systems for effectiveness, functionality, and conformance to UL standards. Systems that are designed, installed, maintained, and tested according to these UL standards by preapproved UL-certified security firms, can become "UL Certificated" if the proper paperwork is filed with and certified UL. Periodically (typically annually) UL system inspectors conduct field visits of the installation firm, the monitoring facility, and the actual installed system to verify conformance with UL standards.

3.15   LEADERSHIP IN ENERGY AND ENVIRONMENTAL DESIGN (LEED)

The energy analysis of building characteristics, the mechanical and electrical components, and all other related energy consumption elements may be an important source of information as to how the building's systems function and its potential vulnerabilities. New construction projects and substantial renovations are often certified through the LEED Green Building Rating System of the U.S. Green Building Council.

Analyses of energy-conserving designs usually include all relevant facets of the building envelope, lighting energy input, domestic water heating, efficient use of local ambient weather conditions, building zoning, efficient part-load performance of all major HVAC equipment and the ability of building automation equipment to automatically adjust for building partial occupancies, optimized start-stop times, and systems resets. Energy analysis should evaluate compliance with the American Society of Heating, Refrigerating, and Air Conditioning Engineers (ASHRAE) Standards 90.1, ASHRAE Standard 62, and 10 CFR 434.

3.16   ILLUMINATIONS ENGINEERING SOCIETY OF NORTH AMERICA (IESNA)

Lack of lighting will substantially increase the potential for risk during a disaster, attack, or after an untoward event. The lighting level standards recommended by the Illuminations Engineering Society of North America (IESNA) should be used. The IESNA Subcommittee on Off-Roadway Facilities standards are the lowest acceptable lighting levels for any parking facility.

Figure 3.1 shows the lighting minimums for various protection levels. A point-by-point analysis should be done in accordance with the IESNA standards. Keeping vehicular entry/exits to a minimum number of locations is beneficial. Long span construction and high ceilings create an effect of openness and aid in lighting the facility. Shear walls should be avoided, especially near turning bays and pedestrian travel paths. Where shear walls are required, large holes in shear walls can help to improve visibility. Openness to the exterior should be maximized. Eliminate dead-end parking areas, as well as other areas with secluded access.

Horizontal Illumination at Pavement Minimum	Low	Low/Med	Medium	Higher
Covered Parking Areas	1.25	1.50	1.75	2.00
Roof and Surface Parking Area	0.25	0.50	0.75	1.00
Stairwells, Elevator Lobbies	2.5	3.5	4.5	5.5
Uniformity Ratio Leverage Minimum	4.1	4.1	4.1	4.1
Uniformity Ratio (maximum, minimum)	20.1	20.1	20.1	20.1
Vertical Illumination 5 Feet above Pavement, Minimum	**Low**	**Low/Med**	**Medium**	**Higher**
Covered Parking Areas	0.625	0.75	0.875	1
Roof and Surface Parking Area	0.125	0.25	0.375	0.5
Stairwells, Elevator Lobbies	1.25	1.75	2.25	2.75

Figure 3.1   Maintained illumination levels.

3.17   STATE AND LOCAL CODES

National building codes are typically the foundation of state and local building codes. However, state and local codes also represent important regional interests and conditions. Compatibility with local plans, zoning compliance, building code compliance, and construction inspections may provide indications of current building status. All requirements (other than procedural requirements) of zoning laws and design guidelines should be evaluated. These include laws relating to landscaping, open space, building setbacks, maximum height of the building, historic preservation, and aesthetics. Access to storm water runoff and erosion control, sanitary sewers and storm drains and water, gas, electrical power and communications, emergency vehicle response organizations, and roads and bridges may all be impacted by some local code restrictions.

Reviews should include drawings and specifications, any on-site inspections, building permits, and compliance with local regulations and compatibility with local fire-fighting practices. When and if these reviews highlight facility vulnerabilities, the interaction between original design or usage intent and local code restrictions should be evaluated.

3.17.1   Water Supply Codes

Regulations of local water authorities include required service connection between building and public water lines. These connections are coordinated with the local water authority. In colder climates, provisions must be made for easy shutoff and drainage during the winter season for water services that extend outdoors. For large buildings or campuses, a loop system fed from more than one source is often required. Dual service for the fire protection systems may be required. Water lines should be located behind curb lines and in unpaved areas if possible, or under sidewalks if not. These lines should not be located under foundations and streets, drives, or other areas where access is severely limited.

Whatever the water connections, the security issues associated with these connections should be considered. In addition, water influx that cannot be remotely controlled or limited during an emergency may be an additional hazard as electrical services are impacted.

3.17.2   Sanitary Sewer: Local Sewer Authority

The regulations of the local sewer authority usually require that all sewer lines be located below unpaved areas if possible. In addition, pipe runs between manholes should be straight lines and manholes should not be located in pedestrian traffic routes. Cleanouts should be provided on all service lines, approximately 1500 mm (5 feet) away from the building, and at all line bends where manholes are not used.

In areas where no public sewers exist, septic tanks and leach fields should be used for sewage discharge. Septic systems will have additional land area (in accordance with local and state code requirements) for future expansion of the discharge system. Cesspools are not permitted in most new construction. Their presence associated with older construction should be evaluated to determine if removal of the cesspool is required.

All sewer systems are potential sources of contaminant influx into buildings. Manholes and cleanouts that are intended to be used to mitigate sewer problems, may present unsecured access routes to dose the sewer system with contaminants.

3.17.3   Storm Water

Storm drains should be separate from sanitary sewers within the property limits, even in cities where separate public systems are not yet available. Most storm drainage systems will be designed for a 25-year minimum storm frequency, unless local criteria are more stringent.

Storm drainage systems should always use gravity flow. Piped systems are preferred. In large campus settings, open ditches or paved channels should be avoided as much as possible. A storm drainage system may consist of an open system of ditches, channels and culverts, or of a piped system with inlets and manholes. Storm drainage pipes should be located in unpaved areas wherever possible. Offset inlets from main trunk lines to prevent clogging.

Detention of storm water on building rooftops must not be permitted. In most cases building roof drainage will be collected by the plumbing system and discharged into the storm drains. Exceptions are small buildings in rural areas where gutters and downspouts may discharge directly onto the adjacent ground surface.

If the disaster situation involves water accumulation, the storm water drainage system may be used to mitigate flooding events. The potential for using this system should be evaluated in all emergency planning and local authorities consulted as to what scenarios will enable storm water drainage system usage in emergencies.

An example of an unplanned resurgence of former storm waters occurred post-Katrina: New Orleans pumped storm water overflow back into Lake Pontchartrain. This water was laced with raw sewage, bacteria, heavy metals, pesticides, and toxic chemicals. In addition, Louisiana officials said that although two large oil spills, from damaged storage tanks, were under control, thousands of other smaller spills continued to coat floodwaters in New Orleans with a rainbow sheen. "It's simply unfeasible" to try and hold the pumped water somewhere to filter out pollution, said Michael D. McDaniel, the Louisiana secretary of environmental quality. "We have to get the water out of the city or the nightmare only gets worse," said Dr. McDaniel, who is a biologist. "We can't even get in to save people's lives. How can you put any filtration in place?" Each of the estimated 140,000 to 160,000 homes that were submerged is a potential source of fuel, cleaners, pesticides, and other potentially hazardous materials found in garages or under kitchen sinks, officials said (*New York Times*, September 2005).

REFERENCES

Water returned to lake contains toxic material, Sewell Chan and Andrew C. Revkin, *New York Times*, September 7, 2005.

ADDITIONAL READING

A Manual for the Prediction of Blast and Fragment Loading on Structures, DOE/TIC 11268. Washington, DC, Headquarters, U.S. Department of Energy, 1992.

Analysis of the Clandestine CB Threat to USAF Strategic Forces (Unclassified), Volumes I and 11. Technical Considerations, Defense Technical Information Center (DTIC) Number AD379465, February 1967.

Antiterrorism Front-End Analysis (Unclassified), Defense Technical Information Center (DTIC) Number AD-C954865, June 1984.

Background Paper, Technologies Underlying Weapons of Mass Destruction, Office of Technology Assessment, U.S. Congress, OTA-BP-ISC-1 15.

Behind the first roar of machinery to drain the city: A tale of pluck and luck, John Schwartz, *New York Times*, September 8, 2005.

Chemical/Biological Hazard Prediction Program, Technical Report, Defense Technical Information Center (DTIC) Number AD-BL 63245, 1991.

Chemical, Biological, and Radiological Incident Handbook, Chemical, Biological and Radiological (CBRN) Subcommittee of the Interagency Intelligence Committee on Terrorism (IICT), October 1998.

Corps closes landfill over asbestos fears EPA says grinding may release fibers, Mark Schleifstein, Staff writer, *New Orleans Times-Picayune*, February 17, 2006.

Department of Homeland Security (DHS), TITLE 6--Domestic Security CHAPTER I—DEPARTMENT OF HOMELAND SECURITY, OFFICE OF THE SECRETARY, Part 27—Chemical Facility Anti-Terrorism Standards and **Appendix to Chemical Facility Anti-Terrorism Standards; Final Rule, US Department Homeland Security, November 20, 2007**.

Department of Transportation (DOT) Regulation 49 CFR, Part 172, Subpart G. 66 FR 45182, Aug. 28, 2001.

Department of Transportation (DOT) Regulation 49 CFR, Part 172, Subpart I. 66 FR 45182, Amdt. 172-127, 59 FR 49133, Sept. 26, 1994.

Department of Transportation (DOT) Regulation 49 CFR, Part 172, Subpart I. 66 FR 45182, Amdt. 172-127, 61 FR 20750, May 8, 1996.

Effects of Terrorist Chemical Attack on Command, Control, Communications, and Intelligence (C³I) Operations, Defense Technical Information Center (DTIC) Number AD-BL 65614, April 1992.

Environmental Considerations in Decision Making, Government Services Administration (GSA), GSA ADM 1095.1F.

Facilities Standards for the Public Building Service, U.S. General Services Administration, Public Buildings Service, Office of the Chief Architect, Washington, DC 20405, March 2003.

Fundamentals of Protective Design for Conventional Weapons, TM 5-855-1.Washington, DC, Headquarters, U.S. Department of the Army, 1986.

Glossary of Terms in Nuclear Science and Technology, Grange Park, IL, American Nuclear Society, 1986.

Guidance for Protecting Building Environments from Airborne Chemical, Biological, or Radiological Attacks, National Institute of Occupational Safety and Health (NIOSH), Department of Health and Human Services (DHHS) (NIOSH) Pub No. 2002-139, May 2002.

HHS awards contract for radiation countermeasures, Global Security Newswire, February 14, 2006.

North American Emergency Response Manual, U.S. Department of Transportation, Transport Canada, Secretariat of Transport and Communications, 1996.

Nuclear Reactor Concepts, U.S. Nuclear Regulatory Commission, May 1993.

Nuclear Terms Handbook, U.S. Department of Energy, Office of Nonproliferation and National Security, 1996.

Protective Construction Design Manual, ESL-TR-87-57. Prepared for Engineering and Services Laboratory, Tyndall Air Force Base, FL. 1989.

Security Engineering, TM 5-853 and Air Force AFMAN 32-1071, Volumes 1, 2, 3, and 4. Washington, DC, Departments of the Army and Air Force, 1994.

Structures to Resist the Effects of Accidental Explosions, Army TM 5-1300, Navy NAVFAC P-397, AFR 88-2. Washington, DC, Departments of the Army, Navy, and Air Force, 1990.

CHAPTER 4

Site and Building Features

Martha Boss, Dennis Day, Charles Allen, and Randy Boss

CONTENTS

Building features should be designed to make security planning more efficient and to lessen vulnerability. A component of all security planning is effective operations and maintenance procedures to include issues such as access, efficacy of decontamination methods, and ongoing detection systems.

4.1   LANDSCAPING

Landscaping design elements that are attractive and welcoming can enhance security. For example, plants can deter unwanted entry, ponds and fountains can block vehicle access, and site grading can also limit access. Avoid landscaping that permits concealment of intruders or obstructs the view of security personnel. Landscaping should be done judiciously so as not to provide hiding places. Hold plantings away from the facility to allow an unobstructed view of all building exteriors.

4.1.1   Slopes

The slopes of planted areas around buildings should permit easy maintenance. Sloping is also used to protect drainage away from the building. Careful design of slope patterns will assure building access during emergency response activities.

Turf areas should have a slope of no more than 3:1 and no less than 1%. A 2% minimum slope is desirable. Areas with slopes steeper than 3:1 must be planted with groundcover or constructed with materials specifically designed to control erosion. Slopes steeper than 2:1 are not acceptable. Terracing may be an appropriate solution for sites with large grade differentials, as long as access for lawn mowers and other maintenance equipment is provided.

The minimum slope for grassy swales and drainage ways is 1% to prevent standing water and muddy conditions. Slopes for walkways will not exceed 5%, unless unavoidable. Slopes greater than 5% may make the construction of special ramps for the disabled necessary. The maximum cross-slope is 2%.

For planted areas adjacent to buildings, the first 3000 mm (10 feet) should be sloped away from the structure to assure no standing water adjacent to exterior basement walls and foundations.

Parking areas or large entrance plazas should have slopes of 1% minimum and 5% maximum. In areas with snowfall, provisions should be made for piling snow removed from roads and parking areas. Drains should be provided at the entrance to ramps into parking structures to minimize the amount of rainwater runoff into the structure.

Paved areas adjacent to buildings should have a minimum 2% slope away from the structure to a curb line, inlet, or drainage way to provide positive drainage of surface water.

4.1.2 Grading

From a cost standpoint, minimizing grading overall and balancing cut and fill is desired. Existing trees or other plant materials to be preserved should be reflected in the grading plan. Where trees are to be preserved, the existing grade within the circle of the tree drip line must not be disturbed by regrading or paving. Snow fencing should be erected at the drip line of the tree to protect existing trees from construction materials or equipment.

4.1.3 Walkways

Preferably, walkways should not have steps. Where steps are necessary, cheek walls enclosing the risers and treads should be used to make a smooth transition to planted areas on the sides of the steps.

No buildings should be built within the 100-year flood plain. If, however, the building location within a flood plain is approved, mechanical and electrical equipment rooms must be located 1500 mm (5 feet) above the level of the 100-year flood plain. No grading should be performed within the boundaries of any wetland without appropriate surveys and permits.

4.1.4 Irrigation

Irrigation systems may not initially seem to be sources of vulnerability. However, potentially interconnected water lines can pose a threat. In addition, properly designed irrigation systems can be a mitigation aid, providing "wash down" waters to decontaminate ground surfaces.

All major components should be installed in protected, accessible locations. Controllers and remote sensing stations should be placed in vandal-proof enclosures. Aboveground components, such as backflow preventers, should be placed in unobtrusive locations and must be protected from freezing. Quick coupling valves should be of two-piece body design and installed throughout the system to allow for hosing down areas and to permit easy access to a source of water. Locate drain valves to permit periodic draining of the system.

Irrigation systems should be zoned so different areas can be watered at different times. Avoid mixing different head or nozzle types (such as a spray head and a bubbler) on the same station. The entire system should be designed to minimize surface runoff. In heavy clay soils, a low application rate may be required. Overspray onto paved surfaces should be avoided.

Non-potable water should be used as a source for the irrigation system when available. Irrigation water should be metered separately from domestic water whenever this would avoid expensive user sewage fees. Rainwater harvesting may be considered as an alternative source for such purposes as irrigation. Rainwater harvesting systems must comply with all local codes and standards.

4.1.5 Lighting

Landscape lighting should be used to enhance safety and security on the site, to provide adequate lighting for nighttime activities, and to highlight special site features. The primary purpose will help determine the requirements for light coverage and intensity. Generally, unobtrusive lighting schemes are preferred.

Where the intent of the lighting is primarily aesthetic, a low-voltage system should be considered. Light fixtures should be placed so people do not look directly at the light source. To avoid plant damage and fire hazard, high intensity or heat-generating fixtures should not be located immediately adjacent to plant material. Fixtures should be resistant to vandalism and easily replaceable from local sources. Landscape lighting and building illumination should be controlled by clock-activated or photocell-activated controllers.

4.2 SITE CIRCULATION

Site circulation design will vary greatly depending on the type of facility being constructed, security issues, and the location within a community. This site design should segregate, at a minimum, pedestrian access, vehicular access (including parking), service vehicle access, and service dock areas. Service dock access may be from an alley, from a below-grade ramp, or from a site circulation drive.

If large trucks are to service the facility, sufficient maneuvering space must be provided, and the service drive should be screened as much as possible. The service drive should always be separate from the access to the parking garage. Where possible, a one-way design for service traffic is preferable to avoid the need for large truck turning areas. The service area of the facility should not interfere with public access roadways.

Confusion over site circulation, parking, and entrance locations can contribute to a loss of site security. Signs should be provided off-site and at entrances. On-site directional, parking, and cautionary signs should be provided for visitors, employees, service vehicles, and pedestrians. Unless required by other standards, signs should generally not be provided that identify sensitive areas.

4.3 ACCESS CONTROLS: BARRIER SYSTEMS

Use various types and designs of buffers and barriers (e.g., walls, fences, trenches, ponds and water basins, plantings, trees, static barriers, sculpture, and street furniture). Design site circulation to prevent high-speed approaches by vehicles. Offset vehicle entrances from the direction of a vehicle's approach to force a reduction in speed.

Barriers to vehicle access should be visually punctuated and as unobtrusive as possible to pedestrians. If addressed skillfully, planters, trees, or sculpted bollards can be employed to provide amenities while meeting vehicle barrier requirements.

Access control and entry screening provide practical means to protect buildings and facilities. To enhance security and still provide effective use, building entrances should be designed to readily permit controlled access.

- Neighboring uses, existing pedestrian patterns, local transit, and the building's orientation should be considered to anticipate pedestrian "desire lines" to and from the building from off-site.
- High blank walls should be avoided; lower walls with sitting edges are preferable, but should be designed to discourage skateboarders or other similar athletic pursuits.
- Buildings may have additional doors to access service areas. Service area doors should not be used as public entrances. Service and maintenance areas of the building should, in general, not be open for public access or traversal.
- A vehicular drop-off area should be located on the street nearest the main entrance. Vehicular drives, parking lots, and service area entrance drives must follow local codes for entrance driveways within the right-of-way limits of city-, county-, or state-maintained roads. Access design analysis should consider approach and vantage points for emergency vehicles.
- Plants (trees, vines) must not be placed in areas where an intruder could use them to climb a wall or reach an upper-story window. Plant material should be selected with appropriate growth patterns to assure that growth over time does not create hiding places for assailants, or a traffic hazard by restricting sightlines.
- To reduce the potential for concealment of devices around screening points, avoid installing features such as trash receptacles and mail boxes that can be used to hide devices. If mail or express boxes are used, locate them well away from checkpoints.

Critical building components should be located no closer than necessary in any direction to any main entrance, vehicle circulation, parking, or maintenance area. If this is not possible, harden as appropriate. Examples of critical building components include: emergency generators including fuel systems, day tank, fire sprinkler, and water supply; fuel storage; main switchgear and critical distribution feeders for emergency power; telephone distribution and main switchgear; building control centers and systems controlling critical functions; main refrigeration systems if critical to building operation; elevator machinery and controls; shafts for stairs, elevators, and utilities; and fire pumps.

Security screening or fencing may be provided at points of low activity to discourage random entrance to the facility on foot, while still maintaining openness and natural surveillance. A system of fencing, grilles, or doors should be designed to completely close down access to the entire facility in unattended hours, or in some cases, all hours. Any ground-level pedestrian exits that open into nonsecure areas should be emergency exits only and fitted with panic hardware and alarmed for exiting movement only.

4.3.1 Garages, Service Areas, and Parking Structures

All garage or service area entrances that are not otherwise protected by site perimeter barriers should be protected by devices capable of arresting a vehicle of the designated threat size at the designated speed. This criterion may be lowered if the access circumstances prohibit a vehicle from reaching this speed.

Mitigating the risks associated with parking requires creative design and planning measures, including parking restrictions, perimeter buffer zones, barriers, structural hardening, and other architectural and engineering solutions. Parking inside the building should be restricted as needed with appropriate ID checks. Similarly, on-site surface or structured parking should be restricted as needed with consideration for minimum stand-off distances to the main occupied structure.

For all stand-alone, aboveground parking facilities, maximizing visibility throughout the parking facility is a key design principle. The preferred parking facility design employs express or non-parking ramps, speeding the user to parking on flat surfaces. For parking inside a facility where the building superstructure is supported by the parking structure, the designer should:

- Protect primary vertical load-carrying members by implementing architectural or structural features that provide a minimum 6-inch standoff.
- Assure that all columns in the parking area are designed for an unbraced length equal to two floors, or three floors where two levels of parking are to be used.

Pedestrian paths should be planned to concentrate activity to the extent possible. For example, bring all pedestrians through one portal rather than allowing them to disperse to numerous access points, which will improve the ability to see and be seen by other users.

Adjacent public parking should be directed to more distant or better protected areas, segregated from employee parking and away from the facility. Parking may be allowed in curb lanes, with a sidewalk between the curb lane and the building. Where distance from the building to the curb provides insufficient setback, parking in the curb lane may require restricted use only.

4.3.2 Stairwells and Elevators

Stairwells required for emergency egress should be located as remotely as possible from areas where blast events or contaminant releases might occur. Wherever possible, stairs should not discharge into lobbies, parking, or loading areas.

Stairways and elevator lobby design should be as open as code permits. The ideal solution is a stair and/or elevator waiting area totally open to the exterior and/or the parking areas. Designs that ensure that people using these areas can be easily seen—and can see out—should be encouraged. If a stair must be enclosed for code or weather protection purposes, glass walls will deter both personal injury attacks and various types of vandalism. Potential hiding places below stairs should be closed off; nooks and crannies should be avoided. Elevator cabs should have glass backs whenever possible. Elevator lobbies should be well lighted and visible to both patrons in the parking areas and the public out on the street.

4.3.3 Loading Docks: Shipping and Receiving Areas

Loading docks and receiving and shipping areas should be separated by at least 50 feet in any direction from utility rooms, utility mains, and service entrances, including electrical, telephone/data, fire detection/alarm systems, fire suppression water mains, and cooling and heating mains. Loading docks should be located so that vehicles will not be driven into or parked under the building. If this is not possible, the service should be hardened for blast.

The entrances and exits at loading docks and service entrances should be provided with a positive means to reduce infiltration and outside debris. Loading docks must be maintained at negative pressure relative to the rest of the building.

The loading dock design should limit damage to adjacent areas and vent explosive force to the exterior of the building. Significant structural damage to the walls and ceiling of the loading dock is acceptable if the areas adjacent to the loading dock do not experience severe structural damage or collapse during an explosive blast. Where peak pressures from the design explosive threats can be shown to be below 1 psi acting on the face of the building, the designer may use the reduced requirements. The floor of the loading dock does not need to be designed for blast resistance if the area below is not occupied and contains no critical utilities.

4.3.4 Mailroom

The mailroom should be located away from facility main entrances, areas containing critical services, utilities, distribution systems, and important assets. In addition, the mailroom should be located at the perimeter of the building with an outside wall or window designed for pressure relief. Adequate space should be provided for explosive disposal containers. An area near the loading dock may be a preferred mailroom location.

Mailrooms where packages are received and opened for inspection, and unscreened retail spaces should be designed to mitigate the effects of a blast on primary vertical or lateral bracing members. Where mailrooms are located in occupied areas or adjacent to critical utilities, walls, ceilings, and floors, mailrooms should be blast and fragment resistant.

4.4 BUILDING PHYSICAL SECURITY

Physical security occurs when a building has adequate protection from structural collapse, untoward intrusion, and electrical safety concerns. Human intrusion may either be in the form of persons or their instruments (machines, surveillance) penetrating a building envelope. Another type of intrusion can occur through the entry of water or air. Water and air contaminants entering the building envelope may significantly change the building's security status.

When evaluating the potential for structural collapse, consideration should be given to materials with inherent ductility. The ductile materials (i.e., cast in place reinforced concrete and steel construction) are better able to respond to load reversals. Careful detailing is required to provide materials such as prestressed concrete, precast concrete, and masonry that will adequately respond to the design loads. Primary vertical load-carrying members should be designed to resist the effects of the specified threat. The designer usually needs to implement architectural or structural features that deny contact with exposed primary vertical load members. A minimum standoff of at least 6 inches from these members is required.

4.4.1 Assessments

The purpose of physical security assessment is to examine and evaluate the systems in place (or being planned) and to identify potential improvements in this area for the sites evaluated. Physical security systems include access controls, barriers, locks and keys, badges and passes, intrusion detection devices and associated alarm reporting and display, closed-circuit television (assessment and surveillance), communications equipment (telephone, two-way radio, intercom, cellular), lighting (interior and exterior), power sources (line, battery, generator), inventory control, postings (signs), security system wiring, and protective force. Physical security systems are reviewed for design, installation, operation, maintenance, and testing.

The physical security assessment should focus on those sites directly related to the critical facilities, including information systems and assets required for operation. Typically included are facilities that house critical equipment or information assets or networks dedicated to the operation of electric or gas transmission, storage, or delivery systems. Other facilities can be included on the basis of criteria specified by the organization being assessed. Appropriate levels of physical security are contingent upon the value of facility or site assets, the potential threats to these assets, and the cost associated with protecting the assets. The focus of the physical security assessment task is determined by prioritizing the facility or site assets; that is, the most critical assets receive the majority of the assessment activity.

At the start of the assessment, survey personnel should develop a prioritized listing of facility or site assets. This list should be discussed with facility or site personnel to identify areas of security strengths and weaknesses. During these initial interviews, assessment areas that would provide the most benefit to the facility or site should be identified; once known, these areas should become the major focus of the assessment activities.

The physical security assessment of each focus area usually consists of the following:

- Physical security program
- Barriers
  - Access controls/badges and locks/keys
  - Intrusion detection systems and communications equipment

The key to reviewing the above topics is to determine the appropriate level that is necessary and consistent with the value of the asset being protected.

4.4.2 Blast Resistance

Risk assessments based on the new construction criteria should be performed on existing structures to examine the feasibility of upgrading the facility. The results, including at a minimum recommendations and cost, should be documented in a written report before submission for project funding.

Progressive collapse analysis must be performed if.

1. The building is to be upgraded for seismic forces.
2. The building structural frame will be exposed as part of the current scope of work making any structural upgrade for preventing progressive collapse appropriate at this time.
3. The exterior façade of the building is to be removed making structural upgrade of the perimeter structural system appropriate at this time.

The three basic approaches to blast-resistant design are:

1. Blast loads can be reduced, primarily by increasing standoff
2. A facility can be strengthened; or
3. Higher levels of risk can be accepted.

The best answer is often a blend of the three.

4.4.2.1 Structural Security

To address blast, the priority for upgrades should be based on the relative importance of a structural or nonstructural element, in the order defined below:

- Primary Structural Elements—essential parts of the building's resistance to catastrophic blast loads and progressive collapse, including columns, girders, roof beams, and the main lateral resistance system
- Secondary Structural Elements—all other load bearing members, such as floor beams, slabs
- Primary Nonstructural Elements—elements (including their attachments) that are essential for life safety systems or elements that can cause substantial injury if failure occurs, including ceilings or heavy suspended mechanical units
- Secondary Nonstructural Elements—all elements not covered in primary nonstructural elements, such as partitions, furniture, and light fixtures

Priority should be given to the critical elements that are essential to mitigating the extent of collapse. Designs for secondary structural elements should minimize injury and damage. Consideration should also be given to reducing damage and injury from primary as well as secondary nonstructural elements.

4.4.2.2 Protection Levels

The entire building structure or portions of the structure will be assigned a protection level according to the facility-specific risk assessment. The following are definitions of damage to the structure and exterior wall systems from the bomb threat for each protection level:

- Low and Medium/Low-Level Protection, Major Damage: The facility or protected space will sustain a high level of damage without progressive collapse. Casualties will occur and assets will be damaged. Building components, including structural members, will require replacement, or the building may be completely unrepairable, requiring demolition and replacement.
- Medium-Level Protection, Moderate Damage, Repairable: The facility or protected space will sustain a significant degree of damage, but the structure should be reusable. Some casualties may occur and assets may be damaged. Building elements other than major structural members may require replacement.
- Higher-Level Protection, Minor Damage, Repairable: The facility or protected space may globally sustain minor damage with some local significant damage possible. Occupants may incur some injury, and assets may receive minor damage.

4.4.2.3 Loads and Stresses

Where required, structures should be designed to resist blast loads. The demands on the structure will be equal to the combined effects of dead, live, and blast loads.

- Blast loads or dynamic rebound may occur in directions opposed to typical gravity loads.
- For purposes of designing against progressive collapse, loads are defined as dead load plus a realistic estimate of actual live load. The value of the live load may be as low as 25% of the code-prescribed live load.

The design should use ultimate strengths with dynamic enhancements based on strain rates. Allowable responses are generally post-elastic.

- Lap splices should fully develop the capacity of the reinforcement.
- Lap splices and other discontinuities should be staggered.
- Ductile detailing should be used for connections, especially primary structural member connections.
- Deflections around certain members, such as windows, must be controlled to prevent premature failure. Additional reinforcement is generally required.
- Balanced design of all building structural components is desired. For example, for window systems, the frame and anchorage should be designed to resist the full capacity of the weakest element of the system.

4.4.2.4 Good Engineering Practice Guidelines

For higher levels of protection from blast, cast-in-place reinforced concrete is normally the construction type of choice. Other types of construction such as properly designed and detailed steel structures are also allowed. Several material and construction types, while not disallowed by these criteria, may be undesirable and uneconomical for protection from blast.

- To economically provide protection from blast, inelastic or post-elastic design is standard. This allows the structure to absorb the energy of the explosion through plastic deformation while achieving the objective of saving lives. To design and analyze structures for blast loads that are highly nonlinear both spatially and temporally, proper dynamic analysis methods must be used. Static analysis methods will generally result in unachievable or uneconomical designs.
- The designer should recognize that components might act in directions that do not match the design decision logic or design intent. This untoward reaction is due to the engulfment of structural members by blast, the negative phase, the upward loading of elements, and dynamic rebound of members. Making steel reinforcement (positive and negative faces) symmetric in all floor slabs, roof slabs, walls, beams, and girders will address this issue. Symmetric reinforcement also increases the ultimate load capacity of the members.
- Special shear reinforcement including ties and stirrups is generally required to allow large post-elastic behavior. The designer should carefully balance the selection of small but heavily reinforced (i.e., congested) sections with larger sections with lower levels of reinforcement.
- Connections for steel construction should be ductile and develop as much moment connection as practical. Connections for cladding and exterior walls to steel frames should develop the capacity of the wall system under blast loads.
- In general, single point failures that can cascade, producing widespread catastrophic collapse, are to be avoided. A prime example is the use of transfer beams and girders that, if lost, may cause progressive collapse and are therefore highly discouraged.
- Redundancy and alternative load paths are generally good in mitigating blast loads. One method of accomplishing this is to use two-way reinforcement schemes where possible.
- In general, column spacing should be minimized so that reasonably sized members can be designed to resist the design loads and increase the redundancy of the system. A practical upper level for column spacing is generally 30 ft for the levels of blast loads described herein.
- In general, floor-to-floor heights should be minimized. Unless an overriding architectural requirement exists, a practical limit is generally less than or equal to 16 ft.
- Designers should use fully grouted and reinforced concrete masonry unit (CMU) construction in cases where CMU is selected.

- Designers should actively coordinate structural requirements for blast with other disciplines including architectural and mechanical.
- Use one-way wall elements spanning from floor to floor to minimize blast loads imparted to columns.

In many cases, the ductile detailing requirements for seismic design and the alternate load paths provided by progressive collapse design assist in the protection from blast. The designer must bear in mind, however, that the design approaches are at times in conflict. These conflicts must be worked out on a case-by-case basis.

4.4.2.5 *Exterior Walls: Limited Loads*

Design exterior walls for the actual pressures and impulses up to a maximum psi and psi-msec as required by the building's inherent design features. The designer should also ensure that the walls are capable of withstanding the dynamic reactions from the windows.

- Shear walls that are essential to the lateral and vertical load-bearing system, and that also function as exterior walls, should be considered primary structures. Design exterior shear walls to resist the actual blast loads predicted from the threats specified.
- Where exterior walls are not designed for the full design loads, special consideration should be given to construction types that reduce the potential for injury.

4.4.2.6 *Progressive Collapse*

Designs that facilitate or are vulnerable to progressive collapse must be avoided. At a minimum, all new facilities should be designed for the loss of a column for one floor above grade at the building perimeter without progressive collapse. This design and analysis requirement for progressive collapse is intended to ensure adequate redundant load paths in the structure should damage occur for whatever reason. Designers may apply static and/or dynamic methods of analysis to meet this requirement. Ultimate load capacities may be assumed in the analyses.

New facilities with a defined threat should be designed with a reasonable probability that, if local damage occurs, the structure will not collapse or be damaged to an extent disproportionate to the original cause of the damage.

In the event of an internal explosion in an uncontrolled public ground-floor area, the design should prevent progressive collapse due to the loss of one primary column, or the designer should show that the proposed design precludes such a loss. That is, if columns are sized, reinforced, or protected so that the threat charge will not cause the column to be critically damaged, then progressive collapse calculations are not required for the internal event. For design purposes, assume no additional standoff from the column (beyond what is permitted by the design) is required.

4.4.2.7 Design for Full Load

Design the exterior walls to resist the actual pressures and impulses acting on the exterior wall surfaces from the threats defined for the facility.

The multidisciplinary team should evaluate the performance requirements for all exterior windows and security-glazing materials proposed for the project. For limited protection, windows do not require design for specific blast pressure loads. Rather, the designer is encouraged to use glazing materials and designs that minimize the potential risks.

- Preferred systems include thermally tempered, heat-strengthened or annealed glass with a security film installed on the interior surface and attached to the frame; laminated thermally tempered, laminated heat-strengthened, or laminated annealed glass; and blast curtains.
- Acceptable systems include thermally tempered glass; and thermally tempered, heat-strengthened or annealed glass with film installed on the interior surface (edge to edge, wet glazed, or daylight installations are acceptable).
- Unacceptable systems include untreated monolithic annealed or heat-strengthened glass; and wire glass.

The minimum thickness of film that should be considered is 4 mil. In a blast environment, glazing can induce loads three or more times that of conventional loads onto the frames. This must be considered with the application of antishatter security film.

The designer should design the window frames to remain intact and not fail prior failure of the glazing under lateral load. Likewise, the anchorage should be stronger than the window frame, and the supporting wall should be stronger than the anchorage. The design strength of a window frame and associated anchorage is related to the breaking strength of the glazing. Thermally tempered glass is roughly four times as strong as annealed, and heat-strengthened glass is roughly twice as strong as annealed.

4.4.2.8 Design Up to Specified Load

Window systems design (glazing, frames, anchorage to supporting walls, etc.) on the exterior facade should be balanced to mitigate the hazardous effects of flying glazing following an explosive event. The walls, anchorage, and window framing should fully develop the capacity of the glazing material selected. When using various design methods, the designer may consider a breakage probability no higher than 750 breaks per 1000 when calculating loads to frames and anchorage.

While most test data use glazing framed with a deep bite, this may not be amenable to effective glazing performance or installation. New glazing systems with a 3/4-inch minimum bite can be engineered to meet the performance standards with the application of structural silicone. However, not much information is available on the long-term performance of glazing attached by structural silicone or with anchored security films.

All glazing hazard reduction products for these protection levels require product-specific test results and engineering analyses performed by qualified independent agents demonstrating the performance of the product under the specified blast loads, and stating that the product meets or exceeds the minimum performance required.

- Window Fenestration: The total fenestration openings are not limited; however, a maximum of 40% per structural bay is a preferred design goal.
- Window Frames: The frame system should develop the full capacity of the chosen glazing up to 750 breaks per 1000, and provide the required level of protection without failure. This can be shown through design calculations or approved testing methods.
- Anchorage: The anchorage should remain attached to the walls of the facility during an explosive event without failure. Capacity of the anchorage system can be shown through design calculations or approved tests that demonstrate that failure of the proposed anchorage will not occur and that the required performance level is provided.

4.4.3 Glazing Alternatives

Glazing alternatives are as follows:

- Preferred systems include thermally tempered glass with a security film installed on the interior surface and attached to the frame; laminated thermally tempered, laminated heat-strengthened, or laminated annealed glass; and blast curtains.
- Acceptable systems include monolithic thermally tempered glass with or without film if the pane is designed to withstand the full design threat.
- Unacceptable systems include untreated monolithic annealed or heat-strengthened glass; and wire glass.

In general, thicker antishatter security films provide higher levels of hazard mitigation than thinner films. Testing has shown that a minimum of a 7-mil-thick film, or specially manufactured 4-mil-thick film, is the minimum to provide hazard mitigation from blast. The minimum film thickness that should be considered is 4 millimeters.

Not all windows in a public facility can reasonably be designed to resist the full forces expected from the design blast threats. As a minimum, design window systems (glazing, frames, and anchorage) to achieve the specified performance conditions for the actual blast pressure and impulse acting on the windows up to a maximum psi and psi-milliseconds. As a minimum goal, the window systems should be designed so that at least a designed percent of the total glazed areas of the facility meet the specified performance conditions when subjected to the defined threats.

4.4.3.1 Interior Windows

Interior glazing should be minimized where a threat exists. The designer should avoid locating critical functions next to high-risk areas with glazing, such as lobbies and loading docks. For lobbies and other areas with specified threats, ballistic windows, if required, should meet the requirements of Underwriters Laboratory (UL) 752 Bullet-Resistant Glazing Level. Glass-clad polycarbonate or laminated polycarbonate are two types of acceptable glazing material. Security glazing, if required, should meet the requirements of American Society for Testing and Materials (ASTM) F1233 or UL 972, Burglary.

4.4.3.2 Resistant Glazing Material

Resistance glazing material should meet the minimum performances specification. However, special consideration should be given to frames and anchorages for ballistic resistant windows and security glazing since their inherent resistance to blast may impart large reaction loads to the supporting walls. Resistance of Window Assemblies to Forced Entry (excluding glazing) is discussed as ASTM F 588 Grade.

4.4.4 Non-Window Openings

Non-window openings such as mechanical vents and exposed plenums should be designed to the level of protection required for the exterior wall. Designs should account for potential in-filling of blast overpressures through such openings. The design of structural members and all mechanical system mountings and attachments should resist these interior fill pressures.

4.4.5 Venting

The designer should consider methods to facilitate the venting of explosive forces and gases from the interior spaces to the outside of the structure. Examples of such methods include the use of blow-out panels and window system designs that provide protection from blast pressure applied to the outside but that readily fail and vent if exposed to blast pressure on the inside.

ADDITIONAL READING

A Manual for the Prediction of Blast and Fragment Loading on Structures, DOE/TIC 11268. Washington, DC, Headquarters, U.S. Department of Energy, 1992.

Analysis of the Clandestine CB Threat to USAF Strategic Forces (Unclassified), Volumes I and 11. Technical Considerations, Defense Technical Information Center (DTIC) Number AD379465, February 1967.

Antiterrorism Front-End Analysis (Unclassified), Defense Technical Information Center (DTIC) Number AD-C954865, June 1984.

Background Paper, Technologies Underlying Weapons of Mass Destruction, Office of Technology Assessment, U.S. Congress, OTA-BP-ISC-1 15.

Behind the first roar of machinery to drain the city: A tale of pluck and luck, John Schwartz, *New York Times*, September 8, 2005.

Biological Effects of Radiation, U.S. Nuclear Regulatory Commission, Office of Public Affairs.

Chemical/Biological Hazard Prediction Program, Technical Report, Defense Technical Information Center (DTIC) Number AD-BL 63245, 1991.

Chemical, Biological and Radiological Incident Handbook, Chemical, Biological and Radiological (CBRN) Subcommittee of the Interagency Intelligence Committee on Terrorism (IICT), October 1998.

Corps closes landfill over asbestos fears EPA says grinding may release fibers, Mark Schleifstein, Staff writer, *New Orleans Times-Picayune*, February 17, 2006.

Desk Guide, Government Services Administration (GSA), GSA PBS NEPA.

Effects of Terrorist Chemical Attack on Command, Control, Communications, and Intelligence (C^3I) Operations, Defense Technical Information Center (DTIC) Number AD-BL 65614, April 1992.

Facilities Standards for the Public Building Service, U.S. General Services Administration, Public Buildings Service, Office of the Chief Architect, Washington, DC 20405, March 2003.

Fundamentals of Protective Design for Conventional Weapons, TM 5-855-1. Washington, DC, Headquarters, U.S. Department of the Army, 1986.

Glossary of Terms in Nuclear Science and Technology, Grange Park, IL, American Nuclear Society, 1986.

Guidance for Protecting Building Environments from Airborne Chemical, Biological, or Radiological Attacks, National Institute of Occupational Safety and Health (NIOSH), Department of Health and Human Services (DHHS) (NIOSH) Pub No. 2002-139, May 2002.

HHS awards contract for radiation countermeasures, Global Security Newswire, February 14, 2006.

North American Emergency Response Manual, U.S. Department of Transportation, Transport Canada, Secretariat of Transport and Communications, 1996.

Nuclear Reactor Concepts, U.S. Nuclear Regulatory Commission, May 1993.

Nuclear Terms Handbook, U.S. Department of Energy, Office of Nonproliferation and National Security, 1996.

Proliferation of Weapons of Mass Destruction, Assessing the Risk, Office of Technology Assessment, U.S. Congress, OTA-ISC-559.

Protective Construction Design Manual, ESL-TR-87-57. Prepared for Engineering and Services Laboratory, Tyndall Air Force Base, FL. 1989.

Replacement pumps don't exist, Officials say damaged equipment may take a week to dry out before it can be repaired. Peter Pae, *Los Angeles Times*, September 3, 2005.

Security Engineering, TM 5-853 and Air Force AFMAN 32-1071, Volumes 1, 2, 3, and 4. Washington, DC, Departments of the Army and Air Force, 1994.

Senate approves disease surveillance bill, Danielle Belopotosky, *National Journal's Technology Daily*, January 6, 2006.

Structures to Resist the Effects of Accidental Explosions, Army TM 5-1300, Navy NAVFAC P-397, AFR 88-2.Washington, DC, Departments of the Army, Navy, and Air Force, 1990.

Weapons of Mass Destruction Terms Handbook, Defense Special Weapons Agency, DSWA-AR-40H, 1 June 1998.

CHAPTER 5

Chemical Threats

Martha Boss, Dennis Day, and Gayle Nicoll

CONTENTS

This chapter primarily discusses chemicals that could be readily used as chemical, biological, or radiological (CBR) agents. However, all chemical stores or potential exposures within a facility should be evaluated during vulnerability assessments.

Some health consequences from CBR agents are immediate, while others may take much longer to appear. CBR agents (e.g., arsine, nitrogen mustard gas, anthrax, radiation from a dirty bomb) can enter the body through a number of routes including inhalation, skin absorption, contact with eyes or mucous membranes, and ingestion.

The amount of a CBR agent required to cause specific symptoms varies among agents. However, these agents are generally much more toxic than common indoor air pollutants. In many cases, exposure to extremely small quantities may be lethal. Symptoms are markedly different for the different classes of agents (e.g., chemical, biological, or radiological). Symptoms resulting from exposure to chemical agents tend to occur quickly. Most chemical warfare agents (gases) are classified by their physiological effects, (e.g., nerve, blood, blister, and choking). Toxic industrial chemicals (TICs) can also elicit similar types of effects (U.S. Army, 1990).

Dilution with outside air is the primary method of maintaining acceptable indoor air quality, whether in a building under normal use or one being used for sheltering-in-place. Exterior airborne sources of contaminants that may be unacceptable for use indoors with respect to odor and sensory irritation would limit the effectiveness of dilution ventilation. Chemical contamination that can become airborne is the most difficult to control. These contaminants can be dispersed as solid particulates or vapor droplets borne by ambient airstreams. More active release through dispersal of compressed gases may also occur.

5.1   ADSORPTION AND ABSORPTION

Chemical contaminants can be transported as a gas, liquid, or a vapor either directly or indirectly. Note: Vapor in this discussion will refer to small droplets of liquid dispersed within a gas phase carrier.

The chemicals may be sorbed or deposited on materials and, thus, sorption may lead to secondary exposures. The term sorbed applies to both adsorption and absorption, which are discussed here relative to the physical phenomena of being sorbed. Something adsorbed to a second thing is attached to the *outside* of the second thing. Something absorbed by a second thing has been taken *inside* the second thing. All surfaces can sorb vapor to some extent. Textiles do so in the greatest quantity because of their large surface areas.

In a cleaner environment, the chemicals may desorb or reaerosolize, causing a potential secondary exposure scenario. Desorption is a slower process than sorption. For example, what may be sorbed in a period of 5 minutes may require an hour or more to desorb in a clean environment, depending upon conditions.

Sorption characteristics are used to determine filtration and neutralization methods for chemical contaminants. Blister and nerve agents that have strong physical adsorption characteristics to carbon and slow hydrolysis after carbon adsorption can be effectively trapped by this adsorbent process in gas masks or filtration systems. However, breakthrough at even the ppb level may cause a severe toxicologic effect, so filtration cannot be relied upon for sustained protection. Most choking and blood agents have weak physical adsorption characteristics and cannot be controlled by the use of carbon filter beds.

All filters will eventually fail. Failure can occur around the perimeter of the filter or via breakthrough of the filter media. Breakthrough occurs when the filter has captured contamination through the majority of its depth. The contaminant can circumvent remaining filter layers via capillary (breakthrough) channels, and be directly emitted on the supposedly clean side of the filter. This direct breakthrough is of concern for any contaminant that can be transmitted as a gas.

5.1.1 Indirect Transport

Liquid agent can be transported into a building or room if items taken in have been exposed to the liquid chemical agent, whether the liquid is sprayed on or through contact with liquid. Liquid contamination on the surfaces of boots, outer garments, gloves, hood, or mask can be transferred to clothing, skin, or hair when outer garments are removed. The agent may then evaporate or desorb from the item over time as desorption temperatures are reached.

5.1.2 Direct Transport

Ambient vapor can enter a building through doorways opened for entry and exit. Airborne contaminants are then diluted and eventually purged as clean air from the filter unit flows through the building. The dilution rate is a function of the airflow exchange rate, percent fresh air, volume of the spaces serviced by the HVAC system, mixing efficiency, and the rate at which agent vapor is sorbed and desorbed by material within the enclosure.

Solid and liquid aerosols transmitted via a gas phase carrier can also enter directly and indirectly. Once deposited on supplies or clothing outdoors, contaminant can become airborne inside the building through movement of the item or the air around the item.

5.2 TOXIC INDUSTRIAL CHEMICALS (TICs)
AND MATERIALS (TIMs)

TICs and toxic industrial materials (TIMs) are commonly categorized by their hazardous properties, such as reactivity, stability, combustibility, corrosiveness, ability to oxidize other materials, and radioactivity (NFPA, 1991). The presence of TICs and TIMs on-site should be evaluated in terms of inherent hazard and also increased hazard should an attack occur.

Building owners and managers should take into account the potential threat posed by large quantities of TICs and TIMs that may be found in the vicinity of their building. Development of structured ventilation filter beds to deal with specific chemicals and impregnation treatments, and to address several high-priority TICs may be required. For the purposes of collection on a sorbent, gaseous agents include:

- Organic vapors (i.e., cyclohexane)
- Acid gases (i.e., hydrogen sulfide)
- Base gases (i.e., ammonia)
- Specialty chemicals (i.e., formaldehyde or phosgene)

TICs that have a combination of high toxicity and ready availability are of principal concern. Those having a volatility of less than 10 torr at room temperature are effectively removed by physical adsorption, including that achieved in filter media. However, a number of high-toxicity TICs, produced industrially on a large scale, have volatilities higher than 10 torr at 20°C and are more difficult to collect.

5.3 CHEMICALS OF INTEREST

The U.S. Chemical Facility Anti-Terrorism Standard (CFATS) identifies a list of chemicals of interest (COIs) and includes the type of risk each chemical poses. Appendix A of CFATS lists 300 COIs. If any of the Appendix A chemicals are identified on-site, the facility may be determined to be "high-risk" by the Department of Homeland Security (DHS). Included COIs are:

- Propane
- Anhydrous ammonia
- Natural gas
- Hydrogen
- Nitric acid
- TNT

- Hydrogen sulfide
- Chlorine
- Ammonium nitrate
- Methane
- Vinyl chloride
- Methyl mercaptan

The rule applies to solids, liquids, and gases, and applies if the chemicals are present in either concentrated form or as a component of a mixture. Chemicals in waste containers are included in addition to process chemicals (stored, handled, used). This list may be an effective starting point to determine both chemicals used on-site, and chemicals stored or used nearby that may be a threat.

5.4 CLASSIC CHEMICAL WARFARE AGENTS

Of particular concern within the macrocosm of CBR are chemical warfare agents. These agents can cause great harm to exposed individuals in a very short period of time. All these agents can be readily transmitted using aerosol dispersion devices.

Compression and subsequent storage in compressed gas tanks is possible. As such, compressed gas cylinders and high-pressure receiving vessels (up to one-ton containers) are suspect. Since toxic effects can also occur with liquid exposures, in particular to mustard agents that do not volatile readily, leakage of unknown liquids should always be suspect.

5.4.1 Blister Agents: Mustard

Mustard agent is a blister agent. Blister agents are irritants that burn the skin and eyes, producing blisters. The earliest symptom will be:

- Red, itchy skin, followed by yellowish blistering
- Irritated, tearing eyes, followed after severe exposure by sensitivity to light, pain, and perhaps blindness
- Runny nose, sneezing, cough
- Diarrhea, fever, nausea, vomiting, and damage to the immune system

Symptom onset can be expected 1 to 6 hours after exposure. However, some symptoms will be immediate. Recovery usually takes up to 4 to 5 weeks. However, if the eyes are severely exposed, permanent blindness may follow. Victims usually suffer large scars due to burns.

Blister agents cause similar tissue effects on the skin and within the lungs. The blistering effect within the lungs, if severe, also causes fluid to build up in the lungs. If the exposed person survives the acute effects, scarring can be expected in the lungs resulting in persistent diminution in lung function.

Mustard is initially supplied as an oily amber liquid. However, to aid dissemination, the agent is often delivered as a gas from compressed gas cylinders. Consequently, any compressed gas cylinders of unknown content may be suspect.

5.4.1.1 *Mustard and Agent HD*

Mustard's chemical formulas include $C_4H_8C_2S$ and $ClCH_2CH_2—S—CH_2CH_2Cl$. The chemical family for mustard agent is chlorinated sulfur. Synonyms for Mustard agent include: sulfur mustard; sulphur mustard gas; sulfide, bis(2-chloroethyl); bis(beta-chloroethyl)sulfide; 1,1'-thiobis(2-chloroethane); 1-chloro-2(beta-chloroethylthio)ethane; beta, beta'-dichlorodiethyl sulfide; 2,2'-dichlorodiethyl sulfide; Di-2-chloroethyl sulfide, Beta, beta'-dichloroethyl sulfide; Iprit S-Lost; S-yperite; Schewefel-lost; Senfgas; Yellow Cross Liquid; Yperite, Y; EA 1033; Kampfstoff Lost.

5.4.1.2 *Agent HD*

Agent HD is the chemical called distilled mustard or bis(2-chloroethyl) sulfide, chemical abstract service registry No. 505-60-2. HD is mustard agent (H) that has been purified by washing and vacuum distillation to reduce sulfur impurities. When released, distilled mustard will dissipate, with dissipation time dependent upon both temperature and air velocity. The vapor pressure of mustard (0.11 mm Hg at 25°C) is moderate, but high enough for mustard gas to remain in the air near liquid sources (*Toxicological Profiles for Mustard*, Agency for Toxic Substances and Disease Registry [ATSDR], 2003).

5.4.2 Nerve Agents

Nerve agents are the most toxic and rapidly acting of the known chemical warfare agents. These agents are similar to organophosphate pesticides in toxicological properties. However, nerve agents are much more potent than organophosphate pesticides.

GB (Sarin) and VX are nerve agents that function as kill agents by blocking the activity of acetyl cholinesterase. In humans, anticholinesterase organophosphates have broadly similar actions to those seen in other species. Acetyl cholinesterase inhibition causes acute effects in humans and other mammals. The symptoms in humans that generally occur when acetyl cholinesterase activity has been reduced by about 50% may include: headache, exhaustion and mental confusion together with blurred vision, sweating, salivation, chest tightness, muscle twitching, and abdominal cramps. The more severe effects can include muscle paralysis leading to severe difficulty in breathing. Convulsions and unconsciousness can occur.

Artificial respiratory support may be required during and after severe exposures. Death may result if the diaphragm cannot relax and breathing stops, essentially the last inspiration is the last breath.

5.4.2.1 GB (Sarin)

GB (Sarin) is the chemical isopropyl methylphosphonofluoridate, Chemical Abstracts Service (CAS) registry number 107-44-8. GB in pure form and in the various impure forms may be found in storage as well as in industrial, depot, or laboratory operations. The chemical family for GB is fluorinated organophosphorous and the chemical formula is $C_4H_{10}FO_2P$. Alternate chemical names include: O-isopropyl methylphosphonofluoridate; phosphonofluoridic acid, methyl-, isopropyl ester; phosphonofluoridic acid, methyl-, and 1-methylethyl ester. Trade names and synonyms include: isopropyl ester of methylphosphonofluoridic acid; methylisopropoxfluorophosphine oxide; isopropyl methylfluorophosphonate; O-isopropyl methylisopropoxfluorophosphine oxide; methylfluorophosphonic acid, isopropyl ester; isopropoxymethylphosphonyl fluoride; isopropyl methylfluorophosphate; isopropoxymethylphosphoryl fluoride; GB; Sarin; and Zarin. Note: The G agents contain fluorine or cyanide, instead of sulfur (which is found in V agents).

Nerve agents with the letter "G" (for German) are classified as nonpersistent chemicals. The "G" agents cause rapid effects among people who inhale or ingest them, resulting in death within minutes. These agents may be vaporized, sprayed, or put into the water supply.

Signs and symptoms of exposure include severe coughing and discomfort in the lungs directly after inhalation of a "G" agent, without fever. Everyone exposed to these agents will have some effect. In moderate to light exposure, symptoms include:

- Runny nose
- Watery eyes, small pinpoint pupils, eye pain, blurred vision
- Drooling
- Excessive sweating
- Cough, chest tightness, rapid breathing
- Diarrhea
- Increased urination
- Confusion, drowsiness, weakness
- Headache
- Nausea, vomiting, and/or abdominal pain
- Slow or fast heart rate
- Abnormally low or high blood pressure

Symptoms may appear seconds to minutes after GB is inhaled or ingested or up to 18 hours later if "G" agent vapor touches the skin. In heavy exposure, symptoms include loss of consciousness, convulsions, paralysis, and respiratory failure possibly leading to death.

5.4.2.2 VX

VX is the chemical phosphonothioic acid, methyl-S-(2-(bis(1-methylethyl)amino) ethyl)0-ethyl ester, CAS registry number 50782-69-9. VX in pure form and in the various impure forms may be found in storage as well as in industrial, depot, or laboratory operations. The chemical formula is: $C_{11}H_{26}NO_2PS$. Alternate chemical names include: methylphosphonothioic acid; S-[2-[bis(1-methylethyl)amino] ethyl]-O-ethyl ester; O-ethyl-S-(2-diisopropylaminoethyl) methylphosphonothioate. Trade names and synonyms include: phosphonothioic acid, methyl-O-ethyl S-(2-bis(1-methylethylamino)ethyl) 0-ethyl ester; S-(2-diisopropylaminoethyl) methylphosphonothioate; S-2-diisopropylaminoethyl O-ethyl methylphosphono-thioate; S-2((2-Diisopropylamino)ethyl) O-ethyl methylphosphonothiolate; O-ethyl S-(2-diisopropylaminoethyl) methylphosphonothioate; O-ethyl S-(2-diisopropyl-aminoethyl) methylthiolphosphonoate; S-(2-diisopropylaminoethyl) o-ethyl methyl phosphonothiolate; ethyl-S-dimethylaminoethyl methylphosphonothiolate; TX60. Also known as EA 1701 (U.S. designation) and T2445 (British designation). Note: The V agents contain sulfur, instead of fluorine or cyanide (which is found in G agents).

VX is the least volatile of the nerve agents, which means VX is the slowest to evaporate from a liquid into a vapor. Because VX evaporates so slowly, VX can be a long-term threat as well as a short-term threat. Surfaces contaminated with VX should therefore be considered a long-term hazard.

VX is the most potent of all nerve agents. Compared with the nerve agent Sarin (also known as GB), VX is considered much more toxic by entry through the skin and somewhat more toxic by inhalation. Symptoms will appear within a few seconds after exposure to the vapor form of VX, and within a few minutes to up to 18 hours after exposure to the liquid form. Visible VX liquid contact on the skin, unless washed off immediately, would be lethal.

Exposure to a low or moderate dose of VX by inhalation, ingestion (swallowing), or skin absorption may cause some or all of the following symptoms within seconds to hours of exposure:

- Runny nose
- Watery eyes, small, pinpoint pupils, eye pain, blurred vision
- Drooling
- Excessive sweating
- Cough, chest tightness, rapid breathing
- Diarrhea
- Increased urination
- Confusion, drowsiness, weakness
- Headache
- Nausea, vomiting, and/or abdominal pain
- Low or fast heart rate
- Abnormally low or high blood pressure

Even a tiny drop of nerve agent on the skin can cause sweating and muscle twitching where the agent has touched the skin. Exposure to a large dose of VX by any route may result in these additional health effects: loss of consciousness, convulsions, paralysis, and respiratory failure possibly leading to death.

Mild or moderately exposed people usually recover completely. Severely exposed people are not likely to survive. Unlike some organophosphate pesticides, nerve agents have not been associated with neurological problems lasting more than 1 to 2 weeks after the exposure—if the victim survives the acute effects.

5.5 NERVE AGENT SAMPLING

In the event that nerve agents are used in a terrorist attack, sampling should be conducted to determine when the areas attacked can be reentered. Outside of military installations, sampling for nerve agents as the attack is happening is not probable. Sampling methodologies are complex and expensive. This text describes basic sampling methods and analytical limitations.

All headspace analysis should take into consideration that GB, VX, and HD will sink in air. Consequently, headspace analysis that does not probe into the lower layers may result in false negatives, especially for VX and HD.

GB will be less likely to remain on any matrices. VX and HD can both be expected to accumulate. This accumulation may exceed the current surrounding agent concentrations, with the equilibrium present at the vapor interphase not predictive of residual agent. Similarly, wipe sampling may not be indicative of agent present beneath the surfaces of containerized components.

Agent	Volatilization Rate mg/m³ at 25°C	Boiling Point	Vapor Pressure (25°C)	Vapor Density	Melting Point
GB	22,000	316°F (158°C)	2.9 mm Hg	4.86	−68.8°F (−57°C)
VX	8.9–10.5	568.4°F (298°C)	0.00063 mm Hg	9.2	−58°F (−50°C)
HD	910	419°F–422°F (215–217°C)	0.11 mm Hg	5.5	57.2°F (13–14°C)

Figure 5.1   Agent physical characteristics.

5.5.1   Vapor Screening and Headspace Monitoring

Headspace monitoring is valid if the monitoring represents all conditions that could reasonably be expected to potentiate off-gassing events. In general, the minimum volatilization temperature for the agents of concern is the required temperature for vapor/headspace monitoring to be a valid indicator of risk. Volatilization (mg/m³) ranges for the various agents at 25°C are listed in Figure 5.1.

GB's volatilization far exceeds that for VX or mustard. As a result GB should be much easier to detect with headspace monitoring. Similarly, mustard is more easily detected than VX. However, other physical characteristics must also be considered.

GB, VX, and HD/HT will all be liquid at room temperature with assumed normal temperature and pressure (NTP). However, if present as vapor, these agents will layer below the surrounding air because of density differentials. Headspace sampling from the top of containers or from near environs versus source points may underestimate agent concentrations. Conversely, mixing may occur both in containers and in the ambient airstream within rooms. This mixing may move the agent into all areas of the container or room.

Since the melting point for HD/HT is 57.2°F, all containers should be monitored only when the containers have been brought to at least 75°F. All air monitoring for waste characterization must be done at or above 75°F; with both the air and the equipment being monitored at that temperature. All container components/contents, including those identified as potential agent residual locations or nodes, must also be brought to this temperature.

5.5.2   Airborne Exposure Limits (AEL)

Airborne exposure limits are the limits at which the general public and workers can be exposed to the agent:

- General population limits (GPL) are 24-hour averaging time for GB and VX and 12-hour averaging time for HD and HT.
- Worker population limits (WPLs) are 8-hour averaging time equivalent to 8-hour time weighted averages (TWAs).

Agent	GPL	WPL	STEL	IDLH	ASC
GB Revised Airborne Exposure Limits (mg/m^3)	0.000001 1×10^{-6}	0.00003 3×10^{-5}	0.0001 1×10^{-4}	0.1 1×10^{-1}	0.0003 3×10^{-4}
VX Revised Airborne Exposure Limits (mg/m^3)	0.0000006 6×10^{-7}	0.000001 1×10^{-6}	0.00001 1×10^{-5}	0.003 3×10^{-3}	0.0003 3×10^{-4}
HD or HT–Interim Airborne Exposure Limits (mg/m^3)	0.00002 2×10^{-5}	0.0004 4×10^{-4}	0.003 3×10^{-3}	0.7 7×10^{-1}	0.03 3×10^{-2}

Figure 5.2 Airborne exposure limits.

- Short-term exposure limits (STELs) are based on 15-minute averaging time for GB and VX, and 5-minute averaging time for HD or HT.
- Immediately dangerous to life and health (IDLH) averaging time is less than or equal to 30 minutes.
- Allowable stack concentration (ASC) is the maximum instantaneous allowable stack concentration (also called the source emission limit [SEL]).

AEL criteria are presented in Figure 5.2.

5.5.3 Monitors

Most agent detection monitors and their associated technologies currently available for nongovernmental uses are focused on the detection of airborne agent concentrations near the Centers for Disease Control's (CDC's) IDLH AELs. An example is open-path air monitoring of gaseous compounds. This monitoring identifies and quantifies gases based on their spectral absorption characteristics. This technology will not sample down to the AEL for most chemical warfare materiel (CWM) agents.

The instrumentation used to detect chemical agents are the Automatic Continuous Air Monitoring System (ACAMS), Miniature Chemical Agent Monitoring System (MINICAMS), and Depot Area Air Monitoring System (DAAMS) reliant systems. GB and HD have traditionally been sampled and detected directly by ACAMS and MINICAMS. VX is also detected using these monitors, however not directly. Because VX has low volatility and high affinity for irreversible adsorption on surfaces, VX is first derivatized by reaction with silver fluoride, which then yields the more volatile and less reactive G analog of VX.

The G analog of VX is then sampled and detected by ACAMS and MINICAMS monitors. One major problem for ACAMS and MINICAMS monitors is the occurrence of false alarms when monitoring at the STEL for VX (equal to the CDC's 1988

TWA value). False alarms may be caused by phosphorus-containing compounds (e.g., pesticides, phosphorus-containing impurities).

5.5.3.1 Automatic Continuous Air Monitoring System (ACAMS)

The ACAMS is a near-real-time (NRT) monitoring system with the ability to detect and report the concentration levels of chemical agent in the air at either low levels or high levels depending on the monitoring configuration in use. Air is sampled during a preset sample period. Agent present in the sample airstream is collected on a solid sorbent bed during the sample period for gas chromatographic (GC) analysis. The results of the GC analysis of the sampled air are displayed on the front panel of the instrument. A permanent record of the chromatogram and the agent concentration is recorded on a strip-chart. The ACAMS produces an audible and visible alarm when the agent concentration level is at or above the preset alarm level.

5.5.3.2 Miniature Chemical Agent Monitoring System (MINICAMS)

MINICAMS are automatic air monitoring systems that:

- Collect compounds on a solid sorbent trap
- Thermally desorb the compounds into a capillary gas-chromatography column for separation
- Detect the compounds with either a halogen-specific detector (XSD) or a flame photometric detector (FPD)

MINICAMS are lightweight, portable, near-real-time, low-level monitors with alarm capability, designed to respond to GB (Sarin), VX; mustard; nitrogen mustard; and Lewisite mustard. The MINICAMS can also be used to detect certain industrial compounds, such as phosgene, chloropicrin, and chloroform.

The halogen-specific detector (XSD) used in some MINICAMS to detect the chlorine in HD is a thermionic device. The target compound is oxidized in a flame to produce halogen ions that react with an alkali-activated, negatively biased platinum electrode. This reaction enhances the platinum electrode's electron emission through a cascade electrical event.

5.5.3.3 Depot Area Air Monitoring System

The Depot Area Air Monitoring System (DAAMS) is a portable air-sampling unit that is used for agent confirmation sampling. The DAAMS typically achieves better chromatographic resolution than do ACAMS or MINICAMS monitors. DAAMS monitors employing FPD detectors may be used to confirm or deny the presence of agent in areas monitored by ACAMS or MINICAMS monitors. However, DAAMS monitors are not NRT instrumentation since a follow-on laboratory analysis is required after the sample collection phase.

DAAMS samples are collected by pulling air through glass sampling tubes packed with a porous polymer for periods of time ranging from a few minutes for NRT confirmation samples to as long as 12 hours for GPL historical monitoring. The sampling apparatus:

- Draws a controlled volume of air through a glass tube filled with a collection material (Tenax GC), and then
- Pulls air through the solid sorbent tube where agent is collected

After sampling for the predetermined period of time and flow rate, the tube is removed from the vacuum line. The tube is transferred for GC analysis. Since DAAMS analyzers with FPD detectors do use the same basic agent detection scheme as FPD-equipped ACAMS or MINICAMS, DAAMS are often subject to the same measurement interferences and errors.

5.5.4   Flame Photometric Detector (FPD)

The Flame Photometric Detector (FPD) may be operated in either the phosphorus- or sulfur-specific mode. Certain pesticides may elicit a false positive and can be used to mask the presence of GB or VX. False positives can also be caused by any phosphorus-containing compounds that:

- Are not agents but have the same GC-column retention time as GB or the G-analog of VX
- Also undergo V-to-G-analog conversion

The FPD is more selective for phosphorus versus carbon than for phosphorus versus sulfur. Thus, despite the selectivity of the FPD, interferences can be caused by:

- Sulfur emissions (resulting from the formation of S2 in the FPD)
- Hydrocarbon emissions (for example, resulting from the formation of CH)

Hydrocarbons can quench (reduce) sulfur and phosphorus emissions, causing false negatives.

5.5.5   Chemical Agent Monitor (CAM)

The CAM is a lightweight, handheld gross-level vapor detector designed to respond to nerve and mustard agent vapors. Vapors of chemical agents are detected by sensing molecular ions of specific mobilities and then using timing and microprocessor techniques to reject interferences. When the CAM detects the presence of a chemical agent vapor, a visual display will indicate the class of agent and the relative concentration of agent. The CAM does not have an audible alarm. Real-time response capability for the detection of GB, VX, and mustard is provided by the CAM; however, the sampler must always remember that only gross-level vapor detection is possible.

5.5.6 Calibration, Certification, and Quality Assurance

Before use, each NRT monitor must be calibrated. Calibration consists of injecting known masses of agent into the inlet of the monitor during successive instrument cycles. Specifically, microliter volumes of a dilute solution of agent are injected. Thus, the response (detector signal) versus the mass of agent can be determined. After calibration, the responses obtained during subsequent cycles of operation of the NRT monitor can be converted to detected masses and to concentration readings that are then reported as found concentrations, that is, agent concentration readings reported by the NRT monitor.

Monitoring systems (and their associated written methods) must be certified before use in accordance with requirements stated in the Chemical Materials Agency's (CMA's) *Programmatic Laboratory and Monitoring Quality Assurance Plan* (U.S. Army, 2004a). Certification generally includes passing a 4-day precision and accuracy (P&A) study. P&A studies are usually conducted over a relatively narrow concentration range, typically 0.5 to 2.0 AEL (as presented in U.S. Army, 2004a). The goals of a P&A study are to:

- Demonstrate that when used for the detection of a true agent concentration of 1.00 AEL, the monitoring system (and its associated written method) is predicted to report a found concentration in the range of 0.75 to 1.25 AEL (that is, 75% to 125% recovery) with a precision of ±25% with 95% confidence
- Document the P&A of the monitoring system at all concentrations used in the study (U.S. Army, 2001)

Monitoring systems and written methods are generally not tested formally outside the concentration range required for the P&A study (U.S. Army, 2004a). Thus, the accuracy of a given monitoring system for concentrations outside the range tested is generally considered to be uncertified. Note: Due to the need to calibrate against actual agent, approval for similar systems outside of military controlled sites may be problematic.

REFERENCES

Programmatic Laboratory and Monitoring Quality Assurance Plan, US Army Chemical Materials Agency, June 2004.

U.S. Army 1990 Field Manual 3–9, *Potential Military Chemical/Biological Agents and Compounds*, Department of the Air Force, Washington, DC, 12 December 1990, PCN 320 00845700.

Toxicological Profiles for Mustard, U.S. Department of Health and Human Services, Agency for Toxic Substances and Disease Registry (ATSDR), 2003.

ADDITIONAL READING

Analysis of the Clandestine CB Threat to USAF Strategic Forces (Unclassified), Volumes I and 11. Technical Considerations, U.S. Department of Defense, Defense Technical Information Center (DTIC) Number AD379465, February 1967.

Antiterrorism Front-End Analysis (Unclassified), U.S. Department of Defense, Defense Technical Information Center (DTIC) Number AD-C954865, June 1984.

Background Paper, Technologies Underlying Weapons of Mass Destruction, Office of Technology Assessment, U.S. Congress, OTA-BP-ISC-1 15.

Behind the first roar of machinery to drain the city: A tale of pluck and luck, John Schwartz, *New York Times*, September 8, 2005.

Biological Effects of Radiation, U.S. Nuclear Regulatory Commission, Office of Public Affairs.

Chemical/Biological Hazard Prediction Program, Technical Report, U.S. Department of Defense, Defense Technical Information Center (DTIC) Number AD-BL 63245, 1991.

Chemical, Biological and Radiological Incident Handbook, Chemical, Biological and Radiological (CBRN) Subcommittee of the Interagency Intelligence Committee on Terrorism (IICT). October 1998.

Effects of Terrorist Chemical Attack on Command, Control, Communications, and Intelligence (C^3I) Operations, U.S. Department of Defense, Defense Technical Information Center (DTIC) Number AD-BL 65614, April 1992.

Fundamentals of Protective Design for Conventional Weapons, TM 5-855-1.Washington, DC, Headquarters, U.S. Department of the Army, 1986.

Glossary of Terms in Nuclear Science and Technology, La Grange, IL, American Nuclear Society, 1986.

Guidance for Protecting Building Environments from Airborne Chemical, Biological, or Radiological Attacks, National Institute of Occupational Safety and Health (NIOSH), U.S. Department of Health and Human Services (DHHS) (NIOSH) Pub No. 2002-139, May 2002.

HHS awards contract for radiation countermeasures, Global Security Newswire, February 14, 2006.

North American Emergency Response Manual, U.S. Department of Transportation, Transport Canada, Secretariat of Transport and Communications, 1996.

Nuclear Reactor Concepts, U.S. Nuclear Regulatory Commission, May 1993.

Nuclear Terms Handbook, U.S. Department of Energy, Office of Nonproliferation and National Security, 1996.

Proliferation of Weapons of Mass Destruction, Assessing the Risk, Office of Technology Assessment, U.S. Congress, OTA-ISC-559.

Protective Construction Design Manual, ESL-TR-87-57. Prepared for Engineering and Services Laboratory, Tyndall Air Force Base, FL. 1989.

Structures to Resist the Effects of Accidental Explosions, Army TM 5-1300, Navy NAVFAC P-397, AFR 88-2.Washington, DC, Departments of the Army, Navy and Air Force, 1990.

Weapons of Mass Destruction Terms Handbook, U.S. Defense Special Weapons Agency, DSWA-AR-40H, June 1, 1998.

CHAPTER 6

Biological Threats

Martha Boss, Dennis Day, and Charles Allen

CONTENTS

Symptoms associated with exposure to biological agents (bacteria, viruses) vary greatly with the agent and may take days or weeks to develop. These agents may result in high morbidity and mortality rates among the targeted population.

The two primary biological threats are:

1. Exposure to biologicals disseminated from disease sources or as a weapon
2. Secondary biological threats as humans sicken from chemical, biological, and nuclear (CBN) agents and environmental fouling of shelters or other localized areas (restrooms) within facilities occurs

Should a site or facility become contaminated with a biological agent, the procedures described in this text should be used to contain and limit the risk. This chapter discusses the types of threats. Chapters 16 through 18 discuss decontamination methods (with mold used as an exemplary threat).

Viruses, bacteria, molds/fungi, yeasts, and protozoa can cause infection in the human body. Viruses and prions are infective proteins that do not exhibit all the characteristics of life forms, and yet are considered a biological threat. Some prion components have even been shown to be infective after incineration.

If a facility becomes vulnerable or is assessed as being vulnerable to biological threat, basic understanding of biologicals is important to determine responses.

6.1 BIOLOGICAL AGENTS

Biological agents such as *Bacillus anthracis* (anthrax), *Variola major* (smallpox), *Yersinia pestis* (bubonic plague), *Brucella suis* (brucellosis), *Francisella tularensis* (tularemia), *Coxiella burnetti* (Q fever), *Clostridium botulinum* (botulism toxin), viral hemorrhagic fever agents, and others have the potential for use in a terrorist attack and may present the greatest hazard. Each of these biological agents may travel through the air as an aerosol.

In nature, biological agents and other aerosols often collide to form larger particles. However, terrorists or other groups may modify these agents to reduce agglomeration, thus increasing the number of biological agents that may be inhaled. An agent's infectivity, toxicity, stability as an aerosol, ability to be dispersed, and concentration influence the extent of the hazard. Other important factors include person-to-person agent communicability and treatment difficulty.

Biological agents have many entry routes and physiological effects. These agents are generally nonvolatile and can normally be removed by appropriately selected particulate filters. However, biological agents may persist on surfaces.

6.2   BIOLOGICAL AGENTS RISK GROUP 2

The primary characteristic of Risk Group 2 is that human derived blood/tissue/cell lines are the sources of the agents or the vectors. This risk group is associated with human disease of varying severity. Under normal circumstances, these pathogens are unlikely to be a serious hazard to healthy workers, community, livestock, or the environment. And exposures rarely cause infection leading to serious disease. In part, the limited severity is due to effective treatment and preventive measures that are available to limit the risk of spread.

Primary hazards relate to:

- Percutaneous or mucous membrane exposures
- Ingestion of infectious materials

6.2.1   Biosafety Level 2: Containment

Even though organisms routinely manipulated at Biosafety Level 2 are not known to be transmissible by the aerosol route, procedures with aerosol or high splash potential that may increase the risk of such personnel exposure must be avoided or contained in either:

- Primary containment equipment, or
- Devices such as a biological safety cabinet (BSC)

A biohazard sign must be posted on the entrance to all contaminated areas. Appropriate information to be posted includes:

- Agent(s) suspected and biosafety level
- Responsible parties' names and phone number(s)
- If entry is anticipated: required immunizations, personal protective equipment (PPE), and exit procedures. Personal protective equipment should include splash shields, face protection, gowns, and gloves.

6.2.2   Sharps

Extreme caution must be taken with any sharps. Rooms, until decontaminated, should not contain sharps of any kind. Identified sharps and suspect fluids/tissues should be assumed to be at least Risk Group 2. Plastic should be used in lieu of glass whenever possible. Containers should be provided for contaminated sharps that may be found. These containers must be leakproof and structurally sound for future transfer either to autoclaves or decontamination.

6.2.3 Examples of Risk Level 2 Organisms

Hepatitis B virus, HIV, the salmonellae, and *Toxoplasma spp.* are representative of microorganisms assigned to this containment level. Other examples of Risk Group 2 agents are as follows:

BACTERIA

CHLAMYDIA

MYCOPLASMA

Actinobacillus: all species

Actinomyces pyogenes (C. pyogenes)

Bacillus cereus

Bartonella bacilliformis, B. henselae, B. quintana, B. elizabethae

Bordetella pertussis, B. parapertussis and B. bronchiseptica

Borrelia recurrentis, B. burgdorferi

Campylobacter spp: C. coli, C. fetus, C. jejuni

Chlamydia pneumoniae, C. psittaci (non-avian strains), *C. trachomatis*

Clostridium botulinum, Cl. chauvoei, Cl. difficile, Cl. haemolyticum, Cl. histolyticum, Cl. novyi, Cl. perfringens, Cl. septicum, Cl. sordellii, Cl. tetani

Corynebacterium diphtheriae, C. haemolyticum, C. pseudotuberculosis, C. pyogenes (A. pyogenes)

Edwardsiella tarda

Erysipelothrix rusiopathae (insidiosa)

Escherichia coli: enterotoxigenic/invasive/hemorrhagic strains

Francisella tularensis Type B, (biovar palaearctica), F. novocida

Fusobacterium necrophorum

Haemophilus influenzae, H. ducreyi

Helicobacter pylori

Legionella spp.

Leptospira interrogans: all serovars

Listeria monocytogenes

Mycobacteria: all species except *M. tuberculosis* and *M. bovis* (non-BCG strain), which are in Risk Group 3

Mycoplasma pneumoniae, M. hominis

Neisseria gonorrhoeae, N. meningitidis

Nocardia asteroides, N. brasiliensis

Pasteurella: all species except *P. multocida type B,* which is in Risk Group 3

Pseudomonas aeruginosa

Salmonella enterica (S. choleraesuis)

Salmonella enterica serovar arizonae (Arizona hinshawii)

Salmonella enterica serovar gallinarum-pullorum (S. gallinarum-pullorum)

Salmonella enterica serovar meleagridis (S. meleagridis)

Salmonella enterica serovar paratyphi B (S. paratyphi B) (Schottmulleri)

Salmonella enterica serovar typhi (S. typhi)

Salmonella enterica serovar typhimurium (S. typhimurium)

Shigella boydii, S. dysenteriae, S. flexneri, S. sonnei

Staphylococcus aureus

Streptobacillus moniliformis

Streptococcus spp: Lancefield Groups A, B, C, D, G

Treponema carateum, T. pallidum (including *T. pertenue), T. vincentii*

Ureaplasma urealyticum

Vibrio cholerae (including El Tor), *V. parahaemolyticus, V. vulnificus*

Yersinia enterocolitica, Y. pseudotuberculosis

FUNGI

Cryptococcaceae

Candida albicans

Cryptococcus neoformans

Moniliaceae

Aspergillus flavus

Aspergillus fumigatus

Epidermophyton floccosum

Microsporum spp.

Sporothrix schenckii

Trichophyton spp.

VIRUSES

Arthropod-borne viruses are identified with an asterisk (*). Only those viruses which may be associated with human or animal disease have been included in this list. Agents listed in this group may be present in blood, CSF, central nervous system and other tissues, and infected arthropods, depending on the agent and the stage of infection.

Adenoviridae

Adenoviruses: all serotypes

Arenaviridae

Lymphocytic choriomeningitis virus: the containment facility adapted strains

Tacaribe virus complex: Tamiami, Tacaribe, Pichinde

Bornaviridae

Borna disease virus

*Bunyaviridae**

Genus Bunyavirus

Bunyamwera and related viruses

California encephalitis group, including LaCrosse, Lumbo and Snowshoe hare virus

Genus Phlebovirus: all species except Rift Valley fever virus

Caliciviridae: all isolates, including Hepatitis E and Norwalk virus

Coronaviridae

Human coronavirus: all strains

Genus Torovirus

Transmissible gastroenteritis virus of swine

Hemagglutinating encephalomyelitis virus of swine

Mouse hepatitis virus

Bovine coronavirus

Feline infectious peritonitis virus

Avian infectious bronchitis virus

Canine, Rat and Rabbit coronaviruses

*Flaviviridae**

Yellow fever virus: 17D vaccine strain

Dengue virus: serotypes 1, 2, 3, 4

Kunjin virus

Hepatitis C virus

Hepadnaviridae

Hepatitis B virus, including Delta agent

Herpesviridae

Alphaherpesvirinae

Genus Simplexvirus: all isolates including HHV 1 and HHV 2, *except Herpes B virus which is in Risk Group 4*

Genus Varicellavirus: all isolates including varicella/zoster virus (HHV 3) and pseudorabies virus

Betaherpesvirinae

Genus Cytomegalovirus: all isolates including CMV (HHV 5)

Genus Muromegalovirus: all isolates

Gammaherpesvirinae

Genus Lymphocryptovirus: Epstein Barr Virus (HHV 4) and EB-like isolates

Genus Rhadinovirus: all isolates except H. ateles and H. saimiri in Risk Group 3

Genus Thetalymphocryptovirus: all isolates

Unassigned Herpesviruses: includes HHV 6 (human B-lymphotrophic virus), HHV 7, HHV 8, etc.

Orthomyxoviridae

Genus Influenzavirus: Influenza virus type A: all isolates

Influenza virus type B: all isolates

Influenza virus type C: all isolates

Papovaviridae

Genus Papillomavirus: all isolates

Genus Polyomavirus: all isolates

Paramyxoviridae

Genus Morbillivirus: all isolates except Rinderpest virus

Genus Paramyxovirus: all isolates

Genus Pneumovirus: all isolates

Parvoviridae

Genus Parvovirus: all isolates

Picornaviridae

Genus Aphthovirus

Genus Cardiovirus: all isolates

Genus Enterovirus: all isolates

Genus Hepatovirus: all isolates (Hepatitis A)

Genus Rhinovirus: all isolates

Poxviridae

Chordopoxvirinae (poxviruses of vertebrates)

Genus Avipoxvirus: all isolates

Genus Capripoxvirus

Genus Leporipoxvirus: all isolates

Genus Molluscipoxvirus

Genus Orthopoxvirus: all isolates except Variola virus and Monkeypox virus which are in Risk Group 4

Genus Parapoxvirus: all isolates

Genus Suipoxvirus: Swinepox virus

Genus Yatapoxvirus

All other ungrouped poxviruses of vertebrates

Reoviridae

Genus Orbivirus: all isolates (with restrictions)

Genus Orthoreovirus: types 1, 2 and 3

Genus Rotavirus: all isolates

Retroviridae

Oncovirinae

Genus Oncornavirus C

Subgenus Oncornavirus C avian: all isolates

Subgenus Oncornavirus C mammalian: all isolates except HTLV-I and HTLV-II

Genus Oncornavirus B: all isolates

Lentivirinae: all isolates except HIV-I and HIV-II

Spumavirinae: all isolates

Rhabdoviridae

Genus Vesiculovirus: all the containment facility adapted strains

Genus Lyssavirus: Rabies virus (fixed virus)

Togaviridae

*Genus Alphavirus**

Semliki forest virus

Sindbis virus

Chikungunya virus: high-passage strains

O'Nyong-Nyong virus

Ross River virus

Venezuelan equine encephalitis virus: only strain TC-83,

no animal inoculation

Genus Rubivirus

Rubella virus

Genus Pestivirus

Bovine diarrhoea virus

Border disease virus

Genus Arterivirus

Equine arteritis virus

Unclassified viruses

Other Hepatitis viruses

Astro viruses

Chronic infectious neuro-pathic agents (CHINAs): Scrapie, BSE

(except Kuru and Creutzfeldt-Jakob Disease agents in Risk Group 3)

PARASITES

Infective stages of the following parasites have caused the contain-ment facility infections by ingestion, skin or mucosal penetration, or accidental injection. Preparations of these parasites known to be free of infective stages do not require this contain-ment level.

PROTOZOA

Babesia microti

Babesia divergens

Balantidium coli

Cryptosporidium spp.

Entamoeba histolytica

Giardia spp. (mammalian)

Leishmania spp. (mammalian)

Naegleria fowleri

Plasmodium spp. (human or simian)

Pneumocystis carinii

Toxoplasma gondii

Trypanosoma brucei, T. cruzi

HELMINTHS

Nematodes

Ancylostoma duodenale

Angiostrongylus spp.

Ascaris spp.

Brugia spp.

Loa loa

Necator americanus

Onchocerca volvulus

Strongyloides spp.

Toxocara canis

Trichinella spp.

Trichuris trichiura

Wuchereria bancrofti

Cestodes

Echinococcus (gravid segments)

Hymenolepis diminuta

Hymenolepis nana (human origin)

Taenia saginata

Taenia solium

Trematodes

Clonorchis sinensis

Fasciola hepatica

Opisthorchis spp.

Paragonimus westermani

Schistosoma haematobium

Schistosoma japonicum

Schistosoma mansoni

6.3   RISK GROUP 3: AGENTS REQUIRING CONTAINMENT LEVEL 3

These agents may cause serious or potentially lethal disease through exposure by the inhalation route. Biosafety Risk Group 3 includes indigenous and/or exotic agents. The risks are high for both individuals who immediately contact these agents and the community (through dispersion). These agents are not ordinarily spread by casual contact, from one individual to another; however a potential for respiratory transmission must always be considered. Serious and potentially lethal infections may result from exposure to these agents.

Primary hazards to personnel working with these agents relate to:

- Auto inoculation
- Ingestion
- Exposure to infectious aerosols

6.3.1   Biosafety Level 3: Containment

At Biosafety Level 3, more emphasis is placed on primary and secondary barriers to protect personnel in contiguous areas, the community, and the environment from exposure to potentially infectious aerosols. Ventilation systems are designed to minimize the spread of these agents and to prohibit release to the general public.

- For buildings that are not designed for this contaminant level, ventilation systems should be assumed to be contaminated and all ventilation exhaust should be stopped. After successful decontamination of both the building proper and the ventilation system, ventilation systems can be restarted.
- If buildings have zoned ventilation systems, sheltering in place in an unaffected zone may be possible if the contaminated areas can be sealed. Sealing would include shutdown of all ventilation systems from contaminated areas, and subsequent monitoring of pressure differentials to assure that occupied zones remain positive relative to contaminated areas.

All penetrations for services in the floors, walls, and ceiling of the containment facility must be sealed. Windows must be sealed and unbreakable (boarded/reinforced as needed). Doors to contaminated areas must be locked. Openings such as those around ducts and the spaces between doors and frames should be sealed.

A biohazard sign must be posted on the entrance to all contaminated areas. Appropriate information to be posted includes:

- Agent(s) suspected and biosafety level
- Responsible parties' names and phone number(s)
- If entry is anticipated: entry procedures, required immunizations, PPE, and exit procedures. Personal protective equipment should include splash shields, face protection, gowns, and gloves.

Entry and exit procedures would include use of airlocks. Prior to exiting, personnel decontamination would include a body shower provided within the containment perimeter. Respiratory and face protection devices should be stored in a "clean" area outside the containment perimeter and accessible from the donning passageway.

6.3.2 Plumbing

- Water supplied to the containment facility must be provided with reduced pressure backflow preventers.
- All vent lines should be provided with high-efficiency particulate air (HEPA) filters or equivalent protection.
- Sink and floor drains should not be used for disposal of infectious materials. Floor drains should be sealed.

6.3.3 Vacuum System

Vacuum lines if used must be protected with liquid disinfectant traps and HEPA filters, or their equivalent. Filters must be replaced as needed, and this replacement must be documented. No vacuum lines may exit through the containment perimeter.

An alternative is to use portable vacuum pumps (also properly protected with traps and filters). Portable vacuum pumps must be fitted with in-line HEPA filters or equivalent equipment.

6.3.4 General Ventilation

A visual monitoring device that indicates and confirms directional inward airflow should be provided at the entry. The Level 3 facility must not become positively pressurized relative to the surrounding area. Audible alarms should be considered to notify personnel of heating, ventilation, and air conditioning (HVAC) system failure. The air supply system to the contaminated area must be interlocked with the air exhaust system. Air must flow inward from both entry and exit areas. The containment facility should be provided with a sealed dedicated air supply and exhaust system. The air discharged must be prevented from recirculating back into the air supply system of the containment facility, the surrounding building, or adjacent buildings.

The outside exhaust must be:

- Dispersed away from occupied areas and air intakes, or
- HEPA filtered

When the air is not exhausted through a dedicated exhaust system, exhaust air must be HEPA filtered before discharge into the main building air exhaust system. The exhaust HEPA filter:

- Plenum must be designed to allow in situ decontamination
- Must pass annual testing and certification by aerosol challenge and scan testing techniques

When the supply air is not provided by a dedicated system, airtight back-draft dampers or HEPA filters must be installed in the air supply system.

6.3.5 Personal Protective Equipment (PPE)

Solid-front clothing is required for entry into a contaminated area. Personal protective clothing includes disposable head covers and dedicated shoes or foot covers.
Gloves must be worn when handling infectious or potentially infectious materials or
contaminated equipment. Frequent changing of gloves accompanied by hand washing is recommended. Disposable gloves are not reused. Appropriate respiratory protection should be used.

6.3.6 Contaminated Material

All potentially contaminated waste materials are decontaminated before disposal or
reuse. A decontamination method should be available in the containment facility and
used, preferably within the containment facility (i.e., autoclave, chemical disinfection, incineration). Waste transported out of the contaminated area must be properly
sealed and transported in secured corridors.

6.3.7 Examples of Risk Level 3 Organisms

Mycobacterium tuberculosis, St. Louis encephalitis virus, and *Coxiella burnetii* are
representative microorganisms assigned to this level. Examples are as follows:

BACTERIA

CHLAMYDIA

RICKETTSIA

Bacillus anthracis

Brucella: all species

*Burkholderia
(Pseudomonas) mallei,
B. pseudomallei*

Chlamydia psittaci: avian
strains only

Coxiella burnetti

Francisella tularensis
type A (biovar tularensis)

Mycobacterium bovis:
non-BCG strains

*Mycobacterium tubercu-
losis 1*

Pasteurella multocida,
type B

Rickettsia: all species

Yersinia pestis

1 Preparation of smears
and primary culture of
M. tuberculosis may be
performed at Level 2
physical containment
using Level 3 opera-
tional procedures and
conditions. All other

manipulations of *M.
tuberculosis* require
Containment Level 3
physical and operational
conditions.

FUNGI

Moniliaceae

*Ajellomyces capsulatum
(Histoplasma capsu-
latum*, including var.
duboisii)

*Ajellomyces derma-
titidis (Blastomyces
dermatitidis)*

Coccidioides immitis

*Paracoccidioides
brasiliensis*

VIRUSES

Arthropod-borne viruses
are identified with an
asterisk (*).

Arenaviridae

Lymphocytic choriomen-
ingitis virus: neurotropic
strains

Bunyaviridae

Unclassified Bunyavirus

Hantaan, Korean hemor-
rhagic fever and epidemic
nephrosis viruses includ-
ing Hantavirus pulmo-
nary syndrome virus

Rift Valley fever virus

*Flaviviridae**

Yellow fever virus: wild
type

St. Louis encephalitis
virus

Japanese encephalitis
virus

Murray Valley
encephalitis virus

Powassan encephalitis
virus

Herpesviridae

Gammaherpesvirinae

Genus *Rhadinovirus*:
Herpesvirus ateles,
Herpesvirus saimiri

Retroviridae

Oncovirinae

Genus *Oncornavirus C*

Human T-cell leukemia/
lymphoma virus 2

Genus *Oncornavirus D*

Mason-Pfizer monkey
virus

Viruses from nonhuman
primates

Lentivirinae

Human immunodefi-
ciency viruses (HIV):
all isolates 2

Rhabdoviridae

Genus *Vesiculovirus*:
wild type strains

Genus *Lyssavirus*

Rabies virus (street virus)

Togaviridae

Genus *Alphavirus**

Eastern equine encepha-
litis virus

Chikungunya virus

Venezuelan equine
encephalitis virus (except
Strain TC-83)

Western equine encepha-
litis virus

Unclassified Viruses

Chronic infectious
neuropathic agents:
Kuru, Creutzfeldt-Jakob
Disease agents (level of
precautions depends on
the nature of the manipu-
lations and the amount of
sera, biopsy/necropsy
materials handled)

PARASITES: none

6.4   RISK GROUP 4: AGENTS REQUIRING CONTAINMENT LEVEL 4

Biosafety Level 4 organisms may be readily transmitted from one individual to another,
or from animal to human or vice versa, directly or indirectly, or by casual contact.

- May be transmitted via the aerosol route
- No available vaccine or therapy is currently available

The primary hazards are:

- Respiratory exposure to infectious aerosols
- Mucous membrane or broken skin exposure to infectious droplets
- Autoinoculation

6.4.1 Biosafety Level 4: Containment

A Biosafety Level 4 facility is generally a separate building or completely isolated zone with complex, specialized ventilation requirements and waste management systems to prevent release of viable agents. In buildings that are not so designed, evacuation may be the only sound alternative. If sheltering in place must occur, all the procedures listed as required for Risk Group 3 should be used in addition to the requirements listed in this section.

All liquid effluent from the contaminated area must be treated prior to discharge. Sewer vents and other service lines should have HEPA filters installed and protection against vermin intrusion.

Redundant exhaust fans are required from the contaminated area. The supply air to and exhaust air from the contaminated area (including vestibules and airlocks to this area) are fitted with HEPA filter(s). The general room exhausts air from the protective suit (donning and doffing) area, decontamination showers, and the decontamination airlock is to be treated by a passage through two HEPA filters in series prior to discharge to the outside. The air is discharged away from occupied spaces and air intakes. The HEPA filter(s) are located as near as practicable to the source in order to minimize the potentially contaminated ductwork.

6.4.2 Decontamination Equipment

Double-door autoclaves are provided to decontaminate materials. Autoclaves that open outside of the containment barrier must be sealed to the containment barrier wall. The autoclave doors are automatically controlled so that the outside door can only be opened after the autoclave *sterilization* cycle has been completed.

Pass-through dunk tanks, fumigation chambers, or equivalent decontamination methods are provided:

- For transfer of materials from room to room
- So that materials and equipment that cannot be decontaminated in the autoclave can be removed from both the Class III BSC(s) and the cabinet room(s)

6.4.3 PPE and Decontamination

Personnel must enter and leave the contaminated area only through the clothing change and shower rooms. Entry is through an airlock fitted with airtight doors. Inner and outer doors to the chemical shower and inner and outer doors to airlocks are interlocked to prevent both doors from being opened simultaneously. Outer and inner change rooms separated by a shower are provided for personnel entering and exiting.

A specially designed suit area is maintained in the containment facility to provide personnel protection. Personnel who enter wear a one-piece positive-pressure suit that is ventilated by a life-support system protected by HEPA filtration. The life support system includes redundant breathing air compressors, alarms, and emergency backup breathing air tanks.

- Entry: Remove and store personal clothing in the outer clothing change room. Complete containment facility clothing, including undergarments, pants and shirts or jumpsuits, shoes, and gloves, is provided and used by all personnel entering the containment facility. Personnel take a decontaminating chemical shower each time the containment facility is exited.
- Exit: When leaving the containment facility and before proceeding into the shower area, personnel remove their containment facility clothing in the inner change room.

Soiled clothing is autoclaved before laundering.

6.4.4 Materials

Supplies and materials needed in the contaminated area are brought in by way of the double-doored autoclave, fumigation chamber, or airlock, which is appropriately decontaminated between each use. After securing the outer doors, personnel within the contaminated area retrieve the materials by opening the interior doors of the autoclave, fumigation chamber, or airlock. These doors are secured after materials are brought in.

Biological materials if removed are:

1. Autoclaved or decontaminated
2. Transferred to a nonbreakable, sealed primary container
3. Enclosed in a nonbreakable, sealed secondary container
4. Removed through a disinfectant dunk tank, fumigation chamber, or an airlock designed for this purpose

Equipment or material that might be damaged by high temperatures or steam may be decontaminated by gaseous or vapor methods in an airlock or chamber designed for this purpose.

6.4.5 Examples of Risk Level 4 Organisms

BACTERIA

None

FUNGI

None

VIRUSES

Arthropod-borne viruses are identified with an asterisk (*).

Arenaviridae

Lassa, Junin, Machupo, Sabia, Guanarito viruses

*Bunyaviridae**

Genus *Nairovirus*

Crimean-Congo hemorrhagic fever virus

Filoviridae

Marburg virus

Ebola virus

*Flaviviridae**

Tick-borne encephalitis complex including Russian Spring-Summer Encephalitis virus

Kyasanur forest virus

Omsk hemorrhagic fever virus

Herpesviridae

Alphaherpesvirinae

Genus *Simplexvirus*: Herpes B virus (*Cercopithecine herpesvirus* 1)

Poxviridae

Genus *Orthopoxvirus*

Variola virus

Monkeypox virus

PARASITES

None

6.5 PRIONS AND DECONTAMINATION

Prions are *infectious protein particles* that lack nucleic acids. The infections caused by prions include a group of similar human and animal neurodegenerative diseases referred to as transmissible spongiform encephalopathies (TSEs) or transmissible degenerative encephalopathies (TDEs). The etiologic agents responsible for TSEs have several unique properties:

- No nucleic acids
- Are impossible to filter
- Cannot be cultured or observed at electron microscopy
- High resistance to chemical and physical sterilization procedures effective against conventional microorganisms

Primary routes of transmission are by ingestion, inoculation, and implantation. Direct contact, droplet, and airborne spread have not been implicated but cannot be ruled out.

6.5.1 Inactivation Issues

Prions are notoriously resistant to most of the physical and chemical methods used for inactivation of pathogens that are conventional microorganisms. Consequently, prion-contaminated items should be segregated from all other wastes to assure effective treatment processes, and to minimize the materials subjected to these more difficult and potentially costly processes. Contact between materials or equipment used and infectious prions must be minimized. Where possible, use disposable items that can be incinerated or treated by alkaline hydrolysis should be used.

Reference treatments recommended by the World Health Organization include:

- Soaking in 1N sodium hydroxide (1 hour at 20°C)
- Soaking in 12.5% bleach (1 hour at 20°C)
- Autoclaving (18 minutes at 134 to 138°C)

Where possible, two or more different methods of inactivation should be combined in any sterilization procedure for these agents. (WHO, 1999).

6.5.2 Inactivation by Autoclaving

The presence of elemental carbon protects infectivity during autoclaving. Infectivity survives autoclaving at 132 to 138°C and the autoclaving effectiveness may actually decline at higher temperatures. Incomplete inactivation by autoclaving may result in the selection for or formation of subpopulations of prions that have biological characteristics that are different from the main population and cannot be inactivated by re-autoclaving. Stabilization of thermostable subpopulations of prions may occur through the smearing and drying of infected tissue on surfaces (Taylor, Fernie, McConnell, and Steel, 1998). Combining sodium hydroxide treatment with autoclaving is extremely effective even at 121°C (Taylor, 1999).

6.5.3 Inactivation by Chemical Sterilants

Chemical inactivation methods are often favored for large or fixed surfaces and fixed equipment. However, prions are extremely resistant to most commonly used disinfectants. Effective sterilants may include using:

- Strong (2 Molar) sodium hydroxide solutions for at least 2 hours
- Strong sodium hypochlorite solutions containing 20,000 ppm of available chlorine
- Proteases-proteinase K, pronase with prolonged digestion

Glycidol may be a potential prion disinfectant. Guanidine thiocyanate (GdnSCN) solutions have been found to be highly effective in inactivating prions, even in complex materials such as whole brain tissue (Yamamotoa, Horiuchi, Ishiguro, Shinagawa, Masuo, and Kaneko, 2001).

Prions are more easily inactivated under alkaline conditions. Higher pH and increasing concentrations of urea or guanidine thiocyanate decreases the protease resistance of prion proteins (Madec, Vanier, Dorier, Bernillion, Belli, and Baron, 1997).

6.5.4 Inactivation by Dry Heating

Thermal inactivation is most efficient with steam under pressure. Under dry conditions, TSEs exhibit extraordinary resistance to thermal inactivation. Specimens of scrapie-infected brain tissue subjected to dry heat at 360°C for one hour still retain a small amount of infectivity. Limited infectivity has also been found in ash from similar tissues heated to 600°C, approaching the operating temperature of medical waste incinerators. A thermally resilient inorganic template of replication has been suggested to explain this pattern of infectivity (Brown, Rau, Bacote, Gibbs, and Gajdusek, 2000; Brown, Liberski, Wolff, and Gajdusek, 1990).

6.5.5 Decontamination of Work Surfaces

Because TSE infectivity persists for long periods on work surfaces:

- Use disposable cover sheets whenever possible to avoid environmental contamination, even though transmission to humans has never been recognized to have occurred from environmental exposure.
- Mechanically clean and disinfect equipment and surfaces that are subject to potential contamination, to prevent environmental buildups.

Surfaces contaminated by TSE agents can be disinfected by flooding, for one hour, with NaOH or sodium hypochlorite, followed by water rinses. Surfaces that cannot be treated in this manner should be thoroughly cleaned.

Cleaning materials must be treated as potentially contaminated.

6.5.6 Personal Protective Equipment

Handling of contaminated instruments during transfers and cleaning should be kept to a minimum. Persons involved in the disinfection and decontamination of instruments or surfaces exposed to the tissues of persons with TSE should wear single-use protective clothing, gloves, mask, and visor or goggles. Disposable gloves and an apron should be worn when removing such spills and should subsequently be disposed of by incineration, together with the recovered waste and cleaning materials.

6.5.7 Waste Disposal

All material classified as prion-exposed waste should be placed in secure leakproof containers and disposed of by incineration at an authorized incineration site. Avoid external contamination of the container to ensure safe handling. TSE infectious waste should be incinerated or treated by a method that is effective for the inactivation of TSE agents. In regions where no incineration facilities are available, these wastes should be either chemically disinfected or burned in pits dedicated to final disposal. Residues should be checked for total combustion (WHO, 1999).

6.5.8 Decontamination of Wastes and Waste-Contaminated Materials

Decontamination of waste liquid and solid residues should be conducted with the same care and precautions recommended for any other exposure to TSE agents.

The work area should be selected for easy containment of contamination and for subsequent disinfection of exposed surfaces.

All waste liquids and solids must be captured and treated as infectious waste. Liquids used for cleaning should be decontaminated in situ by addition of NaOH or hypochlorite or equivalent procedures, and may then be disposed of.

Cleaning tools and methods should be selected to minimize dispersal of the contamination by splashing, splatters and aerosols. Great care is required in the use of brushes and scouring tools. Where possible, cleaning tools such as brushes, toweling, and scouring pads, as well as tools used for disassembling contaminated apparatus, should either be disposable or selected for their ability to withstand the disinfection procedures.

Upon completion of the cleaning procedure, all solid wastes including disposable cleaning materials should be collected and decontaminated. Incineration is highly recommended. The cleaning station should then itself be decontaminated.

6.5.9 Recommended Practices: Disposal of Wastes

Contaminated solid wastes such as table covers, instrument pads, and disposable clothing should be double bagged and incinerated (Baron and Prusiner, 2000).

As a routine procedure, hydroxide-treated wastes should be treated in a gravity-displacement autoclave in sealed, heat-resistant containers that can stand the pressures involved. This practice will make the procedure safer for the operator, and reduce the potential for spills and autoclave damage from the hydroxide solution. Alternatively, hydroxide-treated wastes may be processed in a porous-load autoclave, provided that the containers used can withstand the higher temperatures (134 to 138°C), and the initial and final high vacuum stages that are a features of this process (Taylor, Fernie, and McConnell, 1997).

Some facilities now pretreat wastes or the ash from incineration with sodium hydroxide to assure inactivation of the agent.

6.6 TOXINS

Toxin categories include bacterial (exotoxins and endotoxins), algae (blue-green algae and dinoflagellates), mycotoxins (tricothocenes and aflatoxins), botulinum, and plant- and animal-derived toxins.

Toxins form an extremely diverse category of materials and are typically most effectively introduced into the body by inhalation of an aerosol. Biological toxins are often much more toxic than chemical agents. Their persistency is determined by their stability in water and exposure to heat or direct solar radiation. Under normal circumstances toxins can be collected using appropriately selected particulate filters.

6.7   LEVELS OF CONCERN

Exposure to biological contamination occurs every day. Often the immune system and the lack of concentrated biological contamination in the environment prevents illness from developing. However, if either the immune system functions incorrectly or the biological contamination becomes too large in numbers, allergic reactions and/or other illness symptoms may result.

To protect ourselves we use the following:

- Sterilization
- Disinfection
- Decontamination
- Dilution

Of these, only sterilization attempts to kill all biological contaminants. Disinfection attempts to kill sufficient numbers in order to lessen the infective potential of contaminants.

Decontamination and dilution within interior environments seeks to lessen the numbers of biological contaminants to some defined limit. For molds/fungi and yeasts, various limits are considered acceptable given the use of the buildings or areas, and the health status of the people potentially exposed. Currently, professional opinion and state guidelines are both used to determine acceptable limits.

- In general for molds/fungi and yeasts, the limits for most pathogenic genera are between 100 and 200 colony-forming units in a cubic meter of air. This limit is expressed as cfu/m^3. A cubic meter of air is equivalent to 1000 liters of air. Note: Genera is a classification of biologicals based on their structure and form. For molds/fungi and yeasts, the structure and form of their reproductive structures and spores is also considered during classification.
- For nonpathogenic molds/fungi and yeasts, the limiting factor is the outdoor mold counts versus indoor, and also the potential for amplification to occur indoors. Amplification is the process whereby in certain areas of the building, molds and fungi are increasing in numbers at alarming rates and could ultimately contribute to unsafe airborne spore levels. *In some cases, even if the outdoor counts are higher than indoor counts, amplification if present will be a cause for concern.*
- Some pathogenic molds/fungi are of concern at any level.

Prior to beginning any decontamination effort, the acceptable levels of biological contaminants such as molds/fungi and yeasts must be defined. This acceptable level should be listed in contract documents such as specifications.

6.8   CONTACT SAMPLING AND AIR MONITORING

Air monitoring is accomplished in coordination with contact sampling.

6.8.1 Contact Sampling

Contact sampling may involve the following:

- Collection of bulk samples
- Collection of water samples
- Swabs of surfaces
- Applying agar plates to surfaces
- Vacuuming of small areas
- Tape sampling

All contact sampling requires consistent sampling techniques and ultimately either microscopic analysis or incubation of samples followed by microscopic analysis.

6.8.2 Air Sampling

Air sampling should usually be done in coordination with contact sampling. Sampling events should document the following areas:

- Outdoors
- Control area indoors assumed to be uncontaminated at safe levels
- Area indoors where contamination is suspected and where decontamination is required
- Ventilation systems either taking or supplying air to the indoor area
- Downwind of HEPA negative air filtration units used to exhaust decontamination areas
- Clean rooms used during decontamination for personnel decontamination access

Other areas may need to be sampled depending on the zoning of work and the activity in the building. Air samples may be collected for total viable and/or nonviable airborne components.

- The total spore count samples are collected on special air filtration cassettes (Zefon®).
- Direct impaction onto agar plates or strips is also possible using a variety of impaction devices (Anderson®, RCS®, SAS®). Impaction onto agar plates is used for later laboratory incubation of the sample to determine viable spore presence and numbers.

Air may also be collected in impingers or Summa canisters. Summa canisters are used when microbe-produced volatiles are to be documented. Real-time instrumentation for microbial volatiles is also available that measures volatiles in parts per billion ranges.

Wall checking may be required by probing and drawing air from wall spaces, use of infrared photography, or moisture metering.

6.9   BIOLOGICAL THREAT SUMMARY

Biological threats are expressed as infective agents delivered to prospective hosts. Primary delivery methods within ambient atmospheres may include aerosolization; adsorption and/or absorption to particulates; particulate dispersion of virus, endospores, or prion inorganic homologues; and release of gaseous toxins of biological origin. Secondary delivery can then occur by touching infective surfaces and infection via inhalation/ingestion combinations. Note: Inhalation and ingestion should be viewed as a singular phenomenon since a common pathway is present in the pharynx.

Once released, controlling the spread of biologicals through both primary and secondary delivery means is crucial. For all surfaces, where chemical disinfection procedures are employed, the manufacturer's efficacy statements must be provided prior to disinfectant selection. A regular replacement schedule for decontamination chemicals must be used, based on manufacturer's stocking recommendations. Porous surfaces and those coated with organic greases/lubricants cannot be readily decontaminated and may be ongoing exposure vector assists.

A biohazard sign must be posted at the entrances to areas that are contaminated with infectious agents. Signage must be standardized for both the room and containers within the room that are suspect. An insect and rodent control program must be initiated, since insects and rodents, if uncontrolled, can spread biological contamination.

All wastes and waste containers that are suspect must be either proved negative or decontaminated before disposal. Both decontamination and disposal must be by an approved decontamination method. Materials to be decontaminated outside of the immediate area are:

- Placed in a durable, leakproof container and closed for transport
- Packaged in accordance with applicable regulations before removal from the containment facility

The various risk groups and extra measures that must be taken given infectivity potential are Risk Groups 2 through 4.

REFERENCES

Antloga K, Meszaros J, Malchesky PS, McDonnell GE. (2000). Prion disease and medical devices. *ASAIO J* 46(6):S69–S72.

Bagg J, Sweeney CP, Roy KM, Sharp T, Smith A. (2001). Cross infection control measures and the treatment patients at risk of Creutzfeldt-Jacob disease in UK general dental practice. *British Dental Journal London* 191(2):87–90.

Baron H, Prusiner SB. (2000). Prion diseases. In Fleming DO, Hunt DL, eds., *Biological Safety Principals and Practices*, third edition. Washington, DC, ASM Press, 187–208.

Bosque PJ, Ryou C, Telling G, Peretz D, Legname G, DeArmond SJ, Prusiner SB. (2002). Prions in skeletal muscle. *Proc. Natl. Acad. Sci. USA*, 99(6):3812–3817.

Brown P, Liberski PP, Wolff A, Gajdusek DC. (1990). Resistance of scrapie infectivity to steam autoclaving after formaldehyde fixation and limited survival after ashing at 360°C: practical and theoretical implications. *J Infect Dis* 161(3):467–472.

Brown P, Rau EH, Johnson BK, Bacote AE, Gibbs CJ Jr, Gajdusek DC. (2000). New studies on the heat resistance of hamster-adapted scrapie agent: threshold survival after ashing at 600°C suggests an inorganic template of replication. *Proc Natl Acad Sci USA* 97(7):3418–3421.

Brown P, Wolff A, Gajdusek DC. (1990). A simple and effective method for inactivating virus infectivity in formalin-fixed tissue samples from patients with Creutzfeldt-Jacob disease. *Neurology* 40(6):887–890.

Caspi S, Halimi M, Yanai A, Sasson SB, Taraboulos A, Gabizon R. (1998). The anti-prion activity of Congo red: putative mechanism. *J Biol Chem* 273(6):3484–3489, 1998.

Collins S, Law MG, Fletcher A, Boyd A, Kaldor J, Masters CL. (1999). *Surgical treatment and risk of sporadic Creutzfeldt-Jacob disease: a case-control study.* Lancet 353:693–697.

Darbord JC. Inactivation of prions in daily medical practice. (1999). *Biomed Pharmacother* 53(01):34–38.

DHHS, *Biosafety in Microbiological and Biomedical Laboratories*, 5th ed. Department of Health and Human Services, Centers for Disease Control and Prevention and National Institutes of Health, U.S. Government Printing Office, Washington, DC, 2007.

Fagih B, Eisenberg MJ. (1999). Reuse of angioplasty catheters and risk of Creutzfeldt-Jakob disease. *Am Heart J* 137(6):1173–1178.

Gale P, Young C, Stanfield G, Oakes D. (1998). Development of a risk assessment for BSE in the aquatic environment. *J Appl Microbiol* 84(4):467–477.

HHS Publication No. (CDC) (99-xxxx), *Biosafety in Microbiological and Biomedical Laboratories*, U.S. Department of Health and Human Services, Public Health Service, Centers for Disease Control and Prevention and National Institutes of Health, fifth edition. Washington, DC, Government Printing Office, April 2006.

Laurenson IF, Whyte AS, Fox C. (2001). Iatrogenic prion infection. *N Engl J Med* 345(11):840–841.

Madec JY, Vanier A, Dorier A, Bernillon J, Belli P, Baron T. (1997). Biochemical properties of protease resistant prion protein PrPsc in natural sheep scrapie. *Arch Virol* 142(8):1603–1612.

Manuelidis L. (1997). Decontamination of Creutzfeldt-Jakob disease and other transmissible agents. *J Neurovirol* 3(1):62–65.

McKenzie D, Bartz J, Mirwald J, Olander D, Marsh R, Aiken J. (1998). Reversibility of scrapie inactivation is enhanced by copper. *J Biol Chem* 273(40):25545–25547.

Medical Research Council of Canada (MRC). (1977; revised editions, 1979 and 1980). *Guidelines for the Handling of Recombinant DNA Molecules and Animal Viruses and Cells*, Minister of Public Works and Services, Ottawa.

Patterson P. (2000). Incident may have exposed patients to CJD in surgery. *OR Manager* 16(12):1, 7–8.

Paul J. (1995). Prion, from crazy cows to iatrogenic Creutzfeldt-Jakob disease. Which risk in laboratory or in hospital? *Pathol Biol* (Paris) 43(2):114–120.

Rey JF. (1999). Endoscopic disinfection. A worldwide problem. *J Clin Gastroenterol* 28(4):291–297.

Rutala WA, Weber DJ. (2001). Creutzfeldt-Jakob disease: recommendations for disinfection and sterilization. *Clin Infect Dis* 32(9):1348–1356.

Saborio GP, Permanne B, Soto C. (2001). Sensitive detection of pathological prion protein by cyclic amplification of protein misfolding. *Nature* 14:411(6839):810–813.

Schreuder BE, Geertsma RE, van Keulen LJ, van Asten JA, Enthoven P, Oberthur RC, de Koeijer AA, Osterhaus AD. (1998). Studies on the efficacy of hyperbaric rendering procedures in inactivating bovine spongiform encephalopathy (BSE) and scrapie agents. *Vet Rec* 142(18):474–480.

Taylor DM. (1999). Inactivation of prions by physical and chemical means. *J Hosp Infect* 43 Suppl:S69–S76.

Taylor DM. (2000). Inactivation of transmissible degenerative encephalopathy agents: a review. *Vet J* 159(1):10–17.

Taylor DM, Brown JM, Fernie K, McConnell I. (1997). The effect of formic acid on BSE and scrapie infectivity in fixed and unfixed brain tissue. *Veterinary Microbiology* 58(167–174).

Taylor DM, Fernie K, McConnell I. (1997). Inactivation of the 22A strain of scrapie agent by autoclaving in sodium hydroxide. *Vet Microbiol* 58(2–4):87–91.

Taylor DM, Fernie K, McConnell I, Steele PJ. (1998). Observations on thermostable subpopulations of the unconventional agents that cause transmissible degenerative encephalopathies. *Vet Microbiol* 64(1):33–38.

Taylor DM, Fernie K, McConnell I, Ferguson CE, Steele PJ. (1998). Solvent extraction as an adjunct to rendering: the effect on BSE and scrapie agents of hot solvents followed by dry heat and steam. *Vet Rec* 143(1):6–9.

Timberlin RH, Walker CA, Millison GC, et al. (1983). Disinfect ion studies with two strains of mouse pass aged scrapie agent. *J Neurol Sci* 59:355–369.

University of Toronto Biosafety Committee. (1979). *Guidelines for the Handling of Recombinant DNA Molecules, Animal Viruses and Cells, Microorganisms and Parasites.* Ottawa: Medical Research Council of Canada.

University of Toronto Biosafety Committee. (1996). *Laboratory Biosafety Guidelines*, Second Edition. Ottawa: Medical Research Council of Canada.

University of Toronto Biosafety Committee. (2000). *Biosafety Policies and Procedures Manual 2000*, Ottawa: Medical Research Council of Canada.

Walia JS, Chronister CL. (2001). Possible iatrogenic transmission of Creutzfeldt-Jakob disease via tonometer tips: a review of the literature. *Optometry* 72(10):649–652.

Waste Reduction by Waste Reduction, Inc. Available: http://www.wr2.net/ Cited March 11, 2002.

World Health Organization. (1999) *WHO Infection Control Guidelines for Transmissible Spongiform Encephalopathies. Report of a WHO Consultation*, Geneva, Switzerland, 23–26 March 1999. Available: http://www.who.int/emc-documents/tse/whocdscsraph2003c.html

Yamamoto M, Horiuchi M, Ishiguro N, Shinagawa M, Matsuo T, Kaneko K. (2001). Glycidol degrades scrapie mouse prion protein. *J Vet Med Sci* 63(9):983–990.

Zobley E, Flechsig E, Cozzio A, Enari M. Weissmann C. (1999). Infectivity of scrapie prions bound to a stainless steel surface. *Mol Med* 5(4):240–243.

CHAPTER 7

Radiation and Nuclear Threats

Gayle Nicoll, Martha Boss, and Dennis Day

CONTENTS

7.1   RADIATION SITES

Radionuclides in various chemical and physical forms have become extremely important tools in modern business, industry, research, and teaching. Radioactive materials
are incorporated in many manufactured goods and used in a myriad of industrial services. Use of radioactive materials generates radioactive waste. The ionizing radiations
emitted by these materials and wastes, however, can pose a hazard to human health.
For this reason, special precautions must be observed when radionuclides are used.

The term naturally occurring radioactive materials (NORM) has been used to
define these types of exposures from radioactive materials in their undisturbed (by
human directed enhancement) state.

When radiation is excessive, either from natural sources or man-made dispersal
of radiation sources, a facility and those occupying the facility will be vulnerable.
While filtration systems can capture some incoming radiation particles, in so doing
the filters will become radioactive. Water or other liquid decontamination is also
problematic since the water (or liquid) will become radioactive, too.

In assessing vulnerability, both the limitations to reducing vulnerability and the
very nature of the radiation threats must be considered. In addition, if sheltering-in-
place or other localized confinement is used after radiation exposure has occurred (or
continues to occur), the need for ongoing biological decontamination must be addressed.
Those exposed to the radiation may sicken and even die in a shelter situation.

To monitor exposures and, hopefully, make good choices that delimit future exposures, radiation monitoring equipment should be available and used. Preplanning for
sheltering-in-place given potential nuclear events must include installation of monitors for airborne radiation and availability of handheld real-time instrumentation.

7.1.1 Atomic Structure

The atom, which has been referred to as the "fundamental building block of matter," is itself composed of three primary particles: the proton, the neutron, and the electron. Protons and neutrons are relatively massive compared to electrons and occupy the dense core of the atom known as the nucleus. Protons are positively charged, while neutrons, as their name implies, are neutral. The negatively charged electrons are found in an extended cloud surrounding the nucleus.

The number of protons within the nucleus defines the atomic number, designated by the symbol Z. In an electrically neutral atom (i.e., one with equal numbers of protons and electrons), Z also indicates the number of electrons within the atom. The number of protons plus neutrons in the nucleus is termed the atomic mass, symbol A.

The atomic number of an atom designates its specific elemental identity. For example, an atom with a $Z = 1$ is hydrogen, an atom with $Z = 2$ is helium, while a $Z = 3$ identifies an atom of lithium. Atoms characterized by a particular atomic number and atomic mass are called *nuclides*. A specific nuclide is represented by its chemical symbol with the atomic mass in a superscript (e.g., ^{3}H, ^{14}C, ^{125}I). Nuclides with the same number of protons (i.e., same Z) but a different number of neutrons (i.e., different A) are called isotopes. Isotopes of a particular element have nearly identical chemical properties.

7.1.2 Radioactive Decay

Depending upon the ratio of neutrons to protons within the nucleus, an isotope of a particular element may or may not be stable. Over time, the nuclei of unstable isotopes spontaneously disintegrate or transform in a process known as *radioactive decay* or *radioactivity*. As part of this process, various types of ionizing radiation may be emitted from the nucleus. Nuclides that undergo radioactive decay are called *radionuclides*. This is a general term as opposed to the term *radioisotope*, which is used to describe a particular relationship. For example, ^{3}H, ^{14}C, and ^{125}I are radionuclides. Tritium (^{3}H), on the other hand, is a radioisotope of hydrogen.

Some radionuclides, such as ^{14}C, ^{40}P, and ^{238}U, occur naturally in the environment, while others, such as ^{32}Ph or ^{32}Na, are produced in nuclear reactors or particle accelerators. Any material that contains measurable amounts of one or more radionuclides is referred to as a *radioactive material*.

7.1.3 Activity

The quantity that expresses the degree of radioactivity or radiation-producing potential of a given amount of radioactive material is *activity*. The most commonly used unit of activity is the *curie* (Ci), which was originally defined as that amount of any radioactive material that disintegrates at the same rate as one gram of pure radium. This equals 3.7×10^{10} disintegrations per second (dps). A millicurie (mCi) = 3.7×10^7 dps (2.22×10^9 disintegrations per minute [dpm]) and a microcurie

(μCi) = 3.7×10^4 dps (2.22×10^6 dpm). The activity of a given amount of radioactive material is independent of the mass of the element present and is determined only by the disintegration rate. Thus, two ^{137}Cs one-curie sources might have very different masses depending upon the relative proportion of nonradioactive atoms present in each source.

7.1.4   Types of Ionizing Radiation

Ionizing radiation may be electromagnetic or may consist of high-speed subatomic particles of various masses and charges.

7.1.4.1   *Alpha Particles* (α)

Certain radionuclides of high atomic mass (e.g., ^{226}Ra, ^{238}U, ^{239}Pu) decay by the emission of alpha particles. These are tightly bound units of 2 neutrons and 2 protons each (a helium nucleus). Emission of an α particle results in a decrease of two units of atomic number (Z) and four units of atomic mass (A). Alpha particles can be shielded with gloves or a piece of paper.

7.1.4.2   *Beta Particles* (β)

There are actually two types of beta particles: beta minus (β^-) and beta plus (β^+). A β^- is an electron that is emitted from the nucleus of an atom when a neutron is converted into a proton. A β^+ is a positron (same mass as an electron, but different charge) that is emitted from the nucleus when a proton is converted into a neutron. A nucleus with more neutrons than protons may decay through the emission of a high-speed β^-, while a nucleus with more protons than neutrons may undergo β^+ decay. β particles can be shielded by clothing or glass. For the remainder of this section, the term β will refer to β^-, unless otherwise stated.

7.1.4.3   *Gamma Rays* (γ)

A nucleus that is in an excited state may emit one or more high-energy photons (i.e., particles of electromagnetic radiation) of discrete energies. The emission of these gamma rays does not alter the number of protons or neutrons in the nucleus, but instead has the effect of moving the nucleus from a higher to a lower energy state. γ-ray emission frequently follows β decay, α decay, and other nuclear decay processes.

X-rays and γ rays are electromagnetic radiation, as is visible light. The frequencies of X and γ rays are much higher than that of visible light and so each carries much more energy. Some gamma and X-rays cannot be completely shielded, depending on the amount of energy they carry. Usually lead shielding of various thicknesses can be used.

7.1.4.4 *X-rays*

X-rays are also part of the electromagnetic spectrum and are distinguished from γ rays by their source (i.e., orbital electrons, rather than the nucleus) and energy. X-rays have substantially less energy than γ rays, and can thus be effectively stopped by lead shielding. X-rays are emitted with discrete energies by electrons as they shift orbitals following certain types of nuclear decay processes.

7.1.5   Excitation/Ionization

The various types of radiation (e.g., α particles, β particles, and γ rays) impart their energy to matter primarily through excitation and ionization of orbital electrons. The term *excitation* is used to describe an interaction where electrons acquire energy from a passing particle, but are not removed completely from their atom. Excited electrons may subsequently emit energy in the form of X-rays during the process of returning to a lower energy state. The term *ionization* refers to the complete removal of an electron from an atom following the transfer of energy from a passing particle. Any type of radiation having sufficient energy to cause ionization is referred to as *ionizing radiation*. It is important to note that when an atom is ionized, the electron of interest could have been participating in a chemical bond. Thus, ionizing radiation also has the ability to break bond and cause subsequent biological damage.

7.1.6   Characteristics of Different Types of Ionizing Radiation

Alpha particles have a high specific ionization and a relatively short range. In air, α particles travel only a few centimeters while in tissue, only fractions of a millimeter. For example, an α particle cannot penetrate the dead cell layer of human skin. Therefore, α particles are not dangerous unless they are ingested or inhaled.

Beta particles have a much lower specific ionization than α particles and a considerably longer range. The relatively energetic β from ^{32}P has a range of 6 meters in air or 8 millimeters in tissue. Only 6 millimeters of air or 5 micrometers of tissue, on the other hand, stop the low-energy betas from ^3H.

Gamma and X-rays have the highest energy, and thus the possibility of doing great damage to tissue. However, these particles also have no charge, which means that they undergo only chance encounters with matter. In other words, a given γ ray has a definite probability of passing through any medium of any depth without causing any harm.

7.1.7   Exposure (Roentgen)

Exposure is a measure of the strength of a radiation field at some point. It is usually defined as the amount of charge (i.e., sum of all ions of one sign) produced in a unit mass of air when the interacting photons are completely absorbed in that mass. The most commonly used unit of exposure is the Roentgen (R), which is defined as that amount of X or gamma radiation that produces 2.58×10^{-4} coulombs of charge per kilogram of dry air. In cases where exposure is to be expressed as a rate, the unit would be R/hr or more commonly, milliroentgen per hour (mR/hr). Roentgens are very limited in their use. They apply only to photons, only in air, and only with an energy under 3 MeV.

7.1.8   Absorbed Dose (rad)

Whereas exposure is defined for air, the absorbed dose is the amount of energy imparted by radiation to a given mass of any material. The most common unit of absorbed dose is the radiation absorbed dose (rad), which is defined as a dose of 100 ergs of energy per gram of the material in question. Absorbed dose may also be expressed as a rate with units of rad/hr or millirad/hr.

7.1.9   Dose Equivalent (rem)

Although the biological effects of radiation are dependent upon the absorbed dose, some types of particles produce greater effects than others for the same amount of energy imparted. For example, for equal absorbed doses, α particles may be ten times as damaging as β particles. In order to account for these variations when describing human health risk from radiation exposure, the *dose equivalent* is used. This is the absorbed dose multiplied by a certain *quality* and *modifying* factor indicative of the relative biological damage potential of the particular type of radiation. Note: the term *absorbed*" in the discussion above implies that the radioactive particle is in contact with living tissue. Since α particles are stopped in the dead layer of the skin, α particles would have to be ingested in order to impart their damaging effects.

The unit of dose equivalent is the radiation equivalent in man (rem) or, more commonly, mrem. For gamma or X-ray exposures, the numerical value of the rem is essentially equal to that of the rad.

7.1.10   Biological Effects of Ionizing Radiation

The energy deposited by ionizing radiation as it interacts with matter may result in the breaking of chemical bonds. If the irradiated matter is living tissue, such chemical changes may result in altered structure or function of constituent cells.

Because the cell is composed mostly of water, less than 20% of the energy deposited by ionizing radiation is absorbed directly by macromolecules. More than 80% of the energy deposited in the cell is absorbed by water molecules, with the resultant formation of highly reactive free radicals. A *free radical* is a molecular fragment that contains one unpaired electron (i.e., ·OH). The unpaired electron is highly reactive. These radicals may initiate numerous chemical reactions that result in damage to macromolecules and a corresponding alteration of structure or function.

As a result of the chemical changes in the cell caused by ionizing radiation, large biological molecules may undergo a variety of structural changes that lead to altered function. Some of the more common effects that have been observed are inhibition of cell division, denaturation of proteins and inactivation of enzymes, alteration of membrane permeability, and chromosome aberrations.

7.1.11   Human Health Effects

The effects of ionizing radiation described at the level of the human organism can be divided broadly into one of two categories: stochastic or nonstochastic effects.

7.1.11.1   Stochastic Effects

As implied from the name, *stochastic effects* occur by chance. Stochastic effects caused by ionizing radiation consist primarily of genetic effects and cancer. As the dose to an individual increases, the probability that cancer or a genetic effect will occur also increases. However, at no time, even for high doses, is it certain that cancer or genetic damage will result. Similarly, for stochastic effects, there is no threshold dose below which it is relatively certain that an adversary effect cannot occur.

7.1.11.2   Nonstochastic Effects

Unlike stochastic effects, *nonstochastic effects* are characterized by a threshold dose below which they do not occur. In addition, the magnitude of the effect is directly proportional to the size of the dose. For nonstochastic effects, there is a clear causal relationship between radiation exposure and the effect. Examples of nonstochastic effects include sterility, erythema (skin reddening), and cataract formation. Each of these effects differs from the other in both its threshold dose and in the time over which this dose must be received to cause the effect (i.e., acute vs. chronic exposure).

The range of nonstochastic effects resulting from an acute exposure to radiation is collectively termed, *radiation syndrome*. This syndrome may be subdivided into a:

1. Hemopoietic Syndrome: characterized by depression or destruction of bone marrow activity with resultant anemia and susceptibility to infection (whole body dose of about 200 rads).
2. Gastrointestinal Syndrome: characterized by destruction of the intestinal epithelium with resultant nausea, vomiting, and diarrhea (whole body dose of about 1000 rads).
3. Central Nervous System Syndrome: direct damage to nervous system with loss of consciousness within minutes (whole body doses in excess of 2000 rads). The LD_{50} (i.e., dose that would cause death in half of the exposed population) for acute whole body exposure to radiation in humans is about 450 rads.

7.1.12   Determinants of Dose

The effect of ionizing radiation upon humans or other organisms is directly dependent upon the size of the dose received (although dose rate may also be important).
The dose, in turn, is dependent upon a number of factors including the strength of the
source, the distance from the source to the affected tissue, and the time over which
the tissue was irradiated. The manner in which these factors operate to determine the
dose from a given exposure differs significantly for exposures that are external (i.e.,
resulting from a radiation source located outside the body) and those that are internal
(i.e., resulting from a radiation source located within the body).

7.1.12.1   External Exposures

Exposures to sources of radiation located outside the body are of concern primarily
for sources emitting γ rays, X-rays, or high-energy β particles. External exposures
from radioactive sources that emit α or low-to-medium energy β particles are not significant, since these radiations do not penetrate the dead outer cell layer of the skin.
   As with all radiation exposures, the size of the dose resulting from an external
exposure is a function of:

1. The strength of the source;
2. The distance from the source to the tissue being irradiated; and
3. The duration of the exposure. In contrast to the situation for internal exposures,
    however, these factors can be altered (either intentionally or inadvertently) for a
    particular external exposure situation with a resultant modification of the dose
    received.

The effectiveness of a given dose of external radiation in causing biological damage
is dependent upon the portion of the body irradiated. For example, because of differences in the radiosensitivity of constituent tissues, the hand is far less likely to suffer
biological damage from a given dose of radiation than are the gonads. Similarly, a
given dose to the whole body has a greater potential for causing adversary health
effects than does the same dose to only a portion of the body.

7.1.12.2   Internal Exposures

Exposures to ionizing radiation from sources located within the body are of concern
for sources emitting any and all types of ionizing radiation. Of particular concern
are internally emitted α and β particles that cause significant damage to tissue when
depositing their energy along highly localized paths.
   In contrast to the situation for external exposures, the source to tissue distance,
exposure duration, and source strength cannot be altered for internal radiation
sources. Instead, once a quantity of radioactive material is taken up by the body (e.g.,
by inhalation, ingestion, or absorption) an individual is "committed" to the dose that
will result from the quantities of the particular radionuclide(s) involved.

In general, radionuclides taken up by the body do not distribute equally throughout the body's tissues. Often, a radionuclide concentrates in a critical organ. For example, ^{131}I and ^{125}I concentrate in the thyroid, ^{45}Ca and ^{32}P in the bone, and ^{59}Fe in the spleen.

The dose committed to a particular organ or portion of the body depends, in part, upon the time over which these areas of the body are irradiated by the radionuclide. This, in turn, is determined by the radionuclide's physical and biological half-lives (i.e., the effective half-life). The biological half-life of a radionuclide is defined as the time required for one-half of a given amount of radionuclide to be removed from the body by normal biological turnover.

7.1.13   Sources of Exposure

Exposures to ionizing radiation can be classified broadly according to whether they result from sources within the work environment (i.e., occupational exposures) or from sources outside the work environment (i.e., nonoccupational exposures).

7.1.13.1   Occupational Exposure

Occupational exposures are those received by individuals as a result of working with or near radiation sources (i.e., radioactive material or radiation-producing devices). Occupational exposures differ from nonoccupational exposures in that they are generally received during the course of a 40-hour workweek (exposures from natural background are received continuously over 168 hours per week). Whereas the nonoccupational exposure received by a given individual is largely unknown, an individual's occupational exposure should be closely monitored and controlled.

7.1.13.2   Nonoccupational Exposure

Nonoccupational exposures can be divided into one of two categories: those originating from natural sources or those resulting from man-made sources.

All individuals are continuously exposed to ionizing radiation from various natural sources. These sources include cosmic radiation and naturally occurring radionuclides within the environment and within the human body. The radiation levels resulting from natural sources are collectively referred to as *natural background*. Natural background and the associated dose it imparts vary considerably from one location to another in the United States. It is estimated that the average whole-body dose equivalent from natural background in the United States is about 250 mrem/person/year, which is approximately ¾ of a person's total radiation exposure.

The primary source of man-made nonoccupational exposures is medical irradiation, particularly diagnostic procedures (e.g., X-ray and nuclear medicine examinations). Such procedures, on average, contribute an additional 100 mrem/person/year in the United States. All other sources of man-made nonoccupational exposures, such as nuclear weapons fallout, nuclear power plant operations, and the use of radiation sources in industry and universities contribute an average of less than one mrem/person/year in the United States.

7.1.14   Exposure Limits

Concern over the biological effects of ionizing radiation began shortly after the discovery of X-rays in 1895. From that time to the present, numerous recommendations regarding occupational exposure limits have been proposed and modified by various radiation protection groups, the most important being the International Commission on Radiological Protection (ICRP). These guidelines have, in turn, been incorporated into regulatory requirements for controlling the use of materials and devices emitting ionizing radiation.

7.1.15   Basis of Recent Guidelines

In general, the guidelines established for radiation exposure have had as their principle objectives:

- The prevention of acute radiation effects (e.g., erythema, sterility); and
- The limiting of the risks of late, stochastic effects (e.g., cancer, genetic damage) to "acceptable" levels. Numerous revisions of standards and guidelines have been made over the years to reflect both changes in the understanding of the risk associated with various levels of exposure and changes in the perception of what constitutes an "acceptable" level of risk.

Current guidelines for radiation exposure are based upon the conservative assumption that there is no safe level of exposure. In other words, even the smallest exposure has some probability of causing a late effect such as cancer or genetic damage. This assumption has led to the general philosophy of not only keeping exposures below recommended levels or regulatory limits, but of also maintaining all exposures as low as is reasonably achievable (ALARA). This is a fundamental tenet of current radiation safety practice (Boss and Day, 2001).

7.1.16   Radiation Risk

The risk associated with various levels of exposure to ionizing radiation can be adequately understood only when viewed in the context of other health risks routinely encountered in life.

The risk associated with various activities or situations is most often expressed as an annual mortality rate. For example, the annual mortality rate from home accidents in the United States is 1.1×10^{-4}. Thus, the risk of death from a home accident is approximately one in 10,000/person/year.

Approximately 50% of adult men and 33% of adult women will develop cancer at some time from all possible causes such as smoking, food, alcohol, drugs, air pollutants, and natural background radiation (NCI, 2007). Thus, in any group of 10,000 workers not exposed to radiation on the job, we can expect about 2,500 to develop cancer. If this entire group of 10,000 workers were to receive an occupational radiation dose of 1,000 mrem each, we could estimate that three additional cases might occur, which would give a total of about 2,503. This means that a 1,000-mrem dose to each of 10,000 workers might increase the cancer rate from 25% to 25.03%, an increase of about 3 one-hundredths of one percent.

As an individual, if your cumulative occupational radiation dose is 1,000 mrem, your chances of developing cancer during your entire lifetime may have increased from 25% to 25.03%. If your lifetime occupational dose is 10,000 mrems, we could estimate a 25.3% chance of developing cancer. Using a simple linear model, a lifetime dose of 100,000 mrems may have increased your chances of cancer from 25% to 28%.

7.2 HAZARDS

Three primary scenarios may occur in which radioactive materials could be dispersed by a terrorist:

- Conventional explosives or other means to spread radioactive materials (a dirty bomb)
- Attack on a fixed nuclear facility
- Nuclear weapon

In any of these events, filtration and air-cleaning devices would be ineffective at stopping the blast and radiation itself. However, the air-cleaning or filtration system may be useful in collecting the material from which the radiation is being emitted. Micrometer-sized aerosols from a radiological event are effectively removed from air streams by HEPA filters. This collection could prevent distribution throughout a building. However, subsequent decontamination of the heating, ventilation, and air conditioning (HVAC) system would be required.

Because ionizing radiation cannot be detected by the human senses, detection and quantification must be accomplished by specifically designed instruments. All such methods of measurement employ the use of a substance that responds to radiation in a measurable way, and a system or apparatus to measure the extent of the response. Most radiation detectors operate by one of two methods: ionization or scintillation.

7.3 FEDERAL GUIDANCE

The U.S. Environmental Protection Agency (USEPA) is responsible for developing guidance for all federal agencies as to standards for radiation protection of the public. The USEPA is not authorized to enforce any provisions of the federal guidance, but all federal agencies are expected to comply with the guidance unless compelling reasons (such as specific statutory requirements) require noncompliance.

The federal guidance on radiation protection of the public is intended to apply to all controlled sources of exposure, including sources not associated with operations of the nuclear fuel cycle, but excluding indoor radon and beneficial medical exposures. Therefore, the federal guidance is intended to apply to all exposures of the public to technically enhanced (altered) normally occurring radioactive materials (TENORM) that are no longer in their natural, undisturbed state, but not to NORM. Neither the USEPA nor any other federal agency with responsibilities for radiation protection of the public has developed standards that apply to all exposure situations involving TENORM.

The USEPA has issued proposed federal guidance on radiation protection of the public (FRC) to replace the guidance developed many years ago by the Federal Radiation Council in 1960–1961 (EPA, 1994). The proposed guidance has not been issued in final form and represents the USEPA's current views on the basic, minimal requirements for radiation protection of the public. The USEPA's proposed federal guidance on radiation protection of the public includes:

- No TENORM radiation exposure of the general public (unless justified by an overall benefit from the activity causing the exposure).
- Doses to individuals and populations should be ALARA.
- The *annual effective dose* equivalent to individuals from all controlled sources combined *should not normally exceed 1 mSv (100 mrem)*. This dose equivalency includes sources not associated with operations of the nuclear fuel cycle but excludes indoor radon.
- Annual effective dose equivalents to individuals up to 5 mSv (500 mrem) may be permitted. Approval requires prior authorization with assumed unusual, temporary situations.
- Continued exposure over substantial portions of a lifetime at or near 1 mSv (100 mrem) per year should be avoided.

Authorized limits for specific sources or practices should be established to ensure that the:

- *Primary dose limit of 1 mSv (100 mrem)* per year for all controlled sources combined and the ALARA objective are satisfied, and
- Authorized limit for *any* source or practice normally should be a fraction of the dose limit for all controlled sources combined.

The provisions listed above apply to naturally occurring radionuclides, including TENORM, other than indoor radon. To ensure compliance with these provisions, exposures to human-made radionuclides also need to be taken into account.

7.4   THREAT: NUCLEAR AND RADIOLOGICAL

Unless confirmed by radiation detection equipment, the presence of a radiation haz-ard is difficult to determine. Nuclear bombs, if detonated, pose two threats: physical harm (i.e., demolition of buildings) and biological harm (i.e., radiation exposure). The dispersion of radioactive material will disrupt normal daily activities, create significant long-term health issues, and adversely affect the food supply by contami-nating crops and water sources. Environmental decontamination will be difficult to perform.

While a serious event, such as a plane crash into a nuclear power plant, could result in the release of radioactive material into the air, a nuclear power plant would not explode like a nuclear weapon. However, a radiation danger may exist in the sur-rounding areas, depending on the type of incident, the amount of radiation released, and the current weather patterns.

7.4.1   Nuclear Terrorism

Nuclear terrorism involves the deliberate detonation of a nuclear weapon. Consequences include fatalities and injuries resulting from the initial explosion and subsequent fires, as well as immediate and long-term effects of radiation exposure. An example weapon is an improvised nuclear device (IND), an explosive device designed to cause a nuclear yield using either uranium or plutonium isotopes.

Clearly, an exploded nuclear device could result in a lot of property damage. People would be killed or injured from the blast and might be contaminated by radio-active material. Many people could have symptoms of acute radiation syndrome. After a nuclear explosion, radioactive fallout would extend over a large region far from the point of impact, potentially increasing people's risk of developing cancer over time.

A meltdown or explosion at a nuclear facility could cause a large amount of radioactive material to be released. People at the facility would probably be contami-nated with radioactive material and possibly be injured from the explosion. Those people who received a large dose might develop acute radiation syndrome. People in the surrounding area could be exposed or contaminated.

7.4.2   Radiological Terrorism

Radiological terrorism involves the deliberate contamination of an area using radio-active materials. For example, a terrorist could use conventional explosives to dis-perse a radiation source, such as a spent nuclear reactor fuel rod. Consequences include chronic long-term effects of radiation exposure. Additionally, the chal-lenge of decontaminating those exposed and the contaminated area is of concern. The medical consequences of radiological terrorism are akin to the effects of accidental radiation incidents rather than nuclear bomb detonation. Radiological threats include:

- Radiological dispersal device (RDD): An explosive device used to spread radioactive material (such as that found at a nuclear power plant or in a radiological cargo) upon detonation.
- Simple RDD: A method of spreading radiological material without the use of an explosive. Any nuclear material, including medical isotopes or waste, can be used in this manner.

A dirty bomb is an RDD that uses a conventional explosive, such as dynamite, to spread radioactive materials in the form of powder or pellets. This dispersal does not involve the splitting of atoms to produce the tremendous force and destruction of a nuclear blast, but rather spreads smaller amounts of radioactive material into the surrounding area. The main purpose of a dirty bomb is to frighten people and contaminate buildings or land with radioactive material. A dirty bomb could cause serious injuries from the explosion, and yet the bomb may not have enough radioactive material to cause serious radiation sickness among large numbers of people. However, people exposed to radiation scattered by the bomb could have a greater risk of developing cancer later in life, depending on their dose.

Possible terrorist events involve introducing radioactive material into the food or water supply, using explosives (like dynamite) to scatter radioactive materials, bombing or destroying a nuclear facility, or exploding a small nuclear device. Although introducing radioactive material into the food or water supply would cause great concern or fear, the radioactive material may not cause much contamination or increase the danger of adversary health effects.

7.5   RADIATION DETECTION METHODS

The selection of instruments is based on the type and energy range of the radiation expected on-site. Survey instruments should be chosen for their sensitivity to the type of radiation present in the area to be surveyed. There are three main types of radiation detectors: ionization detectors, scintillation detectors, and semiconductors. Practically speaking, only two are used consistently in the field: ionization and scintillation detectors. Semiconductors, while useful, are expensive and not generally portable. The method of relating the instrument reading to mR/hr should be included.

Some instruments can measure multiple types of radiation and require methods for determining which type of radiation is being measured. For example, Geiger-Mueller detectors respond well to α, β, and γ radiation, although they usually do not discriminate between the type(s) of radiation detected.

Radiation counting systems determine the sample activity (μCi or dpm). However, the radiation detector can never detect 100% of the disintegrations occurring in the sample. The radiation source emits particles in all directions (360°), but the detector, because of its size, location, and sensitivity, only senses a small fraction of the total emitted particles. The counts per minute (cpm) displayed by the counter must therefore be distinguished from the disintegration rate (dpm) of the sample. The ratio of the cpm to the dpm, expressed as a percent, is the efficiency of the counting system.

- Monitoring instruments must be chosen for their sensitivity to the type of radiation to be monitored. A method of relating the instrument reading to microcurie (μCi) must be included.
- Monitoring must be performed slowly and at specified distances. A distance of 1 to 2 millimeters (mm) from the surface will yield a higher count rate than a distance of 1 to 2 meters. This is because of the *inverse square law*, which states that the intensity of the radiation drops off proportional to one over the distance squared from the source.
- Therefore, to determine realistic human radiation exposures, the detector should be placed at the same distance from the source that a normal worker would stand.

Each survey instrument must have an appropriate check source attached or assigned to the instrument. The check source for the instrument must be surveyed immediately after calibration, and the reading must be written onto the calibration sticker on the instrument. Before each use of the instrument, the check source must be monitored, and the reading compared to the reading noted on the calibration sticker. Any meter not measuring within +/−10% of the reading on the calibration sticker:

- Must be tagged as requiring maintenance, and
- Must not be used until maintenance and recalibration have been performed.

Survey instruments must be calibrated periodically using procedures outlined in the instrument manual, within the following guidelines. All sources used to calibrate instruments must be traceable to the National Institute of Standards and Technology (NIST). The survey meter and the probe must be calibrated as a unit. If probes are changed, the unit must be recalibrated prior to use. Instruments must be calibrated

- At least annually
- After every maintenance or repair operation

The date of calibration, the date the next calibration is due, and the initials of the person performing the calibration must be written on a calibration sticker that will be attached to the instrument.

7.6 IONIZATION DETECTORS

Most ionization detectors consist of a gas-filled chamber with a voltage applied between the anode and cathode. The central wire becomes the anode and the chamber wall the cathode. Any radiation entering the chamber ionizes the gas, creating an ion pair. The cation is attracted to the anode while the anion is attracted to the cathode. In the process of migrating toward these electrodes, the ions bump into more gas particles, creating more ion pairs. The result is a cascade effect, amplifying the original signal. At the electrodes, the ions are collected to form a measurable and quantifiable electronic pulse. The pulses are counted, integrated, and displayed on the meter face in Roentgens per hour. Ion chambers provide a nearly linear response to γ and X-radiation above a few kiloelectron-volts (KeV), in energy, and at radiation levels above 0.1 mR/hr. For this reason, an ion chamber is the only instrument for quantifying radiation exposures. Ion chambers may be used to quantify β, γ, or X-ray dose at a location. Depending upon the voltage applied to the chamber, the detector may be considered a gas proportional detector, a Geiger-Muller (GM) detector, or an ion chamber.

The ionizing radiation survey meter is useful for measuring a variety of different radiation sources, such as radon decay products from air samples collected on filters.

- The barometric pressure should be noted for ionizing radiation chambers. The pressure of the gas in the chamber is directly proportional to the amount of gas within the chamber, and thus how much signal amplification will be achieved. A low internal (chamber) pressure will reduce the response of the detector.
- Wipe samples collected on a filter can also be counted with this detector, and general area sampling can be done.

7.6.1 Gas Proportional Detectors and Ion Chambers

Thin-window gas proportional detectors may be used to detect α and β radiation. An extremely thin window, usually made of Mylar or similar material, is necessary because of the low penetrating ability of these particles. Distinction between α and β particles is achieved by adjusting the voltage of the detector. The applied voltage on a proportional detector is limited so that the resulting signal is directly proportional to the type of particle being detected. If the voltage is decreased, different types of particles (and thus particles of different energies) cannot be distinguished, resulting in an ion chamber.

7.6.2 Geiger Muller

Because of its versatility and dependability, the Geiger Muller (GM) detector is the most widely used portable survey instrument. A GM detector with a thin window can detect α, β, and γ radiation. The GM is particularly sensitive to medium- to high-energy β particles (e.g., from ^{32}P), X-rays, and γ rays. The GM detector is fairly insensitive to low-energy X or γ rays such as those emitted from ^{125}I, and to low-energy β particles, such as those emitted by ^{35}S and ^{14}C. The GM cannot detect the weak β from ^{3}H at all.

Unlike the ion chamber, the GM detector does not actually "measure" exposure rate. Instead, the GM "detects" the number of particles interacting in its sensitive volume per unit time. The GM should thus read out in cpm, although the GM can be calibrated to approximate mR/hr for certain situations. With these advantages and limitations, a GM detector is the instrument of choice for initial entry and survey of field radiation sources and radioactive contamination.

GM detectors are calibrated to one energy level of the electromagnetic spectrum, usually 662 KeV, which is the γ energy from the decay of ^{137}Cs. The display is read out in mR/hr. GM detector efficiency for radiation at other energies is not linear. GM detectors may be used to detect the presence of radiation, but only as a rough estimate of the dose rate. Beta shields will allow for blocking betas and reading γ or X-radiation. Subtraction of the γ or X-radiation readings will yield an approximation of the β contribution. Beta contribution should read out in cpm.

7.7 SCINTILLATION DETECTORS

Scintillation detectors are based upon the use of various phosphors (or scintillators) that emit light in proportion to the quantity and energy of the radiation absorbed. As higher energy ionizing radiation enters the scintillation crystal, the electrons within the crystal are excited to higher orbitals. As the electrons return to their ground state, they emit light directly proportional to the amount of energy in the original radiation. The light flashes are converted to photoelectrons that are multiplied in a series of dynodes (i.e., photomultipliers) to produce a large electrical pulse. Because the light output and resultant electrical pulse from a scintillator is proportional to the amount of energy deposited by the radiation, scintillators are useful in identifying the amount of specific radionuclides present (i.e., scintillation spectrometry).

Solid scintillation detectors are particularly useful in identifying and quantifying γ-emitting radionuclides. Sodium iodide (NaI) scintillation detectors, used on survey meters, provide better counting efficiency for γ and X-rays, but have a more limited range of energies, depending on the size of the crystal and the density of the window. NaI detectors may be used to detect the presence of γ radiation, but only if the energy level of the radiation is known and the correct size crystal is used.

The common γ well counter employs a large (e.g., 2″ × 2″ or 3″ × 3″) crystal of NaI within a lead-shielded well. The sample vial is lowered directly into a hollow chamber within the crystal for counting. Such systems are extremely sensitive, but do not have the resolution of more recently developed semiconductor counting systems. Portable scintillation detectors are also widely used for conducting various types of radiation surveys. Of particular use to researchers working with radioiodine is the thin-crystal NaI detector, which is capable of detecting the emissions from ^{125}I with efficiencies nearing 20% (a GM detector is less than 1% efficient for ^{125}I).

The most common means of quantifying the presence of β-emitting radionuclides is through the use of liquid scintillation counting. In these systems, the sample and phosphor are combined in a solvent within the counting vial. The vial is then lowered into a well between two photomultiplier tubes for counting. Liquid scintillation counting has been an essential tool of research involving radiotracers such as ^{3}H, ^{14}C, ^{35}S, and ^{45}Ca.

One problem that occurs with liquid scintillation counting is determining the efficiency of the system. The low-energy photons produced by β particles interacting with the scintillation cocktail are easily shielded from the photomultiplier tubes due to optical and chemical quenching. To account for this, a quench curve must be computed using a set of increasingly quenched standards and a method of determining a quench-indicating parameter (QIP) for each standard. The quench curve is a graph for a certain nuclide of each standard's QIP versus the efficiency of the counter.

The efficiency for an unknown sample is then determined by measuring its QIP and through the graph, determining the counter's efficiency for that sample, then using that and the cpm of the sample to determine the dpm of the sample.

Fortunately, most modern counters compute and store quench curves from a single counting of a standard set, use an external standard to determine the QIP, and automatically output sample counts in dpm.

The other common problem with liquid scintillation counting is chemiluminescence. Certain chemicals, when mixed with some scintillation cocktails, will cause the cocktail to emit photons resulting in a higher count rate than actually exists from the radionuclide itself.

Rule of thumb: Let scintillation vials wait for a few hours after mixing the cocktail to allow for the initial chemiluminescence to be exhausted.

7.8 PERSONNEL MONITORS

A standard method of monitoring routine personnel radiation exposure is with a personal monitor. These monitors take several forms, including rings, badges, and bracelets. The internal detecting device, however, is the same: a photographic plate. Ironically, the first radiation detection was on a photographic plate, and the detection method is still being used today. The photographic film can be doped with specific elements or compounds in order to detect specific radioactive particles. Undoped, the film will detect a variety of common radiation types, with varying responses to α, β, and γ radiation. The badges are generally worn on the arm, hand, or torso. After a month, the badges are collected from the personnel and sent in for analysis: the personnel are issued a new badge each month. The film is developed and quantified for amount of personal exposure. Personal badges are useful for monitoring occupational exposure in industries that routinely handle radioactive material, including medical facilities, cyclotrons and linear accelerators, and nuclear power plants. However, these badges can also be used for prevention and risk calculation.

7.9 POPULATION MONITORING

The term *population monitoring* is made up of immediate monitoring after an incident and long-term monitoring for health effects from the attack.

Within the first hours and days after a radiological attack, people should be monitored with radiation monitors. Public health officials will use the information from the monitoring equipment to find out whether people are contaminated either inside their bodies (internal) or outside their bodies (external) with radioactive materials from the attack. Public health officials will estimate the amount of radiation to which people were exposed, also known as the *dose*, through a process called *dose reconstruction*.

In addition, the health of people who were involved in the incident will be monitored over many years to see whether people are having health effects from the attack. These health effects could include effects related to radiation exposure or effects associated with the stress of being involved in an attack.

In the United States, the Department of Health and Human Services (DHHS) has designated the Centers for Disease Control and Prevention (CDC) as the lead agency for population monitoring. The duties of this designation are described in the Nuclear/Radiological Incident Annex of the *National Response Plan* (NRP). Under the Nuclear/Radiological Incident Annex of the NRP, the CDC is responsible for assisting state, local, and tribal governments in monitoring people for external and internal contamination. The CDC is also responsible for supporting state, local, and tribal governments in decontaminating people who are internally contaminated by providing guidance on giving medicine that can speed up the removal of radioactive material from people's bodies.

The CDC will also help local and state health departments create a registry (list) of people who might have been exposed to radiation from the incident. As part of the work on the registry, the CDC will help the local and state health departments determine how much radiation people were exposed to and follow people for as long as necessary to see whether health effects (from radiation exposure or from the stress of being involved in an attack) develop.

7.10 FACILITY VULNERABILITY

All users and uses of radiation may increase a facility's vulnerability. Vulnerability assessment should focus on how both radioactive materials and the generated wastes (including human wastes) are handled. Storage of radionuclides may have in the past been considered adequate. However, in light of potential terrorist or other criminal activity, this storage may need to be reevaluated. Common facilities with an increased risk due to the use of radioactive materials include medical centers, nuclear research facilities, and nuclear power plants. In addition to the regular topics that should be evaluated during a vulnerability assessment, facilities with radiation sources present should be evaluated for these additional considerations:

- In case of an emergency, could the personnel shelter in place?
- Are potassium iodide (KI) stores warranted and/or adequate?
- Are adequate and appropriate radiation detectors available?
- Are adequate personal stores (i.e., food, water) available?
- Is shielding protection appropriate and adequate?
- Are security measures appropriate?

7.10.1 Medical Centers

Every major hospital in the United States has a nuclear medicine department in which radionuclides are used to diagnose and treat a wide variety of diseases more effectively and safely. Radionuclides are used both to obtain diagnostic information (i.e., a positron [β^+] emission tomography (PET) scan) and to treat illnesses (i.e., by modifying the diseased tissue via exposure to radiation). In a diagnostic scenario, the patient either swallows, inhales, or receives an injection of a tiny amount of a radionuclide. Special cameras reveal where the radioactivity accumulates briefly in the body, providing, for example, an image of the heart that shows normal and malfunctioning tissue. Radionuclides are also used to effectively treat patients with thyroid diseases, including Graves disease—one of the most common forms of hyperthyroidism—and thyroid cancer.

Radionuclides are a technological backbone for much of the biomedical research being done today. These radionuclides are used in identifying and learning how genes work. Much of the research on AIDS is dependent upon the use of radionuclides. Scientists are also "arming" monoclonal antibodies (produced in the laboratory and engineered to bind to a specific protein on a patient's tumor cells) with radionuclides. When such "armed" antibodies are injected into a patient, the antibodies bind to the tumor cells, which are then killed by the attached radioactivity, but the nearby normal cells are spared. So far, this approach has produced encouraging success in treating patients with leukemia. Most new drugs, before being approved by the Food and Drug Administration, have undergone animal studies that use radionuclides to learn how the body metabolizes the compounds.

Another clinical and research tool, PET scanning, involves injecting radioactive material into a person to "see" the metabolic activity and circulation in a living brain. PET studies have enabled scientists to pinpoint the site of brain tumors, identify the source of epileptic activity, and better understand many neurologic diseases.

These are but a few of the many vital uses of ionizing radiation in medicine. About 70 to 80% of all research at the U.S. National Institutes of Health is performed using radiation and radioactive materials.

7.10.2 Nuclear Research Facilities

A wide variety of research is conducted at cyclotrons and linear accelerators around the world. While some of the research is purely academic, some is conducted in the name of medicine and homeland security. These facilities may be independent laboratories or may be affiliated with an academic institution. A unique set of circumstances makes assessment and security of these facilities challenging:

- A variety of different types of research exists within these microcosms.
- Each research group may not be aware of what the others are doing.
- Research often involves transient visitors from other institutions.
- Research is usually conducted round the clock, seven days a week.

- Spent research materials can remain "hot" for extended periods of time.
- Materials used to conduct research, particularly those within the beam, remain "hot" for extended periods of time.
- Each group may be in possession of radioactive materials, so that radioactive sources are distributed throughout the facility.

7.10.3 Nuclear Power Plants

The number of nuclear power plants within the United States is slowly growing as the costs of coal and natural gas increase. Nuclear power is a relatively cheap and renewable energy source. However, nuclear power plants may be enticing targets for terrorist attack. Because of this, the specific address of nuclear power plants is not available to the general public and employees working within a nuclear facility are subject to background checks before being hired. Nuclear power plants are subject to state and federal regulations, as detailed above.

REFERENCES

EPA 1994. *Federal Radiation Protection Guidance for Exposure of the General Public,* Proposed Recommendations, U.S. Environmental Protection Agency, Washington, DC, Federal Register, 59(246): 66414, December 23, 1994.

Probability of Developing or Dying of Cancer, Software version 6.2.1, Statistical Research and Applications Branch, NCI, 2007, http//srab.cancer.gov/devcan (accessed 10/29/08).

ADDITIONAL READING

Biological Effects of Radiation, U.S. Nuclear Regulatory Commission, Office of Public Affairs.

Chemical, Biological and Radiological Incident Handbook, Chemical, Biological and Radiological (CBRN) Subcommittee of the Interagency Intelligence Committee on Terrorism (IICT), October 1998.

Fundamentals of Protective Design for Conventional Weapons, TM 5-855-1. Washington, DC, Headquarters, U.S. Department of the Army, 1986.

Glossary of Terms in Nuclear Science and Technology, La Grange, IL, American Nuclear Society, 1986.

Guidance for Protecting Building Environments from Airborne Chemical, Biological, or Radiological Attacks, National Institute of Occupational Safety and Health (NIOSH), U.S. Department of Health and Human Services (DHHS) (NIOSH) Pub No. 2002-139, May 2002.

Health Effects of Exposure to Low Levels of Ionizing Radiation, BEIR V. International Atomic Energy Agency, National Research Council. Washington, DC, National Academy Press, 1990.

Ionizing Radiation Exposure of the Population of the United States. Report No. 93. National Council on Radiation Protection and Measurements, Bethesda, MD, 1987.

The International Chernobyl Project—Assessment of Radiological Consequences and Evaluation of Protective Measures. Report No. 91-03254. Summary Brochure. Vienna, Austria: 1991.

North American Emergency Response Manual, U.S. Department of Transportation, Transport Canada, Secretariat of Transport and Communications, 1996.

Nuclear Reactor Concepts, U.S. Nuclear Regulatory Commission, May 1993.

Nuclear Terms Handbook, U.S. Department of Energy, Office of Nonproliferation and National Security, 1996.

Radiation Exposure of the U.S. Population from Consumer Products and Miscellaneous Sources. Report No. 95. National Council on Radiation Protection and Measurements, Bethesda, MD, 1987.

Radiation Carcinogenesis—Human Epidemiology. In *The Biological Basis of Radiation Protection Practice*, John D. Boice, Jr., pp. 89–120. Baltimore, MD, Williams and Wilkins, 1992.

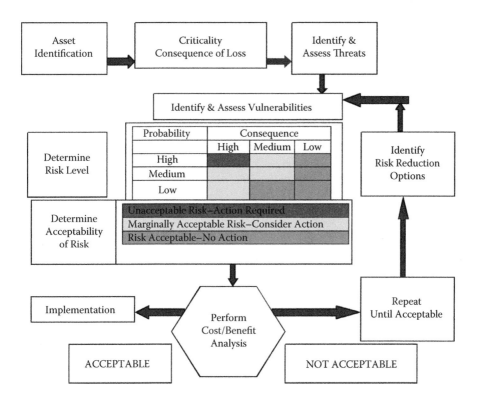

Color Figure 1.1 Example risk management process (DoE, 2002).

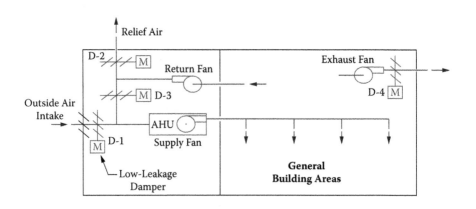

Color Figure 8.1 Emergency air distribution shutoff.

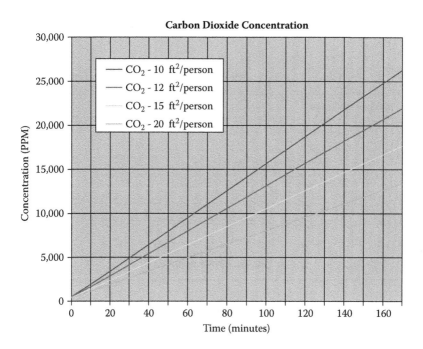

Color Figure 8.4 Carbon dioxide concentrations in tightly sealed shelters.

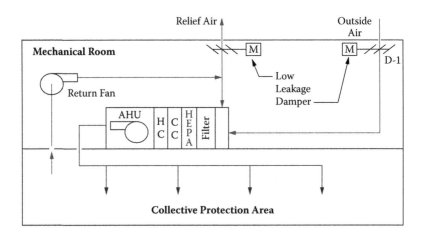

Color Figure 8.5 Central air-handling system with a HEPA filter.

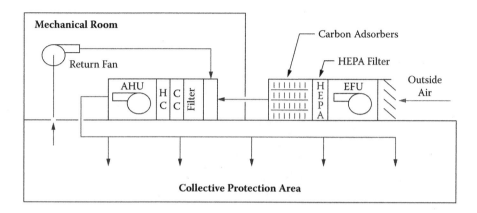

Color Figure 8.6 Air-Handling system with HEPA and carbon filters on the outside air intake.

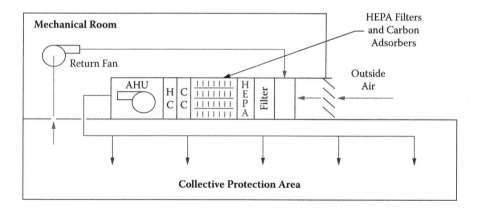

Color Figure 8.7 Air-Handling system with HEPA and carbon filters in the central air-handling unit.

Color Figure 8.8 Blower door assembly.

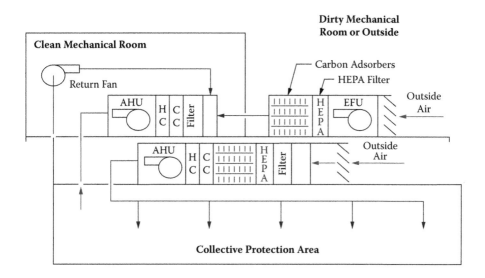

Color Figure 8.9 Blow-through filter unit located outside or in a dirty mechanical room.

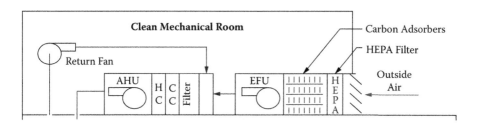

Color Figure 8.10 Draw-through filter unit located in a clean mechanical room.

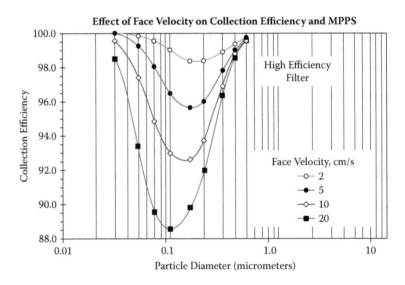

Color Figure 10.2 Effect of face velocity on collection efficiency and MPPS. (NIOSH, 2002.)

CHAPTER 8

Airborne Hazard
Protection Options

Kendall G. Christenson and Donald C. Dittus

CONTENTS

Each year hundreds of chemical incidents occur in the United States that include airborne hazards (such as hazardous fumes, noxious chemicals, or mysterious odors) that permeate buildings and cause illness, injuries, or disruption of activities. Most of these incidents involve accidental releases of chemicals, either inside or outside the building. Some may be the result of malicious acts—vandalism, pranks, or in extreme cases, terrorism. Deliberately releasing chemicals or biological aerosols (fine particles), or a radioactive substance into a building can be life threatening and can create substantial economic impact.

In some cases, these incidents result in building evacuation—the natural response in such emergencies and usually the only practical course of action by which occupants can reach clean air and safety. Even without special protective systems, buildings can provide protection in varying degrees against airborne hazards that originate outdoors. Such protection is limited and effective only under certain conditions, however. Conversely, the hazards produced by a release inside a building can be much more severe than a similar release outdoors. Because buildings allow only a limited exchange of air between indoors and outdoors, not only can higher concentrations occur when a release occurs inside or directly into a building, but hazards persist longer indoors.

This chapter presents a variety of ways to protect building occupants from airborne hazards—to prevent, or mitigate the effects of outdoor and indoor releases of hazardous materials. Prevention relies on building security features and procedures, such as access control and security screening. Mitigation relies on building design and construction features as well as prior planning. Building design and construction features include elevating the outside air intakes, incorporating building sealing measures, modifying ventilation system design, installing a mass notification system, creating chemical, biological, or radiological (CBR) shelters and filtering air.

Types of protection against a CBR attack range from little- or no-cost actions, such as locking mechanical room doors or installing an intercom system, to installing a CBR filtration system similar to those used by the military in high-threat areas.

Since heating, ventilation, and air conditioning (HVAC) systems usage (or shutdown) decisions are so important, HVAC system performance to include flow rates, damper modulation and closure, sensor calibration, filter pressure loss, filter leakage, and filter change-out recommendations must be evaluated. Periodic training of HVAC maintenance staff should be conducted and documented. This training must include the procedures to be followed in the event of a suspected CBR agent release. Development of current, accurate HVAC diagrams and HVAC system labeling protocols should be addressed.

8.1 ACCESS CONTROL

Access control is intended to prevent unauthorized personnel from entering the building. All perimeter doors and windows should have a method to secure them closed from the inside. All secured doors should have an electronic entry or manual lock system that only authorized personnel can activate. Windows should be fixed and of forced-entry-resistant design to prevent a container with a CBR agent from being thrown through a window.

Mechanical room doors should be locked to prevent access by unauthorized or unescorted personnel in order to prevent direct insertion of an agent into the building ventilation system. Roof hatches and roof access doors should be locked and roof access ladders secured to prevent unauthorized access to roof-mounted air-handling equipment and outside air intakes. Likewise, outdoor areas containing building air-handling or ventilation equipment should be fenced and secured. Surveillance cameras, and motion detectors monitored by security personnel can provide additional security.

8.1.1 Security Screening

Entry screening is a two-step process:

1. Detecting a suspect closed container, and
2. Determining whether the contents of the container are hazardous or likely to be hazardous.

Both steps should be established in coordination with the appropriate security personnel with specific operational procedures to detect the presence of hazardous material containers. The first step can include a manual, x-ray, and/or metal detector search of briefcases, handbags, packages, letters, supplies boxes, and food and bottled beverage deliveries. All maintenance personnel and their equipment and authorized deliveries should be checked for hazardous materials. The following are items to consider for further examination or exclusion:

- Aerosol cans or other pressurized containers
- Manual or electric spray devices
- Containers of liquids or powders
- Bottled gases typically used for repair or maintenance within the building
- Pressurized dispensers containing irritating agents, including pepper spray, mace, and tear-producing agents

The second step involves examining the container, its contents, and labels and then determining whether the contents agree with the label. Liquid or powder in an unlabeled container is reason for prohibiting personnel from entering the building. Container labels should also be examined for alterations.

8.2   HVAC SYSTEMS AND PROTECTION OPTIONS

If an airborne hazard is detected inside a building, four possible actions should be considered:

1. Evacuation
2. Sheltering in place
3. Purging with smoke fans
4. Protective mask (i.e., respirator) use

These actions are not applicable for protection on a continuous basis, but can be implemented—singly or in combination—for relatively short periods when a hazard becomes apparent. These measures apply only to chemicals that have a warning property (e.g., odor), agents detectable by automatic detectors, or in response to events resulting in the release of chemicals, such as an accident at a chemical storage facility or on the highway involving a chemical transport vehicle.

To ensure that these actions can be carried out effectively requires a protective-action plan specific to the building, as well as training and familiarization for the occupants.

8.3   OUTDOOR AIR: VENTILATION DESIGN AND AIR INTAKES

Outdoor air design criteria should be based on weather data tabulated in the latest edition of the American Society of Heating, Refrigerating, and Air Conditioning Engineers (ASHRAE) *Handbook of Fundamentals*. Winter design conditions should be based on the 99.6% column dry bulb temperature in the ASHRAE *Fundamentals* volume.

Summer design conditions:

- Sensible heat load calculations should be based on the 0.4% dry bulb temperature with its mean coincident wet bulb temperature.
- Summer ventilation load and all dehumidification load calculations should be based on the 0.4% dew point with its mean coincident dry bulb temperature.

One of the most important steps in protecting a building's indoor environment is the security of the outdoor air intakes. Outdoor air enters the building through these intakes and is distributed throughout the building by the HVAC system. Publicly accessible outdoor air intakes located at or below ground level are at most risk.

Elevation of air intakes and other physical security measures can provide limited protection against external releases that are close to the building. Elevating outside air intakes to a minimum of 10 feet above ground level helps prevent the direct insertion of contaminates and reduces the potential for a contaminate released at ground level to reach the outside air intakes. The best location to mount an intake is high on a wall or on a secure roof. For some building configurations (high-rise buildings), elevation of intakes can also protect against some remote, ground-level releases.

Relocating accessible air intakes to a publicly inaccessible location is preferable. Ideally, the intake should be located on a secure roof or high sidewall. The lowest edge of the outdoor air intakes should be placed at the highest feasible level above the ground or above any nearby accessible level (i.e., adjacent retaining walls, loading docks, handrail). These measures are also beneficial in limiting the inadvertent introduction of other types of contaminants, such as landscaping chemicals, into the building.

If relocation of outdoor air intakes is not feasible, intake extensions can be constructed without creating adversary effects on HVAC performance. The goal is to minimize public accessibility. Design constraints such as excessive pressure loss, and dynamic and static loads on structure must be considered. Extension height should be increased where existing platforms or building features (i.e., loading docks, retaining walls) might provide access to the outdoor air intakes. An extension height of 12 feet (3.7 m) will place the intake out of reach of individuals without some assistance (i.e., use of ladders).

For existing buildings with outside air intakes that are below grade, at ground level, or wall mounted at a low elevation, physical security measures such as fencing around the intakes, surveillance cameras, and motion detectors monitored by security personnel can provide additional security.

The placement and location of outside air intakes is critical to the safety of the occupants inside a building and must be in compliance with the security requirements of the building.

- Roof intakes must be at least 0.2 m (8 inches) above the average maximum snow depth and the potential for drifts at the intake location must be considered.

- Outdoor intakes should be covered by a corrosion-resistant mesh screen to prevent objects from being inserted into them. The following should be followed when placing the screen:
  - The screen openings should be less than 13 mm (0.5 inches).
  - The screen slope should be a minimum of 45 degrees from horizontal to reduce the threat of objects being tossed into the intake and allow objects placed or thrown onto them to roll or slide off.
  - The screen should be located either outside of the intake grille, louver, or rain hood entry or no more than 0.2 m (8 inches) inside of the outside face of the intake grille, louver, or rain hood entry.
- On buildings of more than four stories the outside air supply louvers should be located on the fourth level of the building or higher. On buildings of three stories or less, locate the intakes on the roof or as high as possible.
- Locating intakes high on the exterior wall is preferred to a roof location.
- Outside air intake should not be permitted within 7 meters (25 feet) of a loading dock or any other fume-producing areas.

8.4   VESTIBULES AND REVOLVING DOORS

Designing or retrofitting main building entrances with integral vestibule areas or revolving doors minimizes infiltration caused by wind and stack effect. Vestibules, airlocks, and revolving doors provide a means of controlling infiltration at main building entrances as people enter and exit the building. Vestibules and revolving doors are most beneficial in multistory buildings with large temperature-driven stack-effect pressure differences between indoor and outdoor conditions. At the building entrances, the airflow is inward in the winter and outward in the summer.

8.5   HIGH-RISK AREAS

High-risk areas are portions of a building that have a high risk of an internal CBR release due to the delivery of mail and supplies or due to unsecured public access. These areas include mailrooms, supply and delivery areas, lobby areas, unscreened public access areas, and security screening areas. High-risk areas should be isolated from other portions of the building to the greatest extent possible. Measures to limit the spread of contaminates released in a high-risk area to other portions of the building are as follows:

- Ensure that the envelope of the high-risk area is well sealed and that the perimeter walls are full-height construction, and fully extend and are sealed to the undersides of the roof, to the undersides of the floor above them, or to hard ceilings.
- Provide low-leakage doors between the high-risk area and the rest of the building.

- Provide a separate, dedicated air-handling system to prevent airborne CBR contaminates from migrating from the high-risk area to other areas of the building through the ventilation system.
- Provide an exhaust fan, and maintain the high-risk area at a negative pressure relative to the rest of the building.
- Install weather stripping on all four edges of all perimeter doors or door frames in the high-risk area.
- Consider using recirculating-type filtration units to remove contaminants after a release.

Ventilation requirements for all building spaces should comply with ASHRAE 62.

Note: Utility systems should be located at least 50 feet from loading docks, front entrances, and parking areas. Within building and property lines, incoming utility systems should be concealed and given blast protection, including burial or proper encasement wherever possible.

8.5.1 Atriums

A dedicated air-handling system should be provided to control heat gain/loss in the occupied areas of the atrium. The atrium area should maintain negative pressure relative to adjacent interior and perimeter spaces or zones and positive or neutral pressure relative to adjacent vestibules and lobbies, and positive pressure relative to the outdoors. The design of the HVAC system must be fully coordinated with the smoke control system.

8.5.2 24-Hour Spaces

All areas designated as requiring 24-hour operations should be provided with a dedicated and independent HVAC system. All spaces handling building automation systems (BAS) computer processing of fire alarm monitor and control systems, security monitor and control systems, and/or energy monitoring and control systems should be provided with dedicated HVAC systems to maintain temperature, humidity, and ventilation requirements at all times.

- Twenty-four-hour systems should have dedicated chiller(s), cooling tower(s), boiler(s), and associated pumping systems. However, central system(s) can be used to provide chilled water and hot water during the normal operating hours, or as a backup for the 24-hour system(s).
- Twenty-four-hour systems with a capacity of up to 50 tons should be configured with an air-cooled chiller.

In the event the building's 24-hour operation load, including the dedicated perimeter ventilation system, exceeds 50 tons, the cooling systems may be combined with a central system that includes a dedicated central chilled water supply loop provided along with 24-hour chiller.

8.5.3 High-Occupancy Areas

High-occupancy areas that also have largely variable occupancies, such as conference rooms and lecture theatres are served by dedicated ventilation and air-handling systems. These systems should incorporate a demand-controlled ventilation (DCV) system to minimize energy consumption, while maintaining appropriate levels of ventilation and pressure relationships between spaces and the outdoors. The DCV system devices should be located for ease of maintenance and should provide appropriate operation of the ventilation system being controlled. An enthalpy heat recovery system should be used if economically feasible.

Auditoriums should have dedicated air-handling units equipped with an enthalpy economizer cycle. Units should be designed with an 80% diversity factor to maintain necessary temperature and humidity conditions under partial loads and partial occupancy. Provide dew-point control. The dew point of supply air should not exceed 10°C (50°F) dry bulb.

8.5.4 Food Service Areas

Kitchen areas should be negative in pressure relative to adjacent dining rooms, serving areas, and corridors. Tempered makeup air should be introduced at the kitchen hood and/or the area adjacent to the kitchen hood for at least 80% of exhaust air. Duct air velocity in the grease hood exhaust should be no less than 7.5 to 9 m/s (1500 to 1800 feet per minute [FPM]) to hold particulate in suspension. Dishwashing areas must be under negative pressure relative to the kitchen, and dishwashers should be provided with their own exhaust hoods and duct systems, constructed of corrosion-resistant material.

8.5.4.1 Kitchens and Dishwashing Areas

Kitchens with cooking ranges, steam kettles, ovens, and dishwashers should be provided with dedicated makeup air and exhaust hoods/exhaust systems in accordance with the latest edition of the National Fire Prevention Association (NFPA) Standard 96 ASHRAE Applications Handbook. All components of the ventilation system should be designed to operate in balance with each other, even under variable loads, to properly capture, contain, and remove the cooking effluent and heat, and maintain proper temperature and pressurization control in the spaces efficiently and economically. The operation of the kitchen exhaust systems should not affect the pressure relation between the kitchen and surrounding spaces. Both supply air and makeup air should be exhausted through the kitchen hood heat recovery system while a maximum of 30% of the exhaust air is made up from the space.

Floor drains must be provided at each item of kitchen equipment that requires indirect wastes, where accidental spillage can be anticipated, and to facilitate floor-cleaning procedures. Drains to receive indirect wastes for equipment should be of the floor sink type of stainless construction with a sediment bucket and removable grate.

8.5.5 Mechanical Rooms

The mechanical system should continue the operation of key life safety components following an incident. This system should be located in less vulnerable areas, limiting access to mechanical systems, and providing a reasonable amount of redundancy.

All mechanical rooms must be mechanically ventilated to maintain room space conditions as indicated in ASHRAE 62 and ASHRAE 15. Water lines should not be located above motor control centers or disconnect switches, and should comply with requirements of NEC Chapter 1. Mechanical rooms should have floor drains in proximity to the equipment served to reduce water streaks or drain lines extending into aisles.

8.5.5.1 Chiller Equipment Rooms

All rooms for refrigerant units should be constructed and equipped to comply with ASHRAE Standard 15: Safety Code for Mechanical Refrigeration. Chiller staging controls should be capable of direct digital control (DDC) communication to the central building energy management system.

8.5.5.2 Electrical Equipment Rooms

No water lines are permitted in electrical rooms, except for fire sprinkler piping. Sprinkler piping lines must not be located directly above any electrical equipment.

8.5.5.3 Communications Closets

Communications closets must be cooled in accordance with the requirements of Electronic Industries Alliance/Telecommunications Industry Association (EIA/TIA) Standard 569. Closets that house critical communications components should be provided with dedicated air conditioning systems that should be connected to the emergency power distribution system.

8.5.5.4 Elevator Machine Rooms

A dedicated heating and/or cooling system must be provided to maintain room mechanical conditions required by equipment specifications. In the event the building is equipped throughout with automatic sprinklers, hoistway venting is not required.

8.5.5.5 Emergency Generator Rooms

The environmental systems should meet the requirements of NFPA Standard 110: Emergency and Standby Power Systems combustion air requirements of the equipment.

Rooms must be ventilated sufficiently to remove heat gain from equipment operation. The air supply and exhaust should be located so air does not short circuit.

Generator exhaust should be carried up to roof level in a flue or exhausted in compliance with the generator manufacturer's installation guidelines. Horizontal exhaust through the building wall should be avoided.

8.5.5.6   *Uninterruptible Power Supply (UPS) Battery Rooms*

Battery rooms must be equipped with eye wash, emergency showers, and floor drains.

The battery room must be ventilated and exhausted directly to the outdoors at a rate calculated to be in compliance with code requirements and manufacturer's recommendations, and the exhaust system must be connected to the emergency power distribution system.

Fans should be spark resistant, explosion proof, with the motor outside the air stream; the ductwork should be a negative-pressure system of corrosion-resistant material, with exhaust directly to the outdoors in a dedicated system.

Acoustical enclosures should be provided to maintain a maximum noise criterion (NC) level of 35.

Coordinate with electrical design specifications to include HVAC support equipment in UPS extended servicing agreements.

8.5.6   Toilets and Maintenance Closets

Toilet areas must have segregated exhausts and should be negative in pressure relative to surrounding spaces.

Maintenance or janitorial and housekeeping closets must have segregated exhausts and should be negative in pressure to surrounding spaces.

8.6   VENTILATION PROTECTION STRATEGIES

A building is a system of barriers, and offers some protection for occupants against hazards that originate outdoors. One approach to protection is exclusion—controlling access to the building and an area around the building. Other options include sheltering in place and purging.

Particularly if a large building or large area within the building is used as a shelter, personnel will need to rapidly de-energize all fans, close intake and exhaust dampers, shut all perimeter doors and, if possible, all interior fire doors. Relays or digital control systems can be used to provide easy, one-button control of dampers and hard-wired fans. To close all doors and windows quickly in a large building, a notification system such as a public address system should be used and a protective action plan, in which building occupants have been trained, should be implemented. Even if control systems are used on dampers and hard-wired fans, inhabitants will still need to de-energize personal air-moving devices such as personal fans and portable smoke eaters within the shelter.

Both active and passive protection can be detection based; that is, protection can be initiated upon detecting the presence of an airborne hazard. Only active protection, however, can be employed on a continuous basis, as passive protection requires halting or substantially reducing the intake of fresh air, thus affecting habitability.

8.6.1   Passive Protection: Barriers to Control Air Exchange

The barriers between occupants of the building and the contaminated air provide passive protection that varies inversely with the rate of air exchange between indoors and outdoors. A building's air exchange rate is always greater than zero; therefore, passive protection diminishes as time of exposure to an airborne hazard increases, making the building suitable only for protecting against hazards of short duration. To employ passive protection effectively requires forewarning—the ability to detect the presence of an agent and to temporarily reduce the rate of air exchange.

8.6.2   Active Protection: External or Internal Filtration

Active protection involves air filtration and is typically applied by filtering outside air as the air is forced into a building by fans. This type of application, referred to as *external filtration*, yields far higher levels of protection, is not time dependent, and may be more costly than passive protection. External filtration can provide very high levels of protection, particularly if filtered air is introduced into a building at a rate sufficient to produce an overpressure.

A second approach to active protection is internal filtration. This involves recirculation or drawing air from inside a building and returning the air to the inside of the building. The protection afforded by internal filtration is determined by the normal air exchange rate of the building and the removal rate of the recirculating filter units. Internal filtration is a type of purging.

8.7   EMERGENCY AIR DISTRIBUTION SHUTOFF

If notification is received of a potential external (outdoor) release, an emergency air distribution shutoff switch can be used to minimize the exchange of contaminated outdoor air with clean indoor air. Upon notification of an internal release, activating the shutoff switch can minimize air movement throughout the building thus reducing the contaminant's spread. When an emergency air distribution shutoff switch is activated, air distribution equipment including supply, return, exhaust, and circulation fans throughout the building de-energize and motorized dampers close. The emergency deactivation should also cause elevators, if in motion, to stop at the next floor to further reduce contaminate spread. The elevator doors may open and close, but the elevator is not to move. All portable air-moving devices, such as personal fans and smoke eaters, should also be de-energized. Windows should be closed and

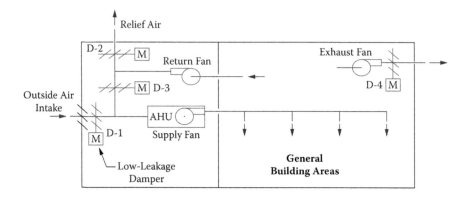

Figure 8.1 **(See color insert following page 34.)** Emergency air distribution shutoff.

locked. Window air conditioners should be de-energized and their outside air dampers, if so equipped, closed.

Figure 8.1 illustrates the emergency air distribution shutoff for a typical central air-handling system. During normal operation, the air-handling unit (AHU) supply air fan, return air fan, and exhaust fan are energized, dampers D-1, D-2, and D-3 are open or modulating, and exhaust damper D-4 is open. When the emergency air distribution switch is activated, the AHU supply air fan, return air fan, and exhaust fan are automatically de-energized, and low-leakage dampers D-1, D-2, and D-4 close to seal the building perimeter and reduce the infiltration of outside air.

As a first response to an unknown release, the facility should operate in the protection mode assuming a release outside of the building, by closing all dampers and de-energizing all fans.

If the release is later determined to be inside the building, then exhaust fans can be energized and exhaust and intake dampers opened in the portion of the building that is affected by the release. By energizing the exhaust system, contaminates in the building may be exhausted to the outdoors, and fresh outdoor air introduced into the building.

8.8   MASS NOTIFICATION OR INTERCOM SYSTEM

A mass notification, public address, or alarm system is critical to CBR protection. Notification alerts building occupants to a CBR release and provides evacuation directions. Alarms and notification systems can also be used to indicate when to evacuate, put on protective masks, or shelter in place. The mass notification, public address, or alarm system may be separate from the fire alarm or integrated with the fire alarm but easily distinguishable from the fire alarm.

8.9 EVACUATION

Evacuation is the most common protective action taken when an airborne hazard is perceived in a building. Existing plans for fire evacuation may be applicable for evacuation in response to a chemical hazard. Two considerations that must be addressed in a nonfire evacuation include whether the: (1) source of the hazard is internal or external, and (2) evacuation may lead to other risks. Evacuation may not be the best action for an external release, particularly one that is a widespread CBR threat. For a suspected internal release, evacuation is a highly recommended option over sheltering in place (SIP). Evacuation due to an internal release ensures building inhabitants quickly come in contact with fresh, uncontaminated outdoor air.

If the area enveloped by the plume of hazardous material is too large to exit rapidly, the use of sheltering in place or protective masks should be considered. The advantage of sheltering in place is that sheltering can be implemented more rapidly than evacuation. The disadvantage is that the protection provided is variable and diminishes with the duration of the exposure.

8.10 SHELTERING IN PLACE

If evacuation is not an acceptable option, employing SIP procedures is the next best option. Along with evacuation, sheltering in place is a conservative and relatively inexpensive response to a perceived airborne threat. Disadvantages of sheltering in place include the fact that a building does not prevent contaminated air from entering; rather, the building envelope minimizes the rate of infiltration. Even for a tightly sealed building, outside air enters slowly; and once the external hazard has passed, the sealed building releases the contaminated air slowly.

Although sheltering is generally for protection against an outdoor release, sheltering in place against an internal release in a large building may be accomplished through HVAC zoning and sealing of building interiors.

SIP can be used as a short-term precautionary measure if a likely, immediate threat is perceived. Advance warning of a potential release or early detection of a release is necessary to apply the protective measures associated with SIP. Therefore, SIP is not practical for protection against any contaminate unless some type of notification accompanies the contaminate release.

The level of protection that can be attained by sheltering in place is substantial, but is much less than that of high-efficiency filtering of the fresh air introduced into the building. The amount of protection varies with the following:

- The building's air exchange rate. The tighter the building and the lower the air exchange rate, the greater the protection the building provides. In most cases, air conditioners and combustion heaters cannot be operated while sheltering in place. Doing so increases the air exchange rate.

- Duration of exposure. Protection varies with time, diminishing as the time of exposure increases; therefore sheltering in place is suitable only for exposures of short duration (roughly an hour or less).
- Purging or period of occupancy. The length of time occupants remain in the building after the hazard has passed affects the level of protection. Because the building slowly releases contaminants that have entered, at some point as the hazard decreases, the concentration inside exceeds the concentration outside. Increasing the air exchange rate after the hazard diminishes or exiting the building into clean air offers maximum protection.
- Natural filtering. Natural filtering occurs as airborne material passes into and out of the building shell, and as chemicals temporarily penetrate building materials. The tighter the building, the greater natural filtering contributes to the protection factor.

In a home, sheltering is readily accomplished by closing windows and doors; and turning off all air conditioners, fans, and combustion heaters. To do so in an office building requires more time and planning. All air-handling units must be turned off and dampers closed.

The area of refuge provided in the event of emergency conditions should be equipped with adequate ventilation energized from the emergency power distribution system and sufficient heating capacity to maintain space temperature of 21°C (70°F) with design winter outdoor temperature. Provide a separate air-handling unit in the refuge area to maintain positive pressure, relative to surrounding spaces, with heating-cooling coils and a differential pressure sensing system.

8.10.1 Actions Required

Sheltering in place requires that two distinct actions be taken without delay to maximize the passive protection a building can provide:

1. The first action is to reduce the building air exchange rate before the hazardous plume arrives by closing all windows and doors and turning off all fans, air conditioners, and combustion heaters.
2. The second action is to increase the building air exchange rate as soon as the hazardous plume has passed, by opening all windows and doors and turning on all fans to aerate the building.

In normal operation, a building's mechanical system is designed to exchange indoor and outdoor air and circulate air within the building. To provide SIP protection, four distinct actions must occur to alter the air exchange rate of a building or safe room with respect to the outdoors:

1. Receipt of a credible warning that an event is likely to occur or has occurred and deciding whether to evacuate or shelter in place.
2. If the entire building is not designated as a shelter, the building occupants go to the designated shelter.
3. Reduce the indoor–outdoor and room-to-room air exchange rates before the airborne hazard reaches the building. To do so, close all windows and doors in the

shelter-in-place envelope; turn off air conditioners, combustion heaters, and all air-moving devices including supply and exhaust fans that induce indoor–outdoor or room-to-room air exchange; and close all outside air intake, relief, and exhaust dampers. Upon a call for fans to de-energize and dampers to begin closing, up to 5 minutes may pass before large fans come to a complete stop and dampers fully close.

4. After all the inhabitants have entered the shelter, lock the shelter perimeter doors from the inside to ensure people on the outside cannot enter and thereby contaminate the shelter.

8.10.2 Selecting and Preparing a Shelter Area

The shelter can be:

- A single-use area that is dedicated for use only during a CBR event, or a multiuse area that could expediently be converted to a shelter given sufficient notice of an event
- Sited to coexist with the building's tornado, hurricane, or other shelter type

Desirable characteristics in selecting a shelter-in-place area include:

- Accessibility: The shelter should be rapidly accessible to its anticipated population including those mentally or physically disabled. Minimize the amount of time people spend outdoors to reach the shelter. Note: No substantial advantage is gained by sheltering in a room on the lower or higher floors of a low-rise building. A location should not be selected based on height above ground level if this height would increase the time for people to reach the shelter in an emergency.
- Shelter size: Provide a minimum of 5 square feet per standing adult, 6 square feet per seated adult, and 10 square feet per wheelchair user for anticipated occupancy up to 2 hours (per FEMA 361 for a tornado shelter). For anticipated durations between 2 and 36 hours provide a minimum of 10 square feet per person (per FEMA 361 criteria for a hurricane shelter).
- Tight construction: Ensure that the shelter area is more tightly constructed than other areas or can be tightly and expediently sealed.

A shelter located in the building's windowless interior is most desirable since a CBR plume must breach two or more architectural barriers before reaching the inhabitants. If only window wall options are available, select the rooms with the fewest number of windows and tightly seal the windows. A room with walls or ceiling in contact with the outdoors can be used as a shelter as long as the exposed building surfaces are adequately sealed from air infiltration.

Stairwells generally make excellent shelters as stairwells have minimal or no ventilation and are usually architecturally separated from the rest of the building by firewalls. The shelter should not have large, uneasily sealable openings such as a kitchen hood or fireplace. If an adequate shelter cannot be defined within the building, consider constructing a building addition.

Provide a CBR Shelter Plan. Include a floor plan that clearly identifies the shelter location. Post this floor plan near the main entrance inside the building, next to the

Fire Evacuation Route Plan and Tornado Shelter Plan. Clearly label the plan as a CBR Shelter Plan.

Architectural, mechanical, and electrical modifications can be made to prepare a safe room, as well as an entire building, to provide greater and more reliable protection when used for sheltering in place. Generally, these modifications have the following four objectives:

1. To seal unintentional openings, thereby reducing the air exchange rate of the shelter envelope.
2. To expedite sheltering by closing all penetrations in the shelter envelope (doors, windows, ducts) and de-energizing all air-moving devices.
3. To add internal, recirculating filtration.
4. To possibly use filtered recirculated air-conditioned air while sheltering, without reducing the level of protection.

8.10.3 Sealing

SIP involves sealing a building, or a portion thereof (safe room), to create a relatively airtight envelope. The envelope protects occupants from airborne contaminants released outside the building. Sealing a building provides additional protection against a potential CBR contaminant plume and reduces the building energy consumption for both heating and cooling by reducing outside air infiltration. Reducing a building's rate of air exchange with the outside air reduces the rate of infiltration of airborne CBR contaminants.

The shelter should have a physical air barrier between the shelter and the rest of the building; perimeter walls should extend up to the roof or floor deck above. Typically, most leakage occurs through the top of the protective area envelope, particularly where suspended lay-in ceilings are used without a hard ceiling or without a well-sealed roof–wall juncture above the lay-in ceiling. In such cases, the ceiling should be replaced with gypsum wallboard.

SIP can use expedient sealing measures or enhanced SIP building components or a combination of expedient and enhanced measures to seal doors and windows and shut down the ventilation system equipment and close ventilation system openings.

Ensure the building or other designated shelter envelope is adequately sealed, including all utility penetrations and building interface joints (i.e., roof–wall interface and floor–wall interface). Use typical weatherizing sealing measures to seal the building or shelter envelope. Caulking and foam sealants are two common materials used for sealing building joints, cracks, and penetrations. Note: Glass fiber insulation does not provide an airtight seal and is, therefore, not acceptable.

- Seal window, window–wall, and wall interfaces and apply weather stripping on exterior doorways and doorways between the shelter and high-risk areas of the building. Ensure door weather stripping is applied to all four sides. Regularly inspect door and window weather stripping and repair or replace as necessary.

- To limit air leakage, ensure the roof–wall interface and floor–wall interface are sealed.
- Seal all pipe, conduit, and duct penetrations at the shelter perimeter.
- In multistory buildings, seal openings or voids in stairwell walls, elevator shafts, mechanical chases, and other openings between floors to reduce the stack effect within the building. Use NFPA-rated materials to seal penetrations through firewalls.

A seal integrity pressure test, with an internal positive and/or negative pressure applied to the building, and smoke tests can be used to locate areas requiring attention.

In many cases, the building's ventilation system also serves the shelter. To help provide an air barrier, either:

- Seal all diffusers and grilles that terminate in the shelter, or
- Install a normally open motorized damper in the ductwork close to the shelter perimeter.

8.10.3.1 Expedient Sealing

Expedient sealing is the least expensive SIP method but requires that inhabitants perform the most labor to adequately seal the shelter perimeter. Shelter inhabitants must close and seal windows and perimeter doors with tape and plastic and de-energize portable and hard-wired air-moving devices. Seal supply and exhaust registers with tape and plastic.

Preplace sealing supplies and materials, including duct tape, sheet plastic, cutting utensils for the plastic, ladders (if necessary to seal ceiling diffusers and registers) in a location accessible during an emergency. To expedite sealing during an emergency, precut to size and label the plastic sheets. Depending on the number of inhabitants and their skill level, the time needed to seal a shelter may exceed the time needed for the contaminate plume to reach the shelter. Particularly in large buildings or shelters, building personnel need to rapidly turn off all fans, close air intake and exhaust dampers, and close all windows and perimeter doors.

8.10.3.2 Enhanced Sealing

For enhanced SIP, inhabitants need only evacuate to the shelter, close perimeter doors and windows, locate the SIP control panel within the shelter, and activate the enhanced SIP system by pressing a button or turning a switch. Relays or digital control systems can be used to provide easy control of fans and dampers. The enhanced SIP system when activated will:

- De-energize all hard-wired air-moving devices serving the building or safe room, including supply, return, and exhaust air fans.

- Close low-leakage isolation dampers installed in all ventilation system openings that penetrate the SIP protective envelope. When considering the entire building as a shelter, the openings include all outside air intakes, relief and exhaust air openings, and elevator shaft vents. For a safe room within a building, the low-leakage dampers can be placed to close off ductwork where the ductwork penetrates the safe room perimeter or in each supply, return, and exhaust air diffuser, register, or grille.

Indicator lights and/or alarms may also be installed on the SIP control panel. These lights and alarms alert the shelter occupants to potential problems in the enhanced SIP system by verifying that all fans are off and isolation dampers and doors are closed. The lights and alarms are activated by differential pressure switches on the fans, end switches on each isolation damper that make on full closure, and balanced magnetic switches (BMS) on each door in the shelter perimeter. Locking the perimeter doors can be performed manually or by magnetic locks that are activated from the control panel.

In an enhanced SIP shelter, the perimeter doors and windows are provided with permanent weather stripping so applying tape and plastic is not necessary. To quickly close all doors in a large building, use a notification system such as a public address system. Develop a protective action plan and train building occupants.

8.10.4 Filtration

8.10.4.1 Internal, Recirculating Filtration

To further increase survivability in SIP shelters, a portable air recirculating and filtering device can be used. This unit filters recirculated air to remove contaminants that enter the shelter over time. This unit contains both high-efficiency particulate air (HEPA) and carbon filters, but does not provide climate control. Ceiling- or wall-mounted units are available as well as free-standing floor or table units. During a CBR event, the unit is energized using a standard 120-volt wall receptacle. Replace filters periodically and based on usage.

8.10.4.2 Internal, Recirculating Filtration with Conditioned Air

Standard air conditioning and heating systems cannot be operated while sheltering in place because fans either directly or indirectly introduce outside air into the shelter. If many people are confined in a shelter without air conditioning, heating, or air movement in hot or very cold weather, conditions may eventually become unbearable and cause inhabitants to prematurely leave. Inhabitants should exit the shelter only after being so directed by local authorities.

To provide an air conditioning and heating system that can operate safely in a CBR emergency mode, the AHU must exclusively serve the shelter. The air-handling unit and associated ductwork are to be located within the shelter to ensure no contact is made with contaminated air. If the unit and associated ductwork were outside of the shelter, contaminated air could potentially be quickly introduced into the shelter via cracks or other openings in the air-handling system. The air-handling unit's outside air intake and exhaust dampers must be reliably operable and should fully close in a CBR emergency. An alternative is to install a split air conditioning system that tempers recirculated air and does not introduce outside air in either the normal or CBR emergency mode.

To provide climate control, carbon and HEPA filters may be retrofitted into the shelter's existing air-handling system. During a chemical emergency, the air-handling system would continue to run but in 100% return air mode, and with its air intake and exhaust dampers fully closed.

8.10.5 Shelter Necessities

Ensure the shelter has adequate restroom facilities for each gender. Generally provide 1 water closet for each 100 persons (or portion thereof) of each gender. Also provide 1 urinal for every 200 males. Portable chemical toilets, such as those most commonly used during extended camping or hunting trips, may also be considered.

As potable water distribution systems may become contaminated during a CBR event, do not drink from water fountains or electric water coolers in the shelter; rather, use bottled water. Contaminated potable water systems may even make hand washing unsafe.

In planning for a potential CBR release, preplace essentials such as portable radios, flashlights, fresh batteries for radios and flashlights, bottled water (assume 1 gallon/person/day), nonperishable food, and personal medicines inside the shelter. Also anticipate seating needs. The shelter should have a means to communicate with local authorities to determine when exiting the facility is safe.

8.11 SIP PROTECTION AND AIR EXCHANGE RATES

The protection provided by SIP varies with the duration of exposure and the shelter air exchange rate. Protection diminishes as the time of exposure increases; therefore SIP is suitable only for exposures of short duration (a few hours or less) depending on factors such as air exchange rate with the shelter, toxicity and concentration, and meteorological conditions.

A tighter, or well-sealed, shelter allows for a lower air exchange rate, therefore providing greater protection for a longer duration. Since shelter construction type and quality vary considerably, determining the air exchange rate for a given shelter requires performing an air leakage test using a tracer-gas test method. In addition, the air exchange rate is also influenced by pressure differentials exerted on the shelter by wind, the stack effect, and air-moving devices, resulting in the air leakage. Figure 8.2 shows air exchange rates, expressed as air changes per hour, for very well-sealed, well-sealed, and poorly sealed shelters under various wind conditions.

Wind Condition	Very Well-Sealed Shelter (air changes/hour)	Well-Sealed Shelter (air changes/hour)	Poorly Sealed Shelter (air changes/hour)
No wind 0 km/hr	0	0.05	0.1
Light 13 km/hr (8 mph)	0.05 to 0.10	0.10 to 0.30	0.30 to 0.50
Moderate 24 km/hr (15 mph)	0.12 to 0.18	0.18 to 0.35	0.35 to 1.00
High 48 km/hr (30 mph)	0.25 to 0.40	0.40 to 0.70	0.70 to 1.5

Figure 8.2 Air exchange rate in air changes per hour for various shelter and wind conditions.

A properly sealed interior room shelter generally provides much greater protection to inhabitants over one with walls or ceiling exposed to the outdoors. Wind helps outdoor air exchange with indoor air through outdoor air infiltration, thereby greatly reducing the protection period. The protection increase for the interior room shelter is due to the minimal amount of air that the shelter exchanges with its neighboring rooms. This minimal air exchange rate delays the contaminated air from reaching the interior room shelter thereby extending the protection period.

SIP Protection is frequently expressed as a protection factor that is the ratio of the contaminant concentration outside the shelter to concentration inside the shelter. Figure 8.3 shows protection factors after sheltering periods of 1, 2, and 4 hours, depending on the air exchange rate.

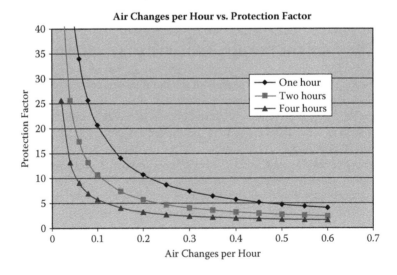

Figure 8.3 Infiltration, expressed as air changes per hour vs. protection factor.

8.12 CARBON DIOXIDE

Increases in carbon dioxide levels and decreases in oxygen are a planning and design concern for sheltering in place with high population densities, 0.93 m^2 (10 ft^2) per person, and tight shelter envelopes for longer durations. Though shelters are generally for short-duration events and a moderate population concentration; carbon dioxide levels increase, and to a lesser extent, oxygen levels decrease within the safe area. The increase in carbon dioxide levels during extended periods of sheltering in place at reduced outside air intake rates is more of a concern than the depletion of oxygen. The elevated concentrations of carbon dioxide far exceed the necessity to replenish the oxygen levels in the safe area. Carbon dioxide buildup makes people lethargic and sleepy, and thus may impair decision-making skills.

The shelter designer should consider carbon dioxide levels when determining the shelter size, and take into account the desired sheltering time and number of people that will be in the shelter.

- The U.S. Occupational Safety and Health Administration (OSHA) has set the carbon dioxide permissible exposure limit (PEL) at 5000 parts per million (ppm) for an 8-hour period and an exposure limit of 30,000 ppm for a 15-minute period.
- The National Institute for Occupational Safety and Health (NIOSH) has set the carbon dioxide threshold limit values (TLV) at 10,000 parts per million (ppm) for a 10-hour period and an exposure limit of 30,000 ppm for a 10-minute period.

The carbon dioxide level should be monitored and displayed on the facility control panel and alarmed at a concentration of 10,000 ppm. At this point the shelter occupants are not in any danger due to the carbon dioxide concentration in the shelter, but should begin to consider the possible need to evacuate the shelter even though an all clear has not been issued by local officials.

Carbon dioxide levels will increase more slowly in shelters that have some outside air exchange. However, for tightly sealed shelters, the effect of outside air exchange is minimal when compared to the effect of population density. Therefore, when designing a shelter, assume no outside air exchange when considering carbon dioxide levels. The carbon dioxide concentration can be calculated for a shelter of any size with no outside air exchange, number of occupants, and sheltering duration using the following formula.

$$C(t) = 15,500 \ P \ (t) \ / \ V + 330$$

where:

C(t) = carbon dioxide concentration, in ppm, in the shelter at time t
  P   = the number of people in the shelter (assumed to be standing or doing light desk work)
  t   = time, in minutes, the occupants have been in the shelter
  V   = shelter volume in ft^3

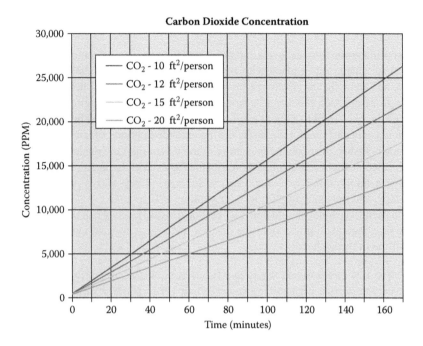

Figure 8.4   (See color insert following page 34.) Carbon dioxide concentrations in tightly sealed shelters.

Figure 8.4 shows a graphical representation of the carbon dioxide level in a shelter with no outside air exchange, a 10-foot ceiling, and population densities of 10, 12, 15, and 20 square feet per person. The shelter with a population density of 10 ft^2 per person reaches a carbon dioxide concentration of 10,000 ppm in 63 minutes and 20,000 ppm in 127 minutes.

8.13   CBR FILTRATION SYSTEMS

Filtration systems to protect shelters from airborne contaminates vary considerably depending on the type of filters that are used and the arrangement selected. All systems use a particulate filter to remove biological and radiological contaminates, and if desired carbon filters are also used to provide protection from chemical contaminates. Systems can be arranged to filter:

- Recirculated air within the shelter as described above for sheltering in place,
- Incoming outside air and create an overpressure in the shelter, or
- Both recirculated and incoming outside air.

An overpressure is created in the shelter by supplying more filtered outside air to the shelter than is exhausted by the shelter's ventilation system. The overpressure prevents external chemical and biological contaminates from infiltrating through the shelter's shell into the protective area. The overpressure can range from 12.5 Pascal (0.05 inches wg) to 75 Pascal (0.3 inches wg). For applications where the threat is a single accidental or intentional release of CBR material, an overpressure in the 12.5 Pascal (0.05 inches wg) to 25 Pascal (0.10 inches wg) range is usually selected.

If the filtration system is intended to run continuously, in most cases, installing a system will increase the facility's energy usage due to higher fan static pressures, increase outside air requirements to pressurize a building or shelter, and possibly eliminate economizer cycles. In addition, operation and maintenance (O&M) costs will increase due to filter replacement and system testing. Designers and facility managers should evaluate these costs to ensure that sufficient funding is available to maintain the system.

8.13.1   High-Efficiency Particulate Air Filters

To be effective against the full range of biological agents, the particulate filter should be a HEPA filter, having at least a 99.97% particle capture efficiency when challenged, at the rated airflow, by aerosol particles having a mean particle diameter of 0.3 microns. The HEPA filters should be installed in sealed holding frames. Less efficient filters and poorly sealed holding frames will filter the air stream to a certain degree, however very small particulate agents may penetrate the filters and frames and contaminate the shelter.

8.13.2   Carbon Filters

If chemical protection is desired, an activated carbon filter is required to remove gaseous chemical contaminates. Unfortunately, an available filter that can remove all chemical contaminates has not been developed; however, an activated carbon media impregnated with reactants will provide the greatest range of protection. Reactants have been engineered to remove multiple groups of similar chemicals, but will allow other groups of chemicals to pass through the filter. Chemicals with a low vapor pressure (less than 10 mm Hg) are physically adsorbed into the pores of the carbon while higher vapor pressure chemicals (greater than 100 mm Hg) are removed from the air stream by a chemical reaction with the reactants impregnated in the carbon. Chemical vapors with an intermediate vapor pressure (greater than 10 mmHg but less than 100 mmHg) can be removed either by adsorption or chemical reaction.

8.14   FILTER AND VENTILATION SYSTEM ARRANGEMENTS

Protection against biological and radiological agents can be provided by installing HEPA filters in the air-handling system. Figure 8.5 shows a schematic of a typical air-handling system with a HEPA filter installed in the central air-handling unit. The HEPA filter will remove biological and radiological contaminants, but not chemical

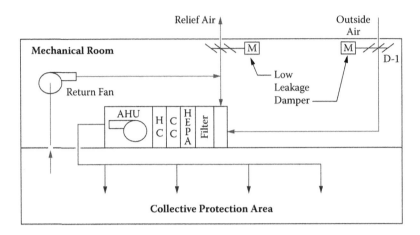

Figure 8.5 **(See color insert following page 34.)** Central air-handling system with a HEPA filter.

contaminants. If the amount of outside air brought into the shelter through the HEPA filters exceeds the air exhausted from the shelter, then a slight overpressure will be created in the shelter. The overpressure may help reduce or even prevent the infiltration of potentially contaminated outside air into the shelter.

Protection against biological and chemical agents that originate outside the building can be provided by installing HEPA filters and carbon filters on the outside air intake. Figure 8.6 shows a schematic of a typical central air-handling system with the outside air passing through an enhanced filtration unit (EFU) that contains both HEPA and carbon filters. The HEPA and carbon filters will remove biological, radiological, and chemical contaminants from incoming outside air. However, if contaminants are released inside the building or safe room, the contaminants will

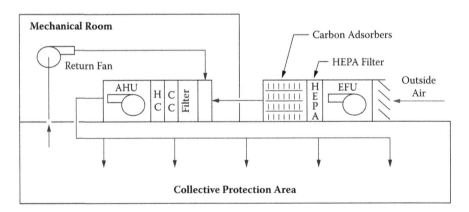

Figure 8.6 **(See color insert following page 34.)** Air-Handling system with HEPA and carbon filters on the outside air intake.

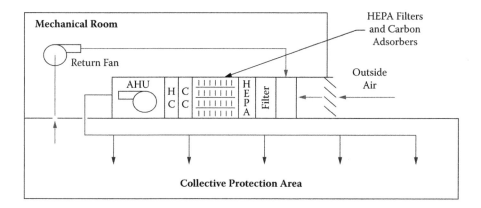

Figure 8.7 **(See color insert following page 34.)** Air-Handling system with HEPA and carbon filters in the central air-handling unit.

not be removed by this filtration scheme. If the amount of outside air introduced into the shelter through the HEPA and carbon filters exceeds the amount exhausted from the shelter, then an overpressure is created in the shelter.

Protection against an external release of CBR contaminants and the removal of recirculated CBR agents can be provided by installing HEPA and carbon filters in the central air-handling unit. Figure 8.7 shows a schematic of a central air-handling system with HEPA and carbon filtration of the return and outside air. In this arrangement, the HEPA and carbon filters will remove biological and chemical agents from incoming outside air and also from the return airstream. The effect of an internal release of agents is not eliminated, but the agents are not spread throughout the building through the ventilation system. If the amount of outside air introduced into the shelter through the HEPA and carbon filters exceeds the air that is exhausted from the shelter, then an overpressure is created in the shelter.

8.15   DESIGN APPROACH

A qualified engineer with experience in HVAC design and collective protection system (CPS) design should be consulted for renovations of existing building and for new building. A complete understanding of the both the HVAC and collective protection system is required.

The engineer must consider the effect a filtration system will have on the heating, cooling, and airflow capacities of the HVAC system and how the filtration system will be integrated into the HVAC control system. The shelter envelope should be sealed and the effect of this sealing on the HVAC system should also be considered.

For collective protection filtration systems, the outside airflow capacity required for the air filtration units is determined by the greater of two components: (1) the sum of the air leakage rate from the shelter at design overpressure and the ventilation rate

Figure 8.8   (See color insert following page 34.) Blower door assembly.

for exhaust requirements, or (2) the amount of outdoor air to satisfy indoor air quality requirements in accordance with ASHRAE Standard 62.1 (The default air intake rate is 15 cfm per shelter occupant).

To determine the air leakage rate from an existing shelter, a pressurization test can be performed using a blower door assembly in accordance with American Society for Testing and Materials (ASTM) E779. If the area is too large to be tested using a blower door, then the facility's air-handling systems can be used to perform the pressure test. A blower door is a calibrated, variable speed fan (or fans) that is temporarily mounted in a perimeter door in a frame that is sealed tightly into the doorjamb (see Figure 8.8). The airflow blowing into or out of the test area and the test area pressure can be adjusted by varying the speed of the fan. The blower door measures both the airflow through the fan and the pressure difference between the test area and the outside. Airflow rates and differential pressures are measured and recorded at a minimum of five points. The data is graphed and an air-flow vs. pressure trend line is drawn through the points. The air leakage from the shelter at design overpressure can then be read from the graph.

For a new shelter, the project architect should specify a maximum leakage rate. The leakage rate is usually expressed in units cubic feet per minute (cfm) per square foot of shelter perimeter area (including walls, roof, and floor) at a given differential pressure. A reasonable and achievable leakage rate is 0.25 cfm/ft^2 at

75 Pascals (0.3 inches wg). The specified leakage rate can then be converted to the leakage rate at the desired overpressure using the following equation:

$$\Delta P_{OP}/\Delta P_{75} = (Q_{OP}/Q_{75})^n$$

Using the following values:

ΔP_{OP} = Shelter's design overpressure (Pascals or inches wg)
ΔP_{75} = Specified test overpressure = 75 Pa (0.30 inches wg)
Q_{OP} = Shelter envelope leakage rate in cfm per square foot at design overpressure
Q_{75} = Specified envelope leakage in cfm per square foot at 75 Pa (0.30 inches wg)
Pressure exponent (n) = 0.65

Ductwork must have the proper configuration or construction to prevent unfiltered air from leaking into the shelter from the ductwork and from an unprotected area into ductwork that is carrying clean filtered air. Ductwork carrying clean filtered air should not be exposed to a contaminated environment (that is, outside the shelter envelope), and ductwork carrying unfiltered air should not be in a clean environment (that is, inside the shelter envelope). If these situations cannot be avoided, ductwork carrying filtered air must be under a positive pressure when located in a contaminated environment, and ductwork carrying unfiltered air must be under a negative pressure when located in a clean environment. If these duct pressures cannot be met or if differential pressures are uncertain, then the ductwork should be sealed by welding seams or other methods and leak tested.

The filtration system unit must have the proper configuration to prevent unfiltered air from leaking into the filter housing and ductwork downstream of the HEPA and carbon filters and entering the shelter, as follows:

If located outside or in a dirty mechanical room (that is, outside the shelter envelope), the filtration system unit must be designed as a blow-through unit with the blower located before the HEPA and carbon filters. Refer to Figure 8.9.

If located in a clean mechanical room (that is, inside the shelter envelope) and drawing in contaminated air through a ductwork system or an intake louver, the air-handling system must be designed as a draw-through arrangement with the blower located after the HEPA and carbon filters. Refer to Figure 8.10. Note that if the filtration system is located in a clean mechanical room, the filters cannot be changed without contaminating the protective area. If the ability to change filters during an event is desired, then the filtration system must be located outside or in a dirty mechanical room. For additional information, refer to UFC 3-340-01.

Clean mechanical rooms should meet the same overpressure requirements as the shelter area. The total static pressure of the filtration system blower must be designed to compensate for dirty filters, ductwork system pressure losses, and the overpressure requirement of the shelter. The filtration system fan can be equipped with a variable frequency drive or inlet vanes to adjust the airflow rate to maintain the desired overpressure in the shelter. The variable airflow will allow the system to adjust for clean or dirty filters and changes in air density from winter to summer

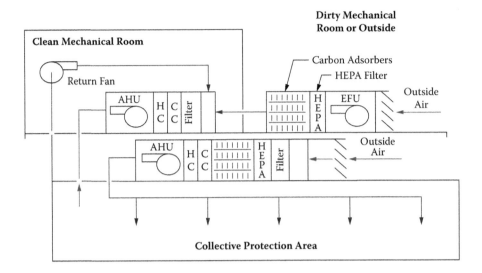

Figure 8.9 **(See color insert following page 34.)** Blow-through filter unit located outside or in a dirty mechanical room.

conditions. The filtration system must be designed for zero bypass around the HEPA filters and carbon absorbers.

A filtration system unit requires a HEPA filter (minimum efficiency reporting value [MERV] 17) and a carbon adsorber in series. In addition, a roughing filter (MERV 8), followed by a prefilter (MERV 13 to 15), are typically used upstream of the HEPA filters to collect large dust particles and extend the life of the HEPA filters. HEPA filters without a prefilter can become fully loaded and require replacement after only 9 months of continuous use. A prefilter extends the life of the HEPA filter to about 2 to 4 years of continuous use. The entire filter train described above, in sequence from upstream to downstream, is as follows:

- Roughing filter (MERV 8)
- Prefilter (MERV 13 to 15)
- HEPA filtration (MERV 17)
- Carbon adsorber

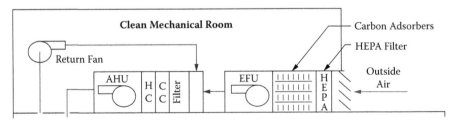

Figure 8.10 **(See color insert following page 34.)** Draw-through filter unit located in a clean mechanical room.

The filter housing should be designed and constructed to allow absolutely no air to bypass the HEPA and carbon filter elements and to accommodate the pressure that will be encountered. Challenge test ports for in-place testing should be incorporated into the system housing or elsewhere in the system ductwork. Pressure test ports for measuring differential pressures across the roughing filter, prefilter, HEPA filter, and carbon adsorber should also be incorporated into the filter housing.

8.16   COLLECTIVE PROTECTION SYSTEM TESTING AND STARTUP

After installation, the collective protection system and affected parts of the HVAC systems should be balanced and adjusted to provide the airflows, pressures, and temperatures specified. If the collective protection system operates in a standby mode, the system should be balanced in both the normal and standby modes of operation. The collective protection system and the HVAC system should be inspected and tested by an experienced engineer to ensure that both systems are operating as designed.

After installation the collective protection filtration system should be inspected and tested to ensure that filters have been installed properly, that no leaks exist between the mounting frame and housing or within the mounting frame, that no bypasses are occurring or will occur, and that the filters have not been damaged. First, visually inspect the gaskets, filters, and components for damage or deterioration. Testing is done by injecting a challenge agent upstream of the filter bank and measuring the challenge agent concentration upstream and downstream of the filters.

For the gas filter (carbon adsorber), the downstream concentration of the test agent should not exceed 0.1% of the upstream concentration. For the HEPA filter, the total test aerosol penetration through the filter should not be greater than 0.03% of upstream concentration. Filter testing is performed in accordance with American National Standard ASME N510, "Testing of Nuclear Air Treatment Systems." For testing the carbon filter, decafluoropentane (HFC 43-10, Vertrel) is a suitable alternative to Freon R-11. For testing the HEPA filter, paraffin hydrocarbon Emery 3004 is a suitable alternative to DOP for use as the upstream challenge agent. Emery 3004 may also be called Ethylflow 164 or Durasyn 164.

8.17   HVAC INSPECTION, OPERATIONS, AND MAINTENANCE

HVAC components serve the core of the building or the periphery or both. Core HVAC systems have a central plant that provides heat or cooling. Larger buildings will use boilers to generate steam for transmittal to heating coils and/or radiators. Cooling systems for larger buildings are usually closed-loop systems that transmit water in parallel to coils carrying Freon mixtures or to cooling towers. The subsequently cooled waters are transmitted to cooling coils.

The heating or cooling coils may be used in various conformations to provide thermal treatment to surrounding airstreams. The treated airstreams may then be delivered by ductwork or plenums to other building areas. In order to move the air around the building, fans are often used. The fans may either push or pull (by creating an upstream vacuum) air from place to place. Large fans are contained within boxes called air-handling units (AHUs).

8.18   INITIAL HVAC INSPECTIONS

To determine vulnerability, a preinspection analysis of HVAC system drawings is recommended prior to an actual on-site inspection. During the analysis of the HVAC system, drawings the use of duct liners within the system may be indicated. Liners used inside ductwork and AHUs may be areas where condensate layer formation has occurred. If liners are present, inspectors should evaluate if the linings show visual evidence of biological growth. AHUs, fan coil units, and ductwork should all be assessed, whether lined or not, to determine the potential for HVAC transmittal of CBR; and conversely the potential for a zoned shutdown.

Plenums are spaces above suspended ceilings. Plenums rather than ductwork are sometimes used to transfer return air to the building. If plenums are used to transfer return air, then all plenums should be inspected. These plenums are very difficult to secure if a CBR contaminant is introduced into a building and may make zoning of the HVAC system impossible. During the design phase specify the use of ductwork rather than plenums for returning air to the thermal treatment systems

Filtration is used to filter both incoming air and air being supplied to the building interiors. Filter bank areas and individual units should be noted and a representative number chosen for visual inspection. Maintenance records for filter cleaning and/ or replacement should be evaluated in terms of industry standards for such maintenance and the later visual evidence of current filter status. The methods used to maintain the filtration systems and the potential for access to filter banks should be evaluated. Uncontrolled access to the filtration system may be an indication that particulate contaminants, if purposely introduced into an HVAC system, would not be moderated by the filtration system if the system is compromised by a parallel disabling of the filtration system.

Air movement throughout a building is controlled by various mechanical, electrical, and plumbing components. Inspectors should note the control systems that determine makeup airflow. In particular, dampers will be examined to determine if the dampers and external HVAC intake air portals are currently operational and if the operational areas can be secured. The inspector should also determine if downdraft from the building or nearby buildings could bring contaminants (from any source) into the makeup airstream. The location of building exhaust systems in relation to building air intake and the potential for exhaust air to be introduced into the building through makeup air intakes is a process called *short-circuiting*.

Visually inspect all cooling towers during the inspection noting the presence of any living organisms (i.e., moss, algae, invertebrates [worms, snails], frogs, and insects). If these organisms are thriving, introduced biological agents or those present due to cooling tower contamination (i.e., *Legionella*) will also thrive.

8.18.1 HEPA Filter Replacement

HEPA filters are usually replaced when the static pressure across the filter reaches two to three inches wg. Replacing the filters at this point limits the overall static pressure of the filtration system and reduces energy usage. The time frame required before filter replacement is required is usually two to four years, depending on the environmental conditions and the quality of the prefilters in the system. An economic analysis that considers the replacement cost of the HEPA filters and the system energy cost should be performed to determine the optimal time to replace the filters.

8.18.2 Carbon Filter Replacement

The service life of an impregnated carbon filter is defined by two components: reactive gas life and physical adsorption life. Filter life against reactive gases is generally the shorter of the two; therefore, the replacement cycle is usually based on the degradation in reactive gas capacity. Filter life is site specific; however, the environmental air quality affects both capacities. The capacity for reactive gasses is typically lost due to exposure to humidity in ambient air; adsorption of organics (air pollutants) reduces the physical adsorption life. The rate at which the impregnant degrades varies with the temperature and the amount of water reacting with the carbon bed. Degradation begins once a filter is opened to the atmosphere. Degradation is more rapid when air is being drawn through the filter because atmospheric pollutants contribute to the degradation. Filter life can also be affected by loosening of the carbon as air is drawn through the filter, a condition that can be determined by conducting an in-place leak test at certain intervals.

The carbon filter should be replaced as indicated by adsorbent life testing conducted on samples removed from the adsorption system. Depending on the local atmospheric pollutants, filter change-out can be expected to be required every two to four years. The filter design must include a means to obtain samples of the adsorbent, either by removing a carbon filter from the filtration system or by incorporating a set of sample pilot canisters in the filter system that represents a sample of the total carbon filter bed thickness. Filters not contaminated by chemical surety materials or by other hazardous materials can be refilled by the manufacturer, but the owner must dispose of any contaminated carbon in accordance with local, state, and federal regulations.

8.19 PURGING WITH SMOKE FANS

Turning on a building's smoke purge fans is a means of reducing internal hazards. Use of smoke purge fans is beneficial for the purging phase of sheltering in place, once the agent is no longer present outside the building.

8.20   PROTECTIVE MASK AND RESPIRATOR USE

New models of universal-fit escape masks have been developed for protection against chemical and biological agents. Such masks form a seal at the wearer's neck and, therefore, fit a wider range of sizes than traditional masks that seal around the face. These masks do not require special fitting techniques or multiple sizes, are designed to store compactly, and can be carried on the belt. However, protective masks and respirators do not protect against all contaminants of concern.

ADDITIONAL READING

Analysis of the Clandestine CB Threat to USAF Strategic Forces (Unclassified), Volumes I and 11. Technical Considerations, Defense Technical Information Center (DTIC) Number AD379465, February 1967.

Antiterrorism Front-End Analysis (Unclassified), Defense Technical Information Center (DTIC) Number AD-C954865, June 1984.

Behind the first roar of machinery to drain the city: A tale of pluck and luck, John Schwartz, *New York Times*, September 8, 2005.

Biological Effects of Radiation, U.S. Nuclear Regulatory Commission, Office of Public Affairs.

Chemical/Biological Hazard Prediction Program, Technical Report, Defense Technical Information Center (DTIC) Number AD-BL 63245, 1991.

Chemical, Biological and Radiological Incident Handbook, Chemical, Biological and Radiological (CBRN) Subcommittee of the Interagency Intelligence Committee on Terrorism (IICT), October 1998.

Effects of Terrorist Chemical Attack on Command, Control, Communications, and Intelligence (C^3I) Operations, Defense Technical Information Center (DTIC) Number AD-BL 65614, April 1992.

Facilities Standards for the Public Building Service, U.S. General Services Administration, Public Buildings Service, Office of the Chief Architect, Washington, DC 20405, March 2003.

Fundamentals of Protective Design for Conventional Weapons, TM 5-855-1. Washington, DC, Headquarters, U.S. Department of the Army, 1986.

Glossary of Terms in Nuclear Science and Technology, La Grange, IL, American Nuclear Society, 1986.

Guidance for Protecting Building Environments from Airborne Chemical, Biological, or Radiological Attacks, National Institute of Occupational Safety and Health (NIOSH), U.S. Department of Health and Human Services (DHHS) (NIOSH) Pub No. 2002-139, May 2002.

HHS awards contract for radiation countermeasures, Global Security Newswire, February 14, 2006.

North American Emergency Response Manual, U.S. Department of Transportation, Transport Canada, Secretariat of Transport and Communications, 1996.

Nuclear Reactor Concepts, U.S. Nuclear Regulatory Commission, May 1993.

Nuclear Terms Handbook, U.S. Department of Energy, Office of Nonproliferation and National Security, 1996.

Proliferation of Weapons of Mass Destruction, Assessing the Risk, Office of Technology Assessment, U.S. Congress, OTA-ISC-559.

Protective Construction Design Manual, ESL-TR-87-57. Prepared for Engineering and Services Laboratory, Tyndall Air Force Base, FL. 1989.

Weapons of Mass Destruction Terms Handbook, Defense Special Weapons Agency, DSWA-AR-40H, 1 June 1998.

CHAPTER 9

Pressure Differentials and Airborne Contaminant Controls

Martha Boss, Dennis Day, and Charles Allen

CONTENTS

Whether for sheltering-in-place considerations or for general design, pressure differentials within a building must be known. This knowledge allows for effective planning should the building's ventilation system be the subject of an attack. In addition, this knowledge will effectively provide information so the heating, ventilation, and air conditioning (HVAC) system can be used to mitigate indoor air contaminant spread.

Calculations should be provided that show the minimum outside airflow rate required for pressurization. Minimum outside airflow rates should be adjusted as necessary to assure building pressurization.

The highest level of protection against an external airborne hazard is achieved by supplying filtered makeup air at a rate sufficient to produce a positive internal pressure in a building. The internal pressure induces an outward flow of clean air through leakage paths to reduce or prevent infiltration, depending upon the level of overpressure relative to external pressures exerted on the leakage paths by wind and stack effect.

To achieve the most effective filtration and air cleaning system against external chemical, biological, and radiological (CBR) threats, outdoor air leakage into the building must be minimized. Dramatically reducing leakage can be impractical for many older buildings. Older buildings may have large leakage areas, operable windows, and decentralized HVAC systems. In these instances, other protective measures should be considered. Typical office buildings are quite porous and may have leakage rates ranging from 0.03 to 0.6 m^3/min per m^2 of floor space (0.1 to 2 cfm/ft^2), at pressures of 50 Pa (U.S. Army Corps of Engineers, 2001). In buildings that have a leaky envelope, maintaining positive indoor pressure may be difficult to impossible.

In cold climates, building design should ensure that an adequate and properly positioned vapor barrier exists before the building is pressurized to minimize condensation, which may cause mold and other problems. All of these factors (leaky envelope, negative indoor air pressure, energy savings mode) influence building air infiltration and must be considered when tightening these buildings. Building pressurization or tracer gas testing can be used to evaluate the air tightness of the building envelope. Information on evaluating building envelope tightness, air infiltration, and water vapor management is described in the American Society of Heating, Refrigerating, and Air Conditioning Engineers (ASHRAE) *Fundamentals Handbook* (ASHRAE, 2001).

9.1   INFILTRATION AND EXFILTRATION

Externally induced infiltration is caused by the wind (forced convection) and inside–outside temperature differences (natural convection). Referred to as the *stack* or *buoyancy effect*, natural convection is caused by air density differences resulting from the temperature difference between indoor and outdoor air.

Stack pressures are highest in the winter, when large indoor–outdoor temperature differences exist. In cooler ambient temperatures, the relatively warm indoor air rises relative to the denser outdoor air. Drawing in air through the lower portions of the building as warmer indoor air exfiltrates from of the building's upper levels, the denser outdoor air replaces the rising indoor air. During the summer, air conditioning maintains indoor temperatures cooler than outdoor temperatures, and the flow is reversed. Because external hazards are more likely to involve a release at ground level, vulnerability due to stack-effect infiltration is greater in winter than in summer.

Infiltration, the exchange of indoor air with outdoor air, is driven by pressure differences across the building envelope and occurs through both unintentional openings in the building shell—cracks, seams, holes, pores—and intentional openings, such as doors, windows, or vents. The pressure differences that cause infiltration can be internally or externally induced. Internally induced infiltration occurs when areas of low pressure develop at the envelope boundary as internal fans circulate air through ducts or plenums of the HVAC system.

9.2 BUILDING AIR TIGHTNESS AND PRESSURIZATION

Significant quantities of air can enter a building by means of infiltration through unintentional leakage paths in the building envelope. The benefit of overpressure is defined by the relative volume of two sources of outdoor air for a building—infiltration and forced ventilation.

The reduction of air leakage is a matter of tight building construction in combination with building pressurization. While building pressurization may be a valuable CBR-protection strategy in any building, in a tight building pressurization will be most effective. However, for pressurization to be effective, filtration of the building supply air must be appropriate for the CBR agent introduced.

Producing an overpressure continuously with filtered makeup air is not practical in most buildings having normal exhaust flows. The flow rates required to prevent infiltration due to convective forces and to make up for exhaust flows necessitates very large filtration systems and greatly increases the heating and air conditioning costs. However, in an emergency situation, overpressurization, if possible, is one more tool available to protect building occupants.

Generally, for concrete buildings, the total exchange of air through unintentional openings is small compared to the volume entering by mechanical ventilation. The more suitable a building is for overpressure, the less important overpressure is for protecting the building interior. The tighter the building shell, the smaller the portion of fresh air that enters the building by infiltration.

9.3 PRESSURE DIFFERENTIALS

Interior/exterior differential air pressures are in constant flux due to wind speed and direction, barometric pressure, indoor/outdoor temperature differences (stack effect), and building operations, such as elevator movement or HVAC system operation. HVAC system operating mode is also important in maintaining positive indoor pressure. For example, many HVAC systems use an energy savings mode on the weekends and at night to reduce outside air supply and, hence, lower building pressurization.

The U.S. Army Corps of Engineers recommends that for external terrorist threats, buildings should be designed to provide positive pressure at wind speeds up to 12 km/hr (7 mph). Designing for higher wind speeds will give even greater building protection (U.S. Army 1999).

9.3.1 Negative Pressure Areas

In areas where exhaust systems are used or an indoor air quality contaminant source is located, a negative pressure should be maintained relative to surrounding spaces. The following spaces should be kept under negative pressure relative to surrounding building areas: smoking lounge, detention cells, toilets, showers, locker rooms, custodial spaces, and battery charging rooms. The air from these spaces must be exhausted directly to the outdoors.

9.3.2 Positive Pressure Areas

To keep dry air flowing through building cavities, systems should be designed with an operations sequence that assures continuous positive pressure with respect to the outdoor environment until the outdoor temperature falls below 4.5°C (40°F), when the building pressure should be brought to neutral. These building HVAC systems should have an active means of measuring and maintaining this positive pressure relationship. Alarms should indicate when the building pressurization drops below a predetermined low limit.

9.4   INTERNAL HEAT GAIN

Internal heat gain from all appliances—electrical, gas, or steam—should be taken into account. When available, manufacturer-provided heat gain and usage schedules should be used to determine the block and peak cooling loads. Typical rates of heat gain from selected office equipment should be based on the latest edition of the *ASHRAE Handbook of Fundamentals*. For preliminary design loads, heat gain given various lighting levels must be calculated. The cooling load estimated for the connected electrical load should be based on the electrical load analysis, estimated receptacle demand load, and anticipated needs.

9.5   STEADY STATE AND DILUTION VENTILATION

Achieving a steady state during a given decontamination/decommissioning event will require controlling the ventilation in the immediate area. Assumptions as to air exchange rates, while valid for whole room ventilation rate averages, are not valid for all areas within a room. Volumetric flow rates (Q) will be unchanging only if the net change in Q is zero (with the assumed net change of mass flow rate being zero). Factors that may cause the net change in Q to be other than zero include:

- Heat transfer
- Introduction of vapor
- Contaminant loading of the ambient air

Consequently, steady state is unlikely for all areas of a room or containment. Steady state may be assumed for incremental areas of a facility and within those areas for equipment on a worst-case basis. Thus, the area or encompassing area being analyzed will be evaluated based on the minimum ventilation rate assumed to be present in any area or encompassing area.

Once the minimum ventilation rate is established, the effects of dilution ventilation should be considered. If dilution ventilation proceeds with ventilation influx equaling off-gassing, the resultant equilibrium may seem to approach steady state. The rate of accumulation is then assumed to equal the rate of generation minus the rate of removal.

$$VdC = Gdt - Q'Cdt$$

or

Volume of the room * derivative of the concentration of gas or vapor = (Generation rate *derivative of time) – (Effective volumetric flow rate * (Concentration * derivative of time)

At steady state:

$$dC = 0$$

and the equation becomes

$$Gdt = Q' Cdt$$

At a constant concentration, C, and uniform generation rate, G

$$G (t_2 - t_1) = Q' C (t_2 - t_1)$$

$$Q' = G/C$$

To account for incomplete mixing, a K factor is introduced.

$$Q' = Q/K$$

Therefore:

$$Q = (G/C) * K$$

The K factor is based on several considerations including:

- Efficacy of mixing
- Distribution of replacement air
- Time
- Toxicity of contaminants
- Location and number of points of generation
- Operational effectiveness of any air-moving devices

The K factor is chosen in a range from 1 to 10. For highly toxic materials (agents), the K factor is 10 solely due to toxicity. *Unfortunately Q' may be overestimated when chemical warfare agents are the contaminants because with a K of 10 solely for toxicity, the other factors are assumed to be optimal (at least in this equation), but may not be.*

9.6 CONTAMINANT CONCENTRATION BUILDUP

Changed conditions after containerization may cause the contaminant concentration to increase over time in the headspace. For example containerized equipment may initially test at acceptable levels (i.e., worker population limits [WPLs]), however shipment of the container when accompanied by heating may cause additional off-gassing with resultant exceedance of formerly acceptable levels (i.e., >WPL).

Contaminant concentration buildup may also occur when desorption occurs from particulate contamination. For example, porous materials are present and absorb agent vapors. Steady state is achieved in the headspace analysis. As conditions change (i.e., heating or exposure to solvent vapors), these particulates may desorb their agent vapor load into the surrounding air.

The concentration of a contaminant can be calculated based on time.

$$dC/G - Q' = dt/V$$

integrated to

$$\ln(G - Q'C_2/G - Q'C_1) = Q'(t_2 - t_1)/V$$

In this equation, subscript 1 is the initial condition and subscript 2 is the resultant condition. The equation can be rearranged to calculate for delta t (Δt).

$$\Delta t = V/Q [\ln(G - Q'C_2/G - Q'C_1)]$$

9.7 PERIMETER ZONE

A dedicated 100% outside air unit should be used to maintain positive pressure.

The ventilation air for the perimeter air handling unit should be sized based on maximum occupancy with diversity and should operate

- Continuously during occupied hours
- At 40% capacity during unoccupied hours

Industrial-grade pressure sensors should be located at several perimeter areas. The internal pressure need only be slightly higher than ambient on average to achieve the goal of excluding humid outdoor air from building cavities.

- Maintain supply air discharge at the unit at no more than 10°C (50°F) dew point when the outside air dew point is above this temperature.
- Maintain neutral pressure (when the outdoor ambient temperature falls below 3°C (37°F) dew point and neutral pressure.

Differential pressure sensors and dew-point sensors should be connected to the building automation system. An alarm should signal if positive or neutral pressures are not maintained, on average, based on multiple samples taken within a five-minute period. Only industrial grade sensors are permitted.

9.8   INTERIOR ZONE

A dedicated outside air handling unit should be used to maintain positive pressure. The unit should:

- Be sized based on the fresh air requirements for maximum occupancy
- Have air monitoring devices and control the exhaust rate during occupied hours at less than 10% of the supply air to ensure positive pressure in the space
- Maintain 10°C (50°F) dew point when the outside air dew point is above that of the outside air
- Use humidification equipment, if necessary, to maintain 3°C (37°F) dew point whenever outside dew point is below 3°C (37°F)
- Maintain neutral pressure by setting the exhaust air quantity to equal the supply air rate. Air monitoring devices should be connected to the building automation system to indicate positive and neutral pressure

9.9   ENTRY/EXIT AND AIRLOCKS

Both hardware (ventilated airlocks) and procedures for transitioning from individual protection equipment to collective protection must be considered. Consider having separate HVAC systems in lobbies, loading docks, and other locations where an event may occur internal to these locations. For buildings having access control, entry zones of concern should be identified. These areas may include the lobby, the mailroom, and the area in which supplies or equipment are received and held temporarily while awaiting distribution. The ventilation system of these entry areas should be isolated from the rest of the building by:

- A separate air handling unit for the entry area
- Exhaust fan(s) to create a slight negative pressure differential in the entry area
- Full-height walls surrounding the entry area
- An airlock or vestibule for the exterior doors to maintain the pressure differential as people enter and exit

Controlling access and locking doors to mechanical rooms where the HVAC initial distribution system is located may be required. When locks are used, personnel who maintain entry must be immediately available 24/7.

A separate variable air volume (VAV) system should serve entrance vestibules and lobby spaces. The VAV system should include a differential pressure control system designed to maintain pressure as needed to protect adjacent areas. The air handling unit and the variable volume dampers at the VAV boxes should have self-contained microprocessor controls capable of connecting to and interoperating with a building's automation system. Air monitoring devices may be required to guarantee efficacy.

9.9.1 Procedures

Depending on the extent of risk, various considerations must be addressed to delimit contaminant movement into and through a building envelope.

- Entry/exit procedures and airlocks can maintain pressures within a protected area to ensure continuous protection during routine operations. Sufficient heating and cooling should be provided to offset the base load plus the infiltration load of the space.
- The entrance design must balance aesthetic, security, risk, and operational considerations. Consider co-locating public and employee entrances. Entrances should be designed to avoid significant queuing. If queuing will occur within the building footprint, the area should be enclosed in blast-resistant construction. If queuing is expected outside the building, a rain cover should be provided.
- The potential for using entrance alcoves or hallways as airlocks should be considered. The entrance vestibule should be positively pressurized relative to atmospheric pressure to minimize infiltration.

9.9.1.1 Entry/Exit

The transition from individual to collective protection is referred to as entry/exit. Entry/exit is an important consideration for all emergency shelters and during building decontamination. For contamination removal, entry/exit involves limited personnel decontamination procedures. These procedures are designed to:

- Minimize contamination transport to safe levels, and are based upon the principle of contamination avoidance.
- Minimize, and ultimately prevent the transport of contaminants

Entry and exit procedures vary according to type of personal protective equipment (PPE) ensemble worn, the type of entry system (airlock or no airlock), and the availability of support personnel and equipment. Such procedures specify the steps of removing (i.e., doffing) protective clothing to prevent contaminant transfer to inner garments or skin. All involve removal of an outer layer of clothing and partial decontamination or exchange of the mask/respirator.

9.9.1.2 Airlocks

An airlock is a transition enclosure, a protected entryway in which people pause for a specified period while entering to allow the purging of airborne contaminants introduced when opening the outer door. The main purpose of the airlock is to prevent direct vapor transport into the enclosure. In doing so, this function of the airlock also ensures that internal pressure is maintained continuously in the enclosure while entries and exits occur.

An airlock is not an air shower; the velocity of the clean air is typically low and has a negligible direct effect upon the rate of desorption or evaporation from the contaminated clothing. Airlocks do, however, provide an indirect effect upon indirect vapor transport. By imposing a delay that occurs in a relatively clean environment, airlocks allow more time for desorption to occur, thus reducing contaminant transport that occurs through adsorption/desorption.

Airlocks and areas where airlocks can be established should be located during emergency planning. Two types of airlocks should be considered: ventilated and unventilated. Ventilated airlocks are transition areas through which people enter and exit the protected air while an airborne hazard is present outside the shelter.

In buildings where security and potential required decontamination efforts may exist simultaneously, several airlock locations should be chosen. Knowledge of the in-place ventilation system is essential to determine which areas should be considered for either airlock or emergency decontamination usage.

One of the advantages of an airlock is its use as a doffing area. If large enough, the airlock can be employed for the removal of the outer clothing, greatly reducing the potential for indirect vapor transport as workers exit a contaminated area. The airlock can be used for chemical monitoring with internal air flows adjusted to be conducive to accurate monitoring. The airlock also can be used to introduce supplies and equipment.

9.9.2 Airflow Requirements

The flow of clean air from the filter unit of a collective protection system serves four purposes:

1. Ventilation (respiratory requirements and comfort of crewmen, 15 to 20 cfm/person)
2. Cooling internal equipment
3. Producing overpressure in the enclosure
4. Purging airborne contaminants from the airlock

The total volume of clean airflow required for a collective protection system is equal to the largest among these four requirements. Adding an airlock creates an additional airflow requirement. If, for example, 500 ft^3/min airflow is required to overcome leakage and establish overpressure, and the requirements for ventilation, cooling, and purging are each less than 500 ft^3/min, then the required total airflow is 500 ft^3/min. If an airlock is a part of the system, the airflow needed to purge the system (for example, 200 ft^3/min) must be added to the airflow needed to achieve an overpressure, raising the total to 700 ft^3/min.

Makeup air that exits through distributed, unintentional leakage paths serves to purge the enclosure, but air to purge an airlock must be directed through the airlock at a rate sufficient for purging the airlock's volume at a required rate. The rate at which an airlock can be purged of airborne contaminants is determined by the:

- Flow rate of clean air
- Flow path of clean air
- Volume of the enclosure
- Sorptive characteristics of the interior surfaces

Purging is most efficient when the path of clean airflow through the enclosure is of maximum length, that is, the clean air sweeps through all areas of the protected enclosure. In simplest terms, this means locating the vent through which air exits the enclosure (the controlled leakage point) at the farthest point from the vents through which clean air enters the enclosure.

The airlock is purged by the flow of clean air. Normally the air flows from the main protected enclosure, through the airlock, and out a discharge vent on the outer airlock door. The flow required for purging can be estimated with the following equation:

$$C/C_o = e^{(-QT/V)}$$

In this equation, C is the concentration at time T, C_o is the initial concentration, Q is the airflow rate in ft^3/min, V is the volume of the enclosure in ft^3, and T is the time in minutes. The purge period or dwell time is commonly no more than 5 minutes. This equation can be used to estimate the necessary flow rate. However, this equation assumes perfect mixing. Depending upon the actual flow path, determined by the distribution of leakage points in the shelter, the actual purge rate may be lower. The equation:

- Is based on simple purging only and does not account for sorption and desorption of vapor on interior surfaces, nor does this equation account for introducing a source of contamination, as occurs with indirect-liquid and indirect-vapor transport. When these become factors, the time required for purging increases.
- Is based on a ratio of beginning and ending concentrations. In practice, the beginning concentration is an unknown, and the ending concentration is the detection threshold of the chemical agent detector used to determine when to unmask.

REFERENCES

Facilities Standards for the Public Building Service, U.S. General Services Administration, Public Buildings Service, Office of the Chief Architect, Washington, DC 20405, March 2003.

Fundamentals of Protective Design for Conventional Weapons, TM 5-855-1. Washington, DC, Headquarters, U.S. Department of the Army, 1986.

Guidance for Protecting Building Environments from Airborne Chemical, Biological, or Radiological Attacks, National Institute of Occupational Safety and Health (NIOSH), U.S. Department of Health and Human Services (DHHS) (NIOSH) Pub No. 2002-139, May 2002.

Protective Construction Design Manual, ESL-TR-87-57. Prepared for Engineering and Services Laboratory, Tyndall Air Force Base, FL. 1989.

Security Engineering, TM 5-853 and Air Force AFMAN 32-1071, Volumes 1, 2, 3, and 4. Washington, DC, Departments of the Army and Air Force, 1994.

CHAPTER 10

Filtration

Martha Boss, Dennis Day, and Charles Allen

CONTENTS

Chemical, biological, and radiological (CBR) agents may travel in the air as a gas or an aerosol. Chemical warfare agents with relatively high vapor pressure are gaseous, while many other chemical warfare agents could potentially exist in either state. Biological and radiological agents are largely aerosols. A diagram of the relative sizes of common air contaminants (e.g., tobacco smoke, pollen, dust) is shown in Figure 10.1. CBR agents could potentially enter a building through either an internal or external release.

Air-filtration and air-cleaning systems can remove a variety of contaminants from a building's airborne environment. The effectiveness of a particular filter design or air-cleaning media will depend upon the nature of the contaminant. Airborne contaminants are gases, vapors, or aerosols (small solid and liquid particles).

Gases, vapors, and mists require adsorbent filters (i.e., activated carbon or other sorbent-type media). Sorbents collect gases and vapors, but not aerosols. The ability of a given sorbent to remove a contaminant depends upon the characteristics of the specific gas or vapor and other related factors. *Air cleaning* refers to the removal of gases or vapors from the air.

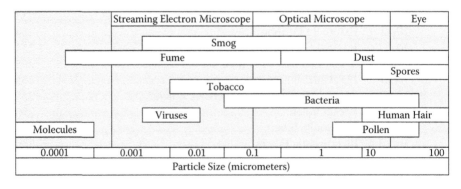

Figure 10.1   Common air contaminants and their relative sizes. (NIOSH, 2002.)

Particulate filters remove aerosols, but not gases and vapors. The efficiency of a particulate filter to remove aerosols depends upon the size of the particles, in combination with the type of filter used and heating, ventilation, and air conditioning (HVAC) operating conditions. Larger-sized aerosols can be collected on lower-efficiency filters, but the effective removal of a small-sized aerosol requires a higher-efficiency filter. Particulate air filters are used for chemical particulates, solid biological, and radiological particles and are not effective for gases and vapors typical of chemical attacks. *Air filtration* refers to removal of aerosol contaminants from the air. Chemical aerosols removed by particulate filters include tear gases and low-volatility nerve agents, such as VX; however, a vapor component of these agents could still exist. Biological agents and radioactive particulates are efficiently removed by HEPA filters.

In most buildings, mechanical filtration systems for particulate and aerosol removal are more common than sorbents for gas and vapor removal. However, the increased protection from CBR aerosols will occur only if the filtration efficiency increase applies to the particle size range and physical state of the CBR contaminant. Decisions regarding collection efficiency levels of particulate filters should be made with respect to ASHRAE Standards 52.1 and 52.2. Selection of the best sorbent or sorbents for gaseous contaminants is more complex. ASZM-TEDA (activated carbon, impregnated with copper, silver, zinc, molybdenum, and triethlyenediamine) carbon is recommended for classical chemical warfare agents. Other sorbents have been developed to collect specific toxic industrial chemicals (TICs).

10.1 SORBENTS AND MOLECULAR FILTRATION

Sorbents capture gas-phase air contaminants by physical adsorption or chemisorption. Physical adsorption results from the electrostatic interaction between a molecule of gas or vapor and a surface. Solid adsorbents—such as activated carbon, silica gel, activated alumina, zeolites, porous clay minerals, and molecular sieves—are useful because of their large internal surface area, stability, and low cost. Many of these sorbents can be regenerated by application of heat or other processes.

The standard for filtering chemicals—that is, molecular filtration—is a packed bed of activated, impregnated carbon granules. A carbon filter employs two different processes to remove molecules from an air stream—physical adsorption and chemical reaction.

Physical adsorption, the trapping of molecules in the micropores of the carbon, is effective against chemicals of low vapor pressure. Activated carbon is a very effective sorbent because its pore sizes vary and have a large surface area. Typically the pores in activated carbon have effective surface areas of about 1,200 square meters per gram. The carbon in a single 200-cubic-ft-per-minute military filter provides more than 5 square miles of active surface area.

Physical adsorption works well only against large molecules. As a rule of thumb, compounds of vapor pressure less than 10 mm Hg at the temperature of the filter bed are strongly adsorbed and retained in the pores of the carbon. To filter chemicals of higher vapor pressure, carbon is impregnated with salts of copper, zinc, silver, molybdenum, and triethylene diamine (TEDA), which react to form products that are innocuous or that can be physically adsorbed.

Carbon filters used for collective protection are designed for high efficiency and capacity. Efficiency is the percentage of agent removed in a single pass, and capacity is the quantity of agent a filter can remove before filtration ceases to be effective at the specified efficiency of 99.999%. A standard carbon filter can be expected to physically adsorb about 20% of its weight in agent or to remove about 5 to 10% of its weight in reactive gas.

The service life of an impregnated carbon filter is defined by both its reactive gas life and physical adsorption life. Filter life is site specific, as both capacities are affected by the environmental air quality. Physical adsorption life is reduced by adsorption of air pollutants. Capacity for reactive gases diminishes gradually over time and is typically lost within about three to four years of exposure to humidity. The rate at which the reactive life degrades varies with the temperature and amount of water adsorbed by the carbon bed, and degradation begins once a filter is opened to the atmosphere.

10.1.1   Gas-Phase Air Cleaning: Sorbent Filters

Some HVAC systems may be equipped with sorbent filters, designed to remove pollutant gases and vapors from the building environment. Sorbents use one of two mechanisms for capturing and controlling gas-phase air contaminants—physical adsorption and chemisorption.

Both capture mechanisms remove specific types of gas-phase contaminants from indoor air. Unlike particulate filters, sorbents cover a wide range of highly porous materials, varying from simple clays and carbons to complexly engineered polymers. Many sorbents, not including those that are chemically active, can be regenerated by application of heat or other processes.

Understanding the precise removal mechanism for gases and vapors is often difficult due to the nature of the adsorbent and the processes involved. While knowledge of adsorption equilibrium helps in understanding vapor protection, sorbent performance depends on such properties as mass transfer, chemical reaction rates, and chemical reaction capacity.

Some of the most important parameters of gas-phase air cleaning include the following:

- *Breakthrough concentration*: the downstream contaminant concentration, above which the sorbent is considered to be performing inadequately. Breakthrough concentration indicates the agent has broken through the sorbent, which is no longer giving the intended protection. This parameter is a function of loading history, relative humidity, and other factors.
- *Breakthrough time*: the elapsed time between the initial contact of the toxic agent at a reported challenge concentration on the upstream surface of the sorbent bed, and the breakthrough concentration on the downstream side of the sorbent bed.

- *Challenge concentration*: the airborne concentration of the hazardous agent entering the sorbent.
- *Residence time*: the length of time that the hazardous agent spends in contact with the sorbent. This term is generally used in the context of superficial residence time, which is calculated on the basis of the adsorbent bed volume and the volumetric flow rate.
- *Mass transfer zone* or *critical bed depth*: interchangeably used terms, which refer to the adsorbent bed depth required to reduce the chemical vapor challenge to the breakthrough concentration.

When applied to the challenge chemicals that are removed by chemical reaction, mass transfer is not a precise descriptor, but is often used in that context. The portion of the adsorbent bed not included in the mass transfer zone is often termed the *capacity zone*.

10.1.2   Sorbent Selection, Installation, and Use

Choosing the appropriate sorbent or sorbents for an airborne contaminant is a complex decision. Removal of gaseous contaminants from a building's air is a less common practice than the installation of particulate filtration. Sorbent filters should be located downstream of the particulate filters. This arrangement will allow the sorbent to collect vapors, generated from liquid aerosols that collect on the particulate filter, and reduce the amount of particulate reaching the sorbent. Gas-phase contaminant removal can potentially be a challenging and costly undertaking; therefore, different factors must be addressed, including the effect of face velocity, as illustrated in Figure 10.2.

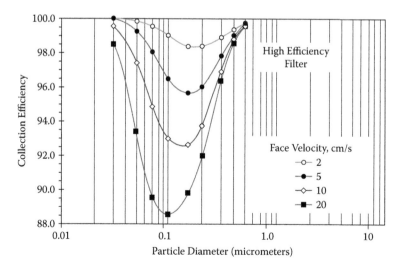

Figure 10.2   (See color insert following page 34.) Effect of face velocity on collection efficiency and MPPS. (NIOSH, 2002.)

10.1.3   Sorbent Properties and Limitations

Sorbents have different affinities, removal efficiencies, and saturation points for different chemical agents, which should be considered when selecting a sorbent. A well-designed adsorption system should have removal efficiencies ranging from 95% to 98% for industrial contaminant concentrations, in the range of 500 to 2,000 ppm; higher collection efficiencies are needed for high-toxicity CBR agents.

Sorbent physicochemical properties, such as pore size and shape, surface area, pore volume, and chemical inertness, all influence the ability of a sorbent to collect gases and vapors. Sorbent manufacturers have published information on the proper use of gas-phase sorbents, based upon contaminants and conditions. Air contaminant concentration, molecular weight, molecule size, and temperature are all important. The activated carbon, zeolites, alumina, and polymer sorbents selected should have pore sizes larger than the gas molecules being adsorbed. This point is particularly important for zeolites because of their uniform pore sizes.

With certain adsorbents, compounds having higher molecular weights are often more strongly adsorbed than those with lower molecular weights. Copper-silver-zinc-molybdenum-triethylenediamine (ASZM-TEDA) carbon is the current military sorbent recommended for collecting classical chemical warfare agents.

10.1.4   Sorbent Types

Activated carbon is the most common sorbent used in HVAC systems, and is excellent for most organic chemicals. Activated carbon is prepared from carbonaceous materials, such as wood, coal, bark, or coconut shells. Activation partially oxidizes the carbon to produce submicrometer pores and channels, which give the high surface area-to-volume ratio needed for a good sorbent.

Activated carbon often has surface areas in the range of 1000 m^2 per gram (m^2/g), but higher porosity materials, i.e., super-activated carbon, are well known. Because activated carbon is nonpolar (does not favorably adsorb water vapor), organic vapors can be captured at relatively high humidity. Activated carbon does not efficiently adsorb volatile, low-molecular-weight gases, such as formaldehyde and ammonia. However, activated carbon is relatively inexpensive and can retain a significant fraction (50%) of its weight in adsorbed material (U.S. Environmental Protection Agency [EPA], 1999).

The surface of activated carbon is highly irregular, and pore sizes range from 0.5 to 50 nm, enabling adsorption of many substances.

Carbons with smaller pore sizes have a greater affinity for smaller high-volatility vapors. Typically, activated carbon prepared from coconut shells has smaller pore sizes, while carbon produced from bituminous coal has larger pores. When the activated carbon has been spent, the carbon may be regenerated thermally or by using solvent extraction. The American Society for Testing and Materials (ASTM) has established standards for determining the quality of activated carbon and addressed issues such as apparent density, particle size distribution, total ash content, moisture, activated carbon activity, and resistance to attrition. The range of vapors that activated carbon will adsorb can be increased by using chemical impregnates, which supplement physical adsorption by an added chemical reaction. Impregnated activated carbon as a removal mechanism has been used since World War I to protect soldiers from chemical warfare agents, such as mustard gas and phosgene.

Chemical impregnates aid activated carbon in removing high-volatility vapors and nonpolar contaminants. Low-vapor-pressure chemicals—such as isopropyl methylphosphonofluoridate (GB), which is a nerve gas (sarin); and bis-(2-chloroethyl) sulfide (HD), which is a vesicant—are effectively removed by physical adsorption. Reactive chemicals have been successfully impregnated into activated carbon to decompose chemically high-vapor-pressure agents, such as the blood agents cyanogen chloride (CK) and hydrogen cyanide (AC).

One type of impregnated activated carbon, ASZM-TEDA carbon, has been used in U.S. military nuclear, biological, and chemical (NBC) filters since 1993. This material is a coal-based activated carbon that has been impregnated with copper, zinc, silver, and molybdenum compounds, in addition to triethylenediamine. ASZM-TEDA carbon provides a high level of protection against a wide range of chemical warfare agents.

Silica gel and alumina are common inorganic sorbents that are used to trap polar compounds. Sorption takes place when the polar functional group of a contaminant molecule is attracted by hydrogen bonding or electron cloud interaction with oxygen atoms in the silica or alumina. Silica gels are inorganic polymers having a variety of pore sizes and surface areas. Silica Gel 100 has a pore size of 10 nm and a surface area of 300 m^2/g. Silica gel 40 has a pore size of 4 nm and surface area of 750 m^2/g.

Impregnate	Chemical Contaminant
Copper/Silver Salts	Phosgene, Chlorine, Arsine
Iron Oxide	Hydrogen Sulfide, Mercaptans
Manganese IV Oxide	Aldehydes
Phosphoric Acid	Ammonia
Potasium Carbonate Potasium Iodide	Acid Gases, Carbon Disulfide
	Hydrogen Sulfide, Phosphine, Mercury, Arsine, Radioactive Methyl Iodide
Potasium Permanganate	Hydrogen Sulfide
Silver	Arsine, Phosphine
Sulfur	Mercury
Sulfuric Acid	Ammonia, Amine, Mercury
Triethylenediamine (TEDA)	Radioactive Methyl Iodide
Zinc Oxide	Hydrogen Cyanide

Figure 10.3 Application of activated carbon impregnates. (NIOSH, 2002.)

Silica gel adsorbs water in preference to hydrocarbons, and wet silica gels do not effectively adsorb hydrocarbons. This property makes silica gel a poor sorbent for humid atmospheres; however, amines and other inorganic compounds can be collected on silica gel. Alumina has pore sizes of approximately 5.8 nm and surface areas as high as 155 m^2/g. By changing the surface pH from acidic to basic, alumina can be modified to sorb a wider polarity range than silica gel.

Zeolites are a large group of naturally occurring aluminosilicate minerals, which form crystalline structures having uniform pore sizes. Zeolites occur in fibrous and nonfibrous forms and may go through reversible selective adsorption. Different molecular structures of zeolites result in pore sizes ranging from 3 to 30 angstroms. Zeolites are hydrophilic and may be chemically impregnated to improve their performance. The zeolites are used for organic solvents and for volatile, low-molecular-weight halides, such as chlorinated fluorocarbons (CFCs). A primary issue related to the effective use of zeolites is the molecular size of the vapor compared to the pore size. Zeolites will not adsorb molecules larger than their pore sizes, nor will zeolites capture compounds for which the zeolites have no affinity.

- Synthetic zeolites are made in crystals from 1 μm to 1 mm and are bonded to large granules, reducing airflow resistance. The zeolites can be manufactured to have large pore sizes and to be hydrophobic for use in high relative humidity. Synthetic zeolites can be designed to adsorb specific contaminants by modification of pore sizes.
- Alumina-rich zeolites have a high affinity for water and other polar molecules while silica-rich zeolites have an affinity for nonpolar molecules (EPA, 1998).

Synthetic polymeric sorbents are designed to collect specific chemical classes based upon their backbone structure and functional groups. Depending on the chemistry, polymeric sorbents can reversibly sorb compounds while others can capture and destroy contaminants. Some commercially available synthetic polymeric sorbents include the following: Ambersorb®, Amberlite®, Carboxen®, Chromosorb®, Hayesep®, and Tenax®.

Chemically impregnated fibers (CIF) are a recently developed technology, using smaller, more active sorbent particles of carbon, permanganate/alumina, or zeolite incorporated into a fabric mat. This design provides a combination of particulate and gas-phase filtration. The smaller sorbent particles are more efficient adsorbers than the larger ones found in typical packed beds. This technology provides the advantages of gas-phase filtration without the associated costs. CIF filters are held in media that range from 1.8 to 2 inches thick. Fibers range in size from 2 to 50 μm in diameter. CIF filters contain less sorbent (as much as 20 times less) than the typical packed beds, resulting in much shorter service life.

10.1.5   Performance Parameters and Breakthrough Prevention

Sorbents are rated in terms of adsorption capacity (i.e., the amount of the chemical that can be captured) for many chemicals. This capacity rises as concentration increases and temperature decreases. The rate of adsorption (i.e., the efficiency) falls as the amount of contaminant captured grows. Information about adsorption capacity (available from manufacturers) allows prediction of the service life of a sorbent bed. Sorbent beds are sized on the basis of challenge agent and concentration, air velocity and temperature, and the maximum allowable downstream concentration.

Gases are removed in the sorbent bed's mass transfer zone. As the sorbent bed removes gases and vapors, the leading edge of this zone is saturated with the contaminant, while the trailing edge is clean, as dictated by the adsorption capacity, bed depth, exposure history, and filtration dynamics. Significant quantities of an air contaminant may pass through the sorbent bed if breakthrough occurs. However, breakthrough can be avoided by selecting the appropriate quantity of sorbent and performing regular maintenance.

A phenomenon known as *channeling* may occur in sorbent beds and should be avoided. Channeling occurs when a greater flow of air passes through the portions of the bed that have lower resistance. Nonuniform packing, irregular particle sizes and shapes, wall effects, and gas pockets cause channeling. Channeling that occurs within a sorbent bed can adversely affect system performance.

10.1.6 Maintenance Schedules Based on Predicted Service Life

Consider service life of the sorbent when determining sorbent bed maintenance schedules and costs. All sorbents have limited adsorption capacities and require scheduled maintenance.

The effective residual capacity of an activated carbon sorbent bed is not easily determined while in use, and saturated sorbents can reemit collected contaminants. Sorbent life depends upon bed volume or mass and its geometric shape, which influences airflow through the sorbent bed. Chemical agent concentrations and other gases (including humidity) affect the bed capacity. Because of differences in affinities, one chemical may displace another chemical, which can then be re-adsorbed downstream or forced out of the bed. Most sorbents come in pellet form and different types can be mixed. Mixed- and layered-sorbent beds permit effective removal of a broader range of contaminants than possible with a single sorbent. Many sorbents can be regenerated, but the manufacturer's guidance must be closely followed to ensure that sorbents are regenerated in a safe and effective manner.

10.1.7 Chemically Active Sorbent Reuse

Some chemically active sorbents are impregnated with strong oxidizers, such as potassium permanganate. The adsorbent part of the bed captures the target gas and gives the oxidizer time to react and destroy other agents. Chemically active sorbents should not be reused because the oxidizer is consumed over time. If the adsorbent bed is exposed to very high concentrations of vapors, exothermic adsorption could lead to a large temperature rise and filter bed ignition.

This risk can be exacerbated by the nature of impregnation materials. Lead and other metals can significantly lower the spontaneous ignition temperature of a carbon filter bed. The risk of sorbent bed fires is generally low and can be further minimized by ensuring that air-cleaning systems are located away from heat sources and that automatic shutoff and warning capabilities are included in the system.

10.1.8   Filter or Sorbent Bypass and Air Infiltration

Ideally, all airflow should pass through the installed filters of the HVAC system. However, filter bypass, a common problem, occurs when air flows around a filter or through some other unintended path. Preventing filter bypass becomes more important as filter collection efficiency and pressure drop increase. Airflow around the filters results from various imperfections, for example, poorly sealed filters, which permit particles to bypass the filters, rather than passing directly into the filter media. Filters can be held in place with a clamping mechanism, but this method may not provide an airtight seal. The best high-efficiency filtration systems have gaskets and clamps that provide an airtight seal. Any deteriorating or distorted gaskets should be replaced and checked for leaks.

The best method of checking for leaks involves a particle counter or aerosol photometer. Visual inspection of filters for major leakage around the edges can also be accomplished by placing a light source behind the filter. No faults or other imperfections should exist within the filter media, and the filter media should be evaluated using a quantitative test, as described in the literature (National Air Filtration Association [NAFA], 2001a; ASHRAE, 2000).

Sample results are time dependent; thus background sampling done during a hot humid day will not be consistent with interior results collected later during a rainstorm. Seasonal and climactic changes must be considered when comparing samples. The relative temperature and humidity should be recorded for the days sampling occurred, particularly if successive sampling days are required.

10.1.9   Chemisorption, Adsorption, and Breakthrough Concentration

In chemisorption the gas or vapor molecules react with the sorbent material or with reactive agents impregnated into the sorbent. The sorbent forms a chemical bond with the contaminant or converts the contaminant into more benign chemical compounds. Potassium permanganate is a common chemisorbent, impregnated into an alumina or silica substrate and used to oxidize formaldehyde into water and carbon dioxide. Other more complex reactions bind the contaminants to the sorbent substrate where the contaminants are chemically altered. Chemisorption is usually slower than physical adsorption and is not reversible.

A number of very toxic vapors (e.g., hydrogen cyanide [AC]) are not retained on activated carbon by physical adsorption due to their high volatility. The traditional approach to provide protection against such materials is to impregnate the adsorbent material with a reactive component to decompose the vapor. Usually the vapor is converted to an acid gas byproduct, which must also be removed by reaction with adsorbent impregnation.

Adsorbent impregnation may potentially lose reactivity over time. Weathering of the impregnate is a particular concern for blood agents, such as AC and cyanogen chloride. Filter replacement schedules have been developed by the U.S. military, based on measurements of CK and AC breakthrough time as a function of environmental conditions, including the most unfavorable (hot and humid conditions).

Many different sorbents are available for various applications. These materials include both adsorbent and chemisorbent materials.

10.2 INITIAL EVALUATION: FILTRATION

Increasing filter efficiency is one of the few measures that can be implemented in advance to reduce the consequences of both an interior and exterior release of a particulate CBR agent. However, the decision to increase efficiency should be made cautiously, with a careful understanding of the protective limitations resulting from the upgrade. The filtration needs of a building should be assessed with a view to implementing the highest filtration efficiency that is compatible with the installed HVAC system and required operating parameters. Relevant issues for HVAC systems and these system's impact on vulnerability after the introduction of CBR agents include:

- HVAC system and components themselves
- Significance of uncontrolled leakage into the building
- Ability of the HVAC system to purge the building
- Efficiency of installed filters
- Capacity of the system relative to potential filter upgrades
- HVAC system controls

Initially, an understanding of the design and operation of the existing building and HVAC system is required. Improving building protection is not an all or nothing proposition. Many CBR agents are extremely toxic and high contaminant removal efficiencies are needed. Complex factors (i.e., agent toxicity, physical and chemical properties, concentration, wind conditions, means of delivery, release location) can influence the human impact of a CBR release.

In responding to potential building contamination, effective air-filtration and air-cleaning systems must be evaluated as to:

- Implementation in an existing HVAC system
- Incorporation into existing buildings when comprehensive renovation occurs
- Maintenance, especially maintenance needed in a contaminant laden-atmosphere

Improved HVAC efficiency should increase building cleanliness, limit effects from accidental releases of contaminants, and generally improve indoor air quality, even in nonemergency situations. These measures may also lessen cases of respiratory infection and reduce exacerbations of asthma and allergies among building occupants.

10.3   HVAC EVALUATION

A quantitative evaluation should be used to determine the total system efficiency. Perform the evaluation for various particle sizes and at the appropriate system flow rate. An evaluation of the results should be used to implement further modifications (e.g., improved filter seals, etc.). Information on quantitative evaluations of HVAC systems and filter performance can be found in the ASHRAE *HVAC Systems and Equipment Handbook* (ASHRAE, 2000) and the NAFA *Guide to Air Filtration* (NAFA, 2001a).

Before selecting a filtration and air-cleaning strategy that includes a potential upgrade in response to perceived types of threats, develop an understanding of the building and its HVAC system. A vital part of this effort will be to thoroughly evaluate the total HVAC system.

Assess how the HVAC system is designed and intended to operate and compare design intent to current operation. In large buildings, this evaluation is likely to involve many different air-handling units and system components. In providing a custom design to fit the needs of the building, consider the following:

- How are the filters in each system held in place and how are the filters and their holding frames or brackets sealed?
- Are the filters simply held in place by the negative pressure generated from downstream fans?
- Do the filter frames (the part of the filter that holds the filter media) provide for an airtight, leak-proof seal with the filter rack system (the part of the HVAC system that holds the filters in place)?
- Does the ductwork and plenum system have integrity, and how is the system activated?
- What types of air contaminants are of concern? Are the air contaminants particulate, gaseous, or both?
  - Are the contaminants TICs, toxic industrial materials (TIMs), or military agents?
  - How toxic are the contaminants? Consider checking with the local emergency or disaster planning body to determine if large quantities of TICs or TIMs are present near the building location or if specific concerns about military, chemical, or biologic agents must be considered.
  - How might the agents enter the building? Are contaminants likely to be released internally or externally to the building envelope, and how can various release scenarios best be addressed?
- What is needed?
  - Are filters or sorbents needed to improve current indoor air quality, provide protection in an accidental or intentional release of a nearby chemical processing plant, or provide protection from a potential terrorist attack using CBR agents?
  - How clean does the air need to be for the occupants, and how much can be spent to achieve that desired level of air cleanliness?
  - What are the total costs and benefits associated with the various levels of filtration?
  - What are the current system capacities (fans, space for filters) and what is desired?
  - What are the minimum airflow needs for the building?
  - Who will maintain these systems and what are their capabilities?

10.3.1 Ducted and Non-Ducted Return Air Systems

Ducted returns offer limited access points to introduce a CBR agent. The return vents can be placed in conspicuous locations, reducing the risk of an agent being secretly introduced into the return system.

Buildings should be designed to minimize mixing between air-handling zones, which can be partially accomplished by limiting shared returns. Non-ducted return air systems commonly use hallways or spaces above dropped ceilings as a return-air path or plenum. CBR agents introduced at any location above the dropped ceiling in a ceiling plenum return system will most likely migrate back to the HVAC unit and, without highly efficient filtration for the particular agent, redistribute to occupied areas. Where ducted returns are not feasible or warranted, hold-down clips may be used for accessible areas of dropped ceilings that serve as the return plenum.

10.3.2 Low-Leakage, Fast-Acting Dampers

When the HVAC system is turned off, the building pressure compared to outdoors may still be negative, drawing outdoor air into the building via many leakage pathways, including the HVAC system. Consideration should be given to installing low-leakage dampers to minimize this flow pathway. Damper leakage ratings are available as part of the manufacturer's specifications and range from ultra-low to normal categories.

Rapid response, such as shutting down an HVAC system, may also involve closing various dampers, especially those dampers controlling the flow of outdoor air (in the event of an exterior CBR release). Assuming that warning prior to a direct CBR release occurs, the speed with which dampers respond to a "close" signal can also be very important. From a protective standpoint, dampers that respond quickly are preferred over dampers that might take 30 seconds or more to respond.

10.4 BUILDING ENVELOPE

The building envelope matters. Filtration and air cleaning affect only the air that passes through the filtration and air-cleaning device. Whether outdoor air, recirculated air, or a mixture of the two, the pattern of air movement through and around the building must be considered. Outside building walls in residential, commercial, and institutional buildings are quite leaky, and the effect from negative indoor air pressures (relative to the outdoors) allows significant quantities of unfiltered air to infiltrate the building envelope.

Unless specific measures are taken to reduce infiltration, as much air may enter a building through infiltration (unfiltered) as through the mechanical ventilation (filtered) system. Therefore, filtration alone will not protect a building from an outdoor contaminant release—especially for systems in which no makeup air or inadequate overpressure is present. Instead, air filtration in combination with other steps, such as building pressurization and envelope air tightness, must be considered to increase the likelihood that the air entering the building actually passes through the filtration and air-cleaning systems.

10.4.1   External Filtration

The size and capacity of a filter system needed for external filtration is determined by the leakage rate of the building. To achieve the highest level of protection, 100% of the air entering the building must be supplied through the filter system. This can be assured only by providing filtered air at a volume great enough to prevent infiltration through all openings, to overcome the forces of wind and stack effect.

Although data are available for estimating the leakage rates of various types of buildings, these have proven to be unreliable for estimating filtration system capacity. The best method is to conduct a fan-pressurization test using commercially available blower-door systems with the building temporarily configured in its protected mode. Fan-pressurization tests require that the openings in the proposed protective envelope be taped to approximate the effect of dampers and seals that are to be added. A fan-pressurization test can be performed at a point in construction prior to installation of filter units or after the installation of the filter units.

The cost of applying high-efficiency filter units is governed by the leakage rate of the building. With very tight buildings, pressurization can be achieved with the same quantity of outdoor air that is required for health and comfort of occupants. Thus, the cost of heating and cooling additional amounts of outside air can be avoided.

10.4.2   Internal Filtration

Internal filtration involves drawing air from inside, rather than outside the building or enclosure and discharging the air inside. Also referred to as *recirculation filtering*, this provides much lower levels of protection against hazards outside the buildings than external filtration. Internal filtration can be achieved by placing filters in the air-handling unit or by using freestanding units—room-type indoor air purifiers or indoor air quality–type filter units.

Once an airborne hazard occurs inside a building, internal filtration can reduce the hazard in two ways: by interrupting the path of transmission to other areas of the HVAC zone, and by purging the agent with filtered air to reduce the concentration more rapidly than would occur through the processes of indoor-outdoor air exchange and deposition.

Applying filters at the air-handling unit also provides external filtration, although this is a relatively inefficient approach to filtering fresh air. To accommodate the large volume of air recirculated by typical air-handling units requires that the carbon and HEPA filters have relatively low resistances. For carbon adsorbers, a low resistance equates to a low efficiency and capacity. Use of freestanding indoor air purifiers having both carbon filters and HEPA filters has been shown to increase the protection factor attainable when sheltering in place.

10.5   MECHANICAL AND ELECTROSTATIC FILTERS FOR PARTICULATES

HVAC filters are critical system components. Filtration efficiency, flow rate, and pressure drop must be considered. Base particulate filter selection on air contaminant sizes, ASHRAE filter efficiency, and performance of the entire filtration system. Upgrading the filtration system may require significant changes in the mechanical components of the HVAC system, depending upon the component capacities. Consider both the direct and indirect impact of upgrading the filtration system.

Where occupancy requirements are likely to generate high levels of airborne particles, special air filtration may be useful on the return air system or dedicated and localized exhaust systems should be used to contain airborne particulates. Particulate air filters are classified as either mechanical filters or electrostatic filters (electrostatically enhanced filters). When selecting particulate air filters, choose between mechanical or electrostatic filters, with emphasis on collection mechanism and pressure drop differences.

Both mechanical and electrostatic filters are fibrous media used extensively in HVAC systems to remove particles, including biological materials, from the air. A fibrous filter is an assembly of fibers that are randomly laid perpendicular to the airflow. The fibers may range in size from less than 1 μm to greater than 50 μm in diameter. Filter packing density may range from 1% to 30%. Fibers are made from cotton, fiberglass, polyester, polypropylene, or numerous other materials (Davies, 1973). Fibrous filters of different designs are used for various applications.

10.5.1   Flat-Panel Filters

Flat-panel filters contain all of the media in the same plane. This design keeps the filter face velocity and the media velocity roughly the same. When pleated filters are used, additional filter media are added to reduce the air velocity through the filter media. This enables the filter to increase collection efficiency for a given pressure drop.

10.5.2   Pleated Filters, Including HEPA Filters

Pleated filters can run the range of efficiencies from a minimum efficiency reporting value (MERV) of 6 up to and including HEPA filter efficiencies. The current standard for filtering biological agents from an air stream is the HEPA filter. A component of all filtration units used by the military and the nuclear industry, the HEPA filter is normally placed upstream of the carbon filter to remove not only biological agents and solid aerosols but also liquid aerosols. The HEPA filter also protects the carbon filter from atmospheric dust.

The minimum filtration efficiency for HEPA filters is specified as 99.97% (a decontamination factor of 3,333) at the 0.3 micrometer (micron) particle size. Most HEPA filters have efficiencies of 99.99% at 0.3 micron and removal efficiency is greater for particles that are outside this size range as long as the filter is operated at or below its rated velocity. HEPA efficiency holds for particles down to about 0.01 micron.

Chemical and biological aerosol dispersions (particulates) are frequently in the 1- to 10-µm range. HEPA filters provide efficiencies greater than 99.99% in that particle size range, assuming no leakage around the filter and no damage to the fragile pleated media. This high level of filtration efficiency provides protection against most aerosol threats.

10.5.3 Pocket Filters

With pocket filters, air flows through small pockets or bags constructed of the filter media. These filters can consist of a single bag or have multiple pockets, and an increased number of pockets increases the filter media surface area. As in pleated filters, the increased surface area of the pocket filter reduces the velocity of the airflow through the filter media, allowing increased collection efficiency for a given pressure drop.

10.5.4 Renewable Filters

Renewable filters are typically low-efficiency media that are held on rollers. As the filter loads, the media are advanced or indexed, providing the HVAC system with a new filter. The spent (and often soiled) filter is present in the inner coil of the media. This type of filter is not recommended for CBR protection.

10.5.5 Electrostatic Filters

Electrostatic filters contain electrostatically enhanced fibers, which actually attract the particles to the fibers, in addition to retaining them. Electrostatic filters rely on charged fibers to dramatically increase collection efficiency for a given pressure drop across the filter. Electrostatically enhanced filters are different from electrostatic precipitators, also known as *electronic air cleaners*. Electrostatic precipitators require power and charged plates to attract and capture particles.

Liquid aerosols are known to cause great reductions in the collection efficiencies of many electrostatic filters, and ambient aerosols may also degrade performance. The degradation is partially related to the stability of the electrostatic charge.

Electrostatic filters may be an acceptable choice for some building protection applications. The filter efficiency rating given by the manufacturer is likely to be substantially higher than the filter will actually achieve when used.

Building owners who cannot feasibly upgrade to traditional high-efficiency mechanical filters may consider extended surface or electrostatic filter systems as an attractive low-cost alternative. Energy costs are minimized by the relatively low pressure drop across these filters, and costly HVAC upgrades (modifications that may be required for higher-efficiency mechanical filters) are frequently avoided.

However, the filtration efficiency of electrostatic filters must be carefully monitored as these filters may substantially degrade with time.

10.6   COLLECTION MECHANISMS

Four different collection mechanisms govern particulate air filter performance: inertial impaction, interception, diffusion, and electrostatic attraction. The first three of these mechanisms apply mainly to mechanical filters and are influenced by particle size.

1. Impaction occurs when a particle traveling in the air stream and passing around a fiber, deviates from the air stream (due to particle inertia) and collides with a fiber.
2. Interception occurs when a large particle, because of its size, collides with a fiber in the filter that the air stream is passing through.
3. Diffusion occurs when the random (Brownian) motion of a particle causes that particle to contact a fiber.
4. Electrostatic attraction, the fourth mechanism, plays a very minor role in mechanical filtration. After fiber contact is made, smaller particles are retained on the fibers by a weak electrostatic force.

Impaction and interception are the dominant collection mechanisms for particles greater than 0.2 μm, and diffusion is dominant for particles less than 0.2 μm. The minimum filter efficiency will shift based upon the type of filter and flow velocity (Lee and Liu, 1980).

10.6.1   Collection Efficiency

Air filters are commonly described and rated based upon their collection efficiency, pressure drop (or airflow resistance), and particulate-holding capacity. Two filter test methods are currently used in the United States:

- ASHRAE Standard 52.1
- ASHRAE Standard 52.2

Standard 52.1 measures arrestance, dust spot efficiency, and dust-holding capacity.

- *Arrestance* means a filter's ability to capture a mass fraction of coarse test dust and is suited for describing low- and medium-efficiency filters. Be aware that arrestance values may be high, even for low-efficiency filters, and do not adequately indicate the effectiveness of certain filters for CBR protection.
- *Dust spot efficiency* measures a filter's ability to remove large particles, those that tend to soil building interiors. *Dust-holding capacity* is a measure of the total amount of dust a filter is able to hold during a dust-loading test.

ASHRAE Standard 52.2 measures particle size efficiency (PSE). This newer standard is a more descriptive test that quantifies filtration efficiency in different particle size ranges for a clean and incrementally loaded filter to provide a composite efficiency value. This gives a better determination of a filter's effectiveness in capturing solid particulate as opposed to liquid aerosols. The 1999 standard rates particle-size efficiency results as a MERV between 1 and 20. A higher MERV indicates a more efficient filter. In addition, Standard 52.2 provides a table showing minimum PSE in three size ranges for each of the MERV numbers, 1 through 16. Thus, if the size of the contaminant is known, an appropriate filter with the desired PSE for that particular particle size can be chosen.

10.6.2 Filter Banks

Multiple filters can extend the life of the more expensive, high-efficiency filters. For example, one or more low-efficiency, disposable prefilters, installed upstream of a HEPA filter, can extend the HEPA filter life by at least 25%. If the disposable filter is followed by a 90% extended surface filter, the life of the HEPA filter can be extended by almost 900% (American Conference of Governmental Industrial Hygienists [ACGIH], 2001).

The cost of prefilter replacement and pressure drop against the extended life of the primary filter must be considered. Overall efficiency may be the same and changing the primary filters more frequently may be a better option than installation of prefilters. This decision is made by weighing the operating cost analysis against the capture efficiencies provided by different systems.

Filter banks often consist of two or more sets of filters; therefore, consider how the entire filtration system will perform—not just a single filter. The outermost filters are coarse, low-efficiency filters (prefilters) that remove large particles and debris while protecting the blowers and other mechanical components of the ventilation system. These relatively inexpensive prefilters are not effective for removing submicrometer particles. Therefore, the performance of the additional downstream filters is critical.

The downstream filters may consist of a single or multiple filters to remove submicrometer particles. Particles in the 0.1 to 0.3 μm size range are the most difficult to remove from the air stream and require high-efficiency filters.

10.7   FILTERS AND PRESSURE CONSIDERATIONS

As mechanical filters load with particles over time, their collection efficiency and pressure drop typically increase. Eventually, the increased pressure drop significantly inhibits airflow, and the filters must be replaced. For this reason, pressure drop across mechanical filters is often monitored and indicates when to replace filters.

Conversely, electrostatic filters, which are composed of polarized fibers, may lose their collection efficiency over time or when exposed to certain chemicals, aerosols, or high relative humidities. Pressure drop in an electrostatic filter generally increases at a slower rate than in a mechanical filter of similar efficiency. Thus, unlike the mechanical filter, pressure drop for the electrostatic filter is a poor indicator of the need to change filters. Other measures, such as collection efficiency or time of use, are more suited for determining electrostatic filter change-out schedules.

To be most effective, filters should be used at their rated pressure drop and face velocity. Filter face velocity refers to the air stream velocity entering the filter. The rated pressure drop for each filter is given for a specific face velocity (typically 1.3 to 2.5 m/s or 250 to 500 fpm), and the pressure drop increases with airflow velocity. With lower efficiency filters, the final (loaded or dirty) pressure drop is often in the range of 125 to 250 Pascals (Pa) (0.5- to 1.0-inch water gauge).

10.7.1 Upgrades and Pressure Loss

Upgrading filtration is not as simple as merely replacing a low-efficiency filter with a higher-efficiency one. Typically, higher-efficiency filters have a higher pressure loss, which will result in some airflow reduction through the system. Pressure loss associated with adsorbent filters can be even greater. The magnitude of the reduction is dependent on the design and capacity of the HVAC system. If the airflow reduction is substantial, inadequate ventilation, reductions in heating and cooling capacity, and reduced energy efficiency may result. To minimize pressure (drop) loss:

- Deep pleated filters or filter banks having a larger nominal inlet area might be feasible alternatives, if space allows.
- Prefilters may be used to avoid high pressure losses.
- More frequent filter change-outs should occur (cleaner filters equal less pressure loss).

Higher-quality filters may have an initial pressure drop higher than 125 Pa (0.5 in. water gauge) and a final pressure drop of as high as 325 Pa (1.5 in. water gauge). Consider the capacity of the existing HVAC system. Many systems (e.g., light commercial, rooftop package units) do not have the fan capacity to handle the higher pressure drop associated with higher-efficiency filters. If the pressure drop of the filters installed in the system is too high, the HVAC system may be unable to deliver the designed volume of air to the occupied spaces.

Higher-capacity fans may be needed to overcome the increased resistance, caused by higher-efficiency filters. Installation of such fans may not be feasible for many HVAC systems because of insufficient physical space or other limitations. In such cases, extended surface filters (i.e., pleated, mini-pleat, or V-bank) or electrostatic filter media, which provide higher efficiency and lower pressure drop, may be an alternative.

High-efficiency filters may experience a significant drop in collection efficiency if operated at too high of a face velocity.

10.7.2 Filter Bypass

Filter bypass is a common problem found in many HVAC filtration systems. Filter bypass occurs when air, rather than moving through the filter, goes around the filter, decreasing collection efficiency and defeating the intended purpose of the filtration system. Filter bypass is often caused by poorly fitting filters, poor sealing of filters in their framing systems, missing filter panels, or leaks and openings in the air-handling unit between the filter bank and blower. Improving filter efficiency requires addressing filter bypass. To optimize effectiveness, the HVAC system should minimize air infiltration and eliminate filter bypass.

10.7.3 Filter Rack or Frame System

The integrity of the HVAC system's filter rack or frame system has a major impact upon the installed filtration efficiency. Reducing the leakage of unfiltered air around filters, caused by a poor seal between the filter and the frame, may be as important as increasing filter efficiency. If filter bypass proves to be significant, corrective actions will be needed.

Some high-efficiency filter systems have better seals and frames constructed to reduce bypass. If upgrading to higher-efficiency filters, the size and shape of the filter rack may need to be changed, in part, to assure appropriate face velocities.

Filters enclosed in metal frames are heavy and may cause problems because of the additional weight placed on the filter racks. The increased weight may require a new filter support system that has vertical stiffeners and better sealing properties to ensure total system integrity.

10.8 OPERATIONS AND MAINTENANCE

Proper maintenance, including monitoring of filter efficiency and system integrity, is critical to ensuring HVAC systems operate as intended. The change-out schedule for various filter types may be significantly different. Filter performance depends on proper selection, installation, operation, testing, and maintenance. The scheduled maintenance program should include procedures for installation, removal, and disposal of filter media and sorbents. The HVAC system must not be operating (locked out/tagged out) to prevent contaminants from being entrained into the moving air stream.

If using mechanical filters, a manometer or other pressure-sensing device should be installed in the mechanical filtration system to provide an accurate and objective means of determining the need for filter replacement.

Ideally determine the change-out schedule for electrostatic filters by using optical particle counters or other quantitative measures of collection efficiency. Collecting objective data (experimental measurements) will allow optimized electrostatic filter life and filtration performance. The data should be particle-size selective to determine filtration efficiencies that are based on particle size (e.g., micrometer, submicrometer, and most penetrating size).

10.8.1 Training and Security

Qualified individuals should be responsible for the operation of the HVAC system. Maintenance personnel must have a general working knowledge of the HVAC system and its function. Maintenance personnel are responsible for monitoring and maintaining the system, including filter change-out schedules, documentation, and record keeping; therefore, maintenance personnel should also be involved in the selection of the appropriate filter media for a given application. Because of the sensitive nature of these systems, appropriate background checks should be completed and assessed for any personnel who have access to the HVAC equipment.

10.8.2 Fragile Mechanical Filters

Mechanical filters, often made of glass fibers, are relatively delicate and should be handled carefully to avoid damage. Polymeric electrostatic filters are more durable and less prone to damage than mechanical filters.

To prevent installation of a filter that has been damaged in storage or one that has a manufacturing defect, maintenance personnel should check all filters before installing them and visually inspect the seams for total integrity. Maintenance personnel should hold the filters in front of a light source and look for voids, tears, or gaps in the filter media and filter frames.

Take special care to avoid jarring or dropping the filter element during inspection, installation, or removal.

10.8.3 After CBR Release

If a CBR release occurs in or near the building, significant hazards may be present, particularly within the building's HVAC system. If the HVAC and filtration systems have protected the building from the CBR release, contaminants will have collected on HVAC system components, on the particulate filters, or within the sorbent bed. These accumulated materials present a hazard to personnel servicing the various systems.

- Maintenance and filter change-out should be performed only when a system is shut down to avoid re-entrainment and system exposure.
- Persons performing maintenance and filter replacement on any ventilation system that is likely to be contaminated with hazardous CBR agents should wear appropriate personal protective equipment (respirators, gloves, etc.).
- Decontaminating filters exposed to CBR agents requires knowledge of the type of agent, safety-related information concerning the decontaminating compounds, and proper hazardous waste disposal procedures. Before removal and where feasible, particulate filters should be disinfected in a 10% bleach solution or other appropriate biocide. Maintenance personnel must shut down the HVAC system when using disinfecting compounds and ensure that the disinfecting compounds are compatible with the HVAC system components contacted.
- Place old filters in sealed plastic bags upon removal.

10.9 LIFE-CYCLE ANALYSIS

Life-cycle cost should be considered (initial installation, replacement, operating, maintenance). Not only are higher-efficiency filters and sorbent filters more expensive than the commonly used HVAC system filters, but also fan units may need to be changed to handle the increased pressure drop associated with the upgraded filtration systems. Although improved filtration will normally come at a higher cost, many of these costs can be offset by the accrued benefits, such as cleaner and more efficient HVAC components and improved indoor environmental quality.

Life-cycle analysis will ensure that filtration and air-cleaning options meet the needs of the building maintenance and operations staff while providing protection to the building occupants. Incremental improvements to the removal efficiency of a filtration or air-cleaning system may substantially lessen the impact of a CBR attack to a building environment and its occupants while generally improving indoor air quality.

While higher filtration efficiency is encouraged and should provide indoor air quality benefits beyond an increased protection from CBR terrorist events, the overall cost of filtration should be evaluated. Improved filtration generally keeps heating and cooling coils cleaner, and thus may reduce energy costs through improvements in heat transfer efficiency. However, when HEPA filters and activated carbon adsorbers are used, the overall costs will generally increase substantially.

Filtration costs include the periodic cost of the filter media, the labor cost to remove and replace filters, and the fan energy cost required to overcome the pressure loss of the filters. Maintenance of ventilation systems and training of staff are critical for controlling exposure to airborne contaminants, such as CBR agents.

Filter cost, always a consideration, is directly related to efficiency, duration of effectiveness, and collection mechanism. Higher-efficiency filters tend to have a higher life-cycle cost than lower-efficiency filters. With some higher-efficiency filter systems, higher acquisition and energy costs can be offset by longer filter life and a reduced labor cost for filter replacements. Costs can be minimized by using an appropriate filter change out schedule, as illustrated in Figure 10.4.

Mechanical filters (pleated glass fiber) are quite likely to be more expensive than electrostatic (polymeric media) filters, but both may have the same initial fractional collection efficiency. However, over time the two types of filters will perform differently.

10.10 AN ALTERNATIVE: COLLECTIVE PROTECTION FILTER UNITS

Since 1932, when the U.S. Army standardized the first collective protection filter unit, known as the M1 Collective Protector, many collective protection systems have been developed for both mobile and fixed applications. Among the filter systems available for application to buildings—systems having flow rates of about 1,000 cfm and higher—are the FFA 580 filter units in the 600-cfm and 1000-cfm sizes; the new Multi-Cell Radial Flow (MCRF) filter unit in an 8,400-cfm configuration; the M49 Filter Unit as two 15,000-cfm units; and commercial V-bed filter units.

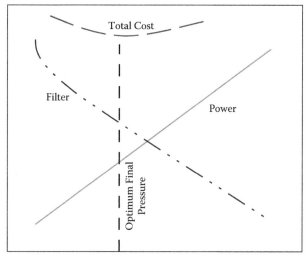

By selecting the appropriate change out schedule based upon the
optimum final pressure drop, the total cost can be minimized.

Figure 10.4 Relationship: cost, filter life, power needed. (NIOSH, 2002.)

The military has employed collective protection against airborne hazards—
chemical, biological, and radiological agents—since World War I. The concept is
simple, requiring merely an enclosure and a supply of clean air.

10.10.1 Class I

Supplying a continuous flow of clean air to an enclosure yields far higher levels of protection than unventilated shelters, particularly if the flow rate is sufficient to achieve
an overpressure and establish an outward flow of clean air through all intentional
and unintentional openings. This approach—developing a positive pressure in the
protected zone—achieves extremely high protection factors and is defined as Class
I collective protection.

10.10.2 Class II

An intermediate class, or Class II collective protection, involves a flow of purified
air, but at a rate that achieves little or no overpressure.

10.10.3 Class III

In its simplest form, the supply of clean air is the air retained in a building when the
occupants close windows, doors, and other openings to shelter in place. This type

of collective protection—the unventilated shelter—is similar to the first protective shelters, the gas-proof dugouts used in World War I. Today, unventilated shelters are defined as Class III collective protection.

REFERENCES

ANSI/ASHRAE (American National Standards Institute/American Society of Heating, Refrigerating, and Air-Conditioning Engineers). (2001). *ASHRAE Standard 62: ventilation for acceptable indoor air quality*. Atlanta, GA: American Society of Heating, Refrigerating, and Air-Conditioning Engineers, Inc., p. 27.

Davies CN. (1970). The clogging of fibrous aerosol filters. *Aerosol Sci* 1:35–39.

Lee KW, Liu BYH. (1980). On the minimum efficiency and the most penetrating particle size for fibrous filters. *J Air Pollut Control Assoc* 30:377–381.

NAFA (National Air Filtration Association). (2001). *NAFA position statement on bioterrorism*. Washington, DC, National Air Filtration Association. NFPA (1991).

ADDITIONAL READING

Alves BR, Clark AJ. (1986). An examination of the products formed on reaction of hydrogen cyanide and cyanogen with copper, chromium (6+) and copper-chromium (6+) impregnated activated carbons. *Carbon* 24:287–294.

American Institute of Architects (AIA). (2001). *Building security through design*. Washington, DC: American Institute of Architects, p. 4.

Avery RH. (1978). Energy effective air filtration. In *Plant Engineering*, Barrett LW, Rousseau AD.

BOMA (Building Owners and Managers Association). (2002). *Emergency preparedness*. Washington, DC: Building Owners and Managers Association International, p. 170.

Brink JA, Burgrabe WF, Greenwell LE. (1968). Mist eliminators for sulfuric acid plants. *Chem Eng Prog* 64(11):82–86.

Brown PN. (1989). Effect of aging and moisture on the retention of hydrogen cyanide by impregnated charcoals. *Carbon* 27:821–833.

Brown RC, Wake D, Gray R, Blackford DB, Bostock GJ. (1988). Effect of industrial aerosols on the performance of electrically charged filter material. *Ann Occup Hyg* 32(3):271–294.

Burroughs HE. (2002). Taking action against the new threat. *Refrigeration Service Eng Soc J (RSES)* 74(1):32–36.

CDC (Centers for Disease Control and Prevention). (2001). Update: investigation of anthrax associated with intentional exposure and interim public health guidelines. *MMWR* 50(41):889–897.

Chemical/Biological Hazard Prediction Program, Technical Report, Defense Technical Information Center (DTIC) Number AD-BL 63245, 1991.

Chemical, Biological and Radiological Incident Handbook, Chemical, Biological and Radiological (CBRN) Subcommittee of the Interagency Intelligence Committee on Terrorism (IICT), October 1998.

Cooper DW. (1980). Mechanisms for electrostatic enhancement of fabric filter performance. *Filtration and Separation* 17(6):520–523.

Dennis WL. (1973). Effect of humidity on the efficiency of particulate air filters. *Filtration and Separation* 10(2):149–150.

Dixon TC, Meselson M, Guillemin, J, Hanna PC. (1999). Anthrax. *N Engl J Med* 341(11): 815–826.

Doughty DT. (1991). *Development of a chromium-free impregnated carbon for adsorption of toxic agents.* Pittsburgh, PA: Calgon Carbon Corporation.

Facilities Standards for the Public Building Service, U.S. General Services Administration, Public Buildings Service, Office of the Chief Architect, Washington, DC, March 2003.

Filtration and Air-Cleaning Systems to Protect Building Environments from Airborne Chemical, Biological, or Radiological Attacks, Department of Health and Human Services, Centers for Disease Control and Prevention, National Institute for Occupational Safety and Health (NIOSH), U.S. Department of Health and Human Services (DHHS) (NIOSH) Publication No. 2003-136, Washington DC, April 2003.

Fire protection handbook. Quincy, MA, National Fire Protection Association.

Friedlander SK. (1977). *Smoke, dust and haze fundamentals of aerosol behavior.* New York, John Wiley & Sons.

Fundamentals of Protective Design for Conventional Weapons, TM 5-855-1. Washington, DC, Headquarters, U.S. Department of the Army, 1986.

Glossary of Terms in Nuclear Science and Technology, La Grange, IL, America Nuclear Society, 1986.

GSA (U.S. General Services Administration). (2001). *Security in the workplace—informational material.* Washington, DC, U.S. General Services Administration.

Guidance for Protecting Building Environments from Airborne Chemical, Biological, or Radiological Attacks, National Institute of Occupational Safety and Health (NIOSH), U.S. Department of Health and Human Services (DHHS) (NIOSH) Pub No. 2002-139, May 2002.

Gupta A, Novick VJ, Biswas P, Monson PR. (1993). Effect of humidity and particle hygroscopicity on the mass loading capacity of high efficiency particulate air (HEPA) filters. *Aerosol Sci Technol* 19(1):94–107.

Henderson DA. (1998). Bioterrorism as a public health threat. *Emerging Infectious Diseases* 4(3):1–7.

Henderson DA. (1999). The looming threat of bioterrorism. *Science* 283:1279–1282.

HHS awards contract for radiation countermeasures, Global Security Newswire, February 14, 2006.

Inglesby TV, Henderson DA, Bartlett JG, Ascher MS. (1999). Anthrax as a biological weapon: medical and public health management. *JAMA* 281(18):1735–1745.

Karwacki CJ, Jones P. (2000). *Technical report: toxic industrial chemicals: assessment of NBC filter performance.* Aberdeen Proving Ground, MD, Edgewood Chemical Biological Center.

Lee KW, Liu BYH. (1981). Experimental study of aerosol filtration by fibrous filters. *Aerosol Sci Technol* 1(1):35–46.

Lee KW, Liu BYH. (1982). Theoretical study of aerosol filtration by fibrous filters. *Aerosol Sci Technol* 1(2):147–162.

Liu BYH, Pui DYH, Rubow KL. (1983). Characteristics of air sampling filter media. In *Aerosols in the mining and industrial work environments,* Marple, VA, Liu BYH, eds. Ann Arbor, MI, *Ann Arbor Science* 3:989–1038.

Miller JD. (2002). Defensive filtration. *ASHRAE Journal* 44(12):18–23.

North American Emergency Response Manual, U.S. Department of Transportation, Transport Canada, Secretariat of Transport and Communications, 1996.

Noyes WA. (1946). *Military problems with aerosols and nonpersistent gases, summary technical report of division 10*. Washington, DC, National Defense Research Committee.

NRC (National Research Council). (1988). *Protection of federal office buildings against terrorism*. Washington, DC, Committee on the Protection of Federal Facilities Against Terrorism, Building Research Board, National Research Council, p. 60.

Protective Construction Design Manual, ESL-TR-87-57. Prepared for Engineering and Services Laboratory, Tyndall Air Force Base, FL, 1989.

Vincent JH. (1995). *Aerosol science for industrial hygienists*. Oxford, UK, Pergamon.

Weapons of Mass Destruction Terms Handbook, Defense Special Weapons Agency, DSWA-AR-40H, 1 June 1998.

Willeke K, Baron P, eds. (1993). *Aerosol measurement: principles, techniques, and applications*. New York, Van Nostrand Reinhold.

CHAPTER 11

VAC Control Systems

Martha Boss, Charles Allen, and Donald C. Dittus

CONTENTS

Many central heating, ventilation, and air conditioning (HVAC) systems have energy management and control systems that can regulate airflow and pressures within a building on an emergency response basis. Some modern fire alarm systems may also provide useful capabilities during chemical, biological, and radiological (CBR) events. In some cases, the best response option (given sufficient warning) might be to shut off the building's HVAC and exhaust systems, thus, avoiding the introduction of a CBR agent from the outside. In other cases, interior pressure and airflow control may prevent the spread of a CBR agent released in the building and ensure the safety of egress pathways.

The decision to install emergency HVAC control options should be made in consultation with an HVAC professional qualified to evaluate the ramifications of various HVAC operating modes on building operation and safety systems.

- Depending upon the design and operation of the HVAC system and the nature of the CBR agent release, HVAC control may not be appropriate in all emergency situations. Lobbies, loading docks, and mailrooms might be provided with manually operated exhaust systems, activated by trained personnel to remove contaminants in the event of a known release, exhausting air to an appropriate area.
- In other instances, manipulation of the HVAC system could minimize the spread of an agent. If an HVAC control plan is pursued, building personnel should be trained to recognize a terrorist attack quickly and to know when to initiate the control measures. For example, emergency egress stairwells should remain pressurized (unless known to contain the CBR source). Other areas, such as laboratories, clean rooms, or pressure isolation rooms in hospitals, may need to remain ventilated. All procedures and training needs associated with the control of the HVAC system should be addressed in the building's emergency response plan.

11.1   METERS, GAUGES, AND FLOW MEASURING DEVICES

Each piece of mechanical equipment should be provided with the instrumentation or test ports to verify critical parameters, such as capacity, pressures, temperatures, and flow rates.

Thermometers and pressure gauges are required on the suction and discharge of all pumps, chillers, boilers, heat exchangers, cooling coils, heating coils, and cooling towers. To avoid pressure gauge tolerance errors, a single pressure gauge may be installed, and valved to sense both supply and return conditions. For coils with less than 10 gpm flow, provisions for use of portable instruments to check temperatures and pressures should be made.

Duct static pressure gauges should be provided for the central air-handling unit air supply fan discharge, branch take-offs of vertical supply risers, and at all duct locations at which static pressure readings are being monitored to control the operation of a variable air volume (VAV) system.

Differential static pressure gauges should be placed across filters in air-handling units and to measure building pressure relative to the outdoors. A thermometer is required at the outside air intake to each air-handling unit.

11.1.1   Flow-Measuring Devices

Airflow measuring grids are required for all central air-handling units. Measuring grids should be provided at the supply air duct, return air duct, and the outside air duct. Airflow measuring grids must be sized to give accurate readings at minimum flow. Reducing the duct size at the station may be required to permit accurate measurement.

Water flow or energy-measuring devices are required for each chilled water refrigeration machine, hot water boiler, pump, and connections to district energy plants. Individual water flow or energy-measuring devices should be provided for chilled water lines serving computer rooms and chilled water and hot water lines to out-leased spaces.

Flow-measuring devices should be capable of communicating with the central building automation systems (BAS). Water flow– and air flow–measuring devices should confirm or validate requirements.

11.1.2 Testing Stations

Permanent or temporary testing stations should be provided for startup and testing of building systems. Connections should be designed so temporary testing equipment can be installed and removed without shutting down the system.

11.2 AUTOMATIC TEMPERATURE AND HUMIDITY CONTROLS

A direct digital control (DDC) system with host computer-controlled monitoring and control should be provided. Control systems should be BACnet or LonsWork, conforming to American Society of Heating, Refrigerating, and Air-Conditioning Engineers (ASHRAE) BACnet Standard 135.

11.2.1 Controls

Preprogrammed stand-alone single- or multiple-loop microprocessor proportional-integral-derivative (PID) controllers should be used to control all HVAC and plumbing subsystems. PID loops should be used.

All chillers, boilers, terminal units, and air-handling units (AHUs) should have self-contained BACnet or LonsWork controllers, capable of communicating with the building automation system.

11.2.2 Temperature Controls

Heating and cooling energy in each zone should be controlled by a thermostat or temperature sensor located in that zone. Independent perimeter systems must have at least one thermostat or temperature sensor for each perimeter zone. A 1.5°C (35°F) dead band should be used between independent heating and cooling operations within the same zone.

Night setback and setup controls must be provided for all comfort conditioned spaces, even if initial building occupancy plans are for 24-hour operation. Morning warm-up or cool-down must be part of the control system. Controls for the various operating conditions must include maintaining pressurization requirements.

11.2.3 Humidity Controls

Indoor and outdoor humidity sensors should be calibrated in-place during system startup and at least annually thereafter. Dew point control is preferred to provide more stable humidity levels. However, the sensors are acceptable if calibrated in-place, and co-located with dry bulb sensors, so that the BAS can convert these two signals to a dew point value for control purposes.

11.2.4 Temperature Reset Controls Air Systems

Systems supplying heated or cooled air to multiple zones must include controls that automatically reset supply air temperature required by building loads or by outside air temperature.

11.3 HYDRONIC SYSTEMS

Systems supplying heated and/or chilled water to comfort conditioning systems must also include controls that automatically reset supply water temperatures required by temperature changes responding to changes in building loads (including return water temperature) or by outside air temperature.

11.4 VEHICLE GARAGES OR INDOOR USE AREAS

Vehicle garage exhaust fans should generally be activated based upon carbon monoxide sensors within the garage. Carbon monoxide sensors should also be located in all floor areas where vertical shafts penetrate the garage areas.

11.5 BUILDING AUTOMATION SYSTEMS

Building automation systems (BAS) should be evaluated based on the size of the building. Buildings of 100,000 gross square feet (gsf) and more should have a BAS. The size of the building, number of pieces of equipment, expected energy savings, and availability of trained staff should all be considered before a decision is made. BAS is required and considered part of the system on large facilities (above 9,300 gross square meters [100,000 gross square feet), both new facilities and major modernizations.

BAS should be DDC for providing lower operating costs and ease of operation. Microprocessor controllers monitor and adjust building systems to optimize their performance and the performance with other systems such as HVAC and lighting in order to minimize overall power and fuel consumption of the facility.

The system should consist of a series of direct digital controllers interconnected by a local area network. The BAS should be accessible through a Web browser. The system should have a graphical user interface and must offer trending, scheduling, downloading memory to field devices, real-time "live" graphic programs, parameter changes of properties, set point adjustments, alarm/event information, confirmation of operators, and execution of global commands.

11.5.1 Level of Integration

Since the advent of microcomputer BAS systems, integration of many systems has become possible. However, caution is advised when planning BAS systems with a high level of integration. The more integration, the more complex the system becomes and the more training is required for the operating staff. Also, reliability requirements for the different systems may vary. Lighting control systems should not be connected to the BAS except for monitoring of the lighting system. Fire alarm systems, security systems, and elevator systems should not be controlled by a BAS. These systems should have independent control panels and networks. The BAS system should monitor the status of these systems only, in order to prompt emergency operating modes of HVAC and lighting systems.

The BAS should utilize *open* communication protocols per ASHRAE Standard 135, to minimize the costs of providing integration and to allow interoperability between building systems and control vendors. Compatibility with overall regional and/or central monitoring and central strategies must be assured. Functional design, hardware, software, operation, and maintenance manuals should be provided both in hard copy and electronic formats. The BAS should have energy management and monitoring software.

11.5.2 Energy Conservation

The best targets for energy conservation in building systems are the HVAC system and the lighting system. HVAC control algorithms should include optimized start/stop for chillers, boilers, air-handling units and all associated equipment, and feedforward controls based on predicted weather patterns. Lighting control should be accomplished by use of separate control equipment, which allows BAS monitoring and reporting and control settings. Optimal start/stop calculates the earliest time systems can be shut down prior to the end of occupancy hours and the latest time systems can start up in the morning with the aim of minimizing equipment run time without letting space conditions drift outside comfort set points. Weather prediction programs store historic weather data in the processor memory and use this information to anticipate peaks or part-load conditions. Programs also run economizer cycles and heat recovery equipment. The BAS should include programs for control that switch pumps and compressors from operating equipment to standby on a scheduled basis. Also, programs that provide maintenance schedules for equipment in every building system should be included, complete with information on what parts and tools are needed to perform each task.

11.5.3 System Design Considerations

The BAS requires measurements at key points in the building system to monitor part-load operation and adjust system set points to match system capacity to load demands. Controls cannot correct inadequate source equipment, poorly selected components, or mismatched systems. Energy efficiency requires a design that is optimized by realistic prediction of loads, careful system selection, and full control provisions. System ability must include logs of data created by user selectable features. In new buildings and major renovations, the BAS should have approximately 20% spare capacity for future expansion. The system must provide for stand-alone operation of subordinate components. The primary operator workstation should have a graphical user interface. Stand-alone control panels and terminal unit controllers can have text-based user interface panels that are handheld or fixed.

11.5.4 Energy Measurement Instrumentation

The BAS should have the capability to allow building staff to measure energy consumption and monitor performance that is critical to the overall success of the system. Electrical values, such as V, A, kW, KVAR, KVA, PF, kWh, KVARH, frequency and percent total harmonic distortion (THD), should be measured. Energy management measurements should be totalized and trended in both instantaneous and time-based numbers for chillers, boilers, air-handling units, and pumps. Energy monitoring data should be automatically converted to standard database and spreadsheet format and transmitted to a designated personal computer (PC). Energy points are those points that are monitored to ensure compliance with ASHRAE Standard 90.1.

11.6 DETECTION SYSTEMS

Detection-based approaches for building protection offer substantially lower levels of protection than those of continuous, external filtration. Against outdoor releases, the level of protection attainable with air-exchange-rate control approaches that of the unventilated shelter. Against an internal release, the benefit (in reduced exposure to occupants) is governed by the increase in air exchange rate that can be achieved with fans. The level of protection in both cases is also strongly limited by the response time of the system.

Using detectors to initiate protective actions—changing air exchange rates with dampers, turning ventilation fans or smoke purge fans on or off, or activating standby filter units—requires real-time detection capabilities of a broad range for both chemicals and aerosols. Automatic detectors that are currently available provide real-time capability for only a very limited number of chemicals. Other protective actions, aside from collective protection, are to apply protective masks and to evacuate the building.

Response time, a limiting factor in a detector-based system, has three components: the time of transport of the agent to the detector, the internal response time of the detector, and the time for completing the protective action, such as closing dampers and turning fans on or off. Protective capability diminishes as the time for response increases from zero. All response-time components must be considered in employing a detection-based protective system.

11.7 STARTUP, TESTING, AND BALANCING EQUIPMENT AND SYSTEMS

Factory representatives should be present for startup of all major equipment, such as boilers, chillers, and automatic control systems. Testing, adjusting, and balancing parameters at startup should be documented. The testing and balancing contractor should have up-to-date certification by the Associated Air Balance Council (AABC), the National Environmental Balance Bureau (NEBB), or the Testing, Adjusting, and Balancing Bureau (TABB).

Documentation of performance testing of all equipment and systems for part load and full load during seasonal variations and assumed zoning alternatives should include comparison to initial design parameters. HVAC tests should be conducted at static pressures equal to the maximum design pressure of system. The maximum leakage allowable should not exceed 50% of that allowed in SMACNA's *HVAC Air Duct Leakage Manual*.

The initial or baseline indoor air quality (IAQ) testing for CO, CO_2, volatile organic compounds, NO_2, O_3, and tobacco smoke should be conducted when the system is stabilized. Operational tests for each air and hydronic system should also be conducted at this time. Tests should provide measurements to validate that energy and efficiency requirements of ASHRAE 90.1 and 62 have been met. During the course of system usage, both peak summer and winter energy consumption and performance should be measured and documented.

REFERENCES

Desk Guide, Government Services Administration (GSA), GSA PBS NEPA.

Facilities Standards for the Public Building Service, U.S. General Services Administration, Public Buildings Service, Office of the Chief Architect, Washington, DC 20405, March 2003.

Fundamentals of Protective Design for Conventional Weapons, TM 5-855-1. Washington, DC, Headquarters, U.S. Department of the Army, 1986.

North American Emergency Response Manual, U.S. Department of Transportation, Transport Canada, Secretariat of Transport and Communications, 1996.

Protective Construction Design Manual, ESL-TR-87-57. Prepared for Engineering and Services Laboratory, Tyndall Air Force Base, FL. 1989.

CHAPTER 12

VAC Systems Design Features

Martha Boss, Charles Allen, and Donald C. Dittus

CONTENTS

In order to understand the vulnerabilities of heating, ventilation, and air conditioning (HVAC) systems, the preferred design features must be understood. Even these preferred design systems have inherent vulnerabilities and may constitute preventative maintenance challenges. This chapter illustrates the basis of preferred HVAC design for most government buildings as a basis for understanding HVAC design principles for all buildings.

Mechanical systems are designed to function at full load and part load associated with all projected occupancies and modes of operation. System solutions should also be planned to accommodate future occupancies and modes of operation.

All mechanical systems should be designed to automatically respond to the local climatic conditions and heat-recovery opportunities. These designs will provide cost-effective energy conservation measures while assuring set point control. Designs must optimize program areas to the extent possible, assuring attainment of all critical performance goals.

For office spaces, the average density of the occupied floor area of a building should be one person per 9.3 usable square meters (100 usable square feet). Within areas occupied by workstations, the occupancy load can be as dense as one person per 7 usable square meters (75 usable square feet) in local areas. Block loads and room loads should be calculated accordingly. Sensible and latent loads per person should be based on the latest edition the American Society of Heating, Refrigeration, and Air Conditioning Engineers (ASHRAE) *Handbook of Fundamentals*.

For dining areas, auditoriums and other high-occupancy spaces, occupancy loads should represent the number of seats available. Areas such as storage rooms or mechanical rooms do not have occupancy loads.

Assessors should, at a minimum, ask the following questions:

- What is the mechanical condition of the equipment within the building? Is the equipment still operating as designed? Is all equipment appropriately connected and controlled? Are equipment access doors and panels in place and appropriately sealed?
- How is the HVAC system controlled, and what is the response time should shutdown or emergency zoning be required? Are all dampers—outdoor air, return air, bypass, fire and smoke—functioning? How does the HVAC system respond to manual fire alarm, fire detection, or fire-suppression device activation? Are outdoor air louvers in use and observable by the general public?
- How is the building zoned? Where are the air handlers for each zone? Is the system designed for smoke control? How does air flow through the building? What are the pressure relationships between zones? Which building entryways are positively or negatively pressurized? Is the building connected to other buildings by tunnels or passageways? Are utility chases and penetrations, elevator shafts, and fire stairs significant airflow pathways? Is obvious air infiltration present? Localized?
- Are water filtration systems in place? What are their efficiencies?
- Do adjacent structures or landscaping allow access to the building roof?

What fire protections and life safety systems must remain operational should building security be compromised? Assure that all facility managers understand the key components of these systems.

12.1   NEW BUILDING SYNDROME

When a building is new, volatile compounds (VOC) may be released in large quantities from materials such as adhesives, vinyl, and carpets. All products to be incorporated into the building, including finishes and furniture, should be researched regarding characteristics of off-gassing and noxious odors that will affect indoor air quality. An outside air purge cycle should be provided to air-handling equipment enabling evening removal of VOC buildups during the first weeks of occupancy.

12.2   ASHRAE STANDARD 62

The outside air and ventilation rates of ASHRAE Standard 62 are the minimum acceptable criteria. Instrumentation and controls should be provided to assure outdoor air intake rates are maintained within 90% of required levels during occupied hours. Dilution with outside air is the primary method of maintaining acceptable air quality.

Type of Area	Summer		Winter	
	DB[1]	RH[2]	DB[1]	RH[2]
General Office	24 (75)		22(72)	
ADP Rooms[9]	22(72)	45[4]	22(72)	
Corridors	24(75)		22(72)	
Building Lobbies[10]	24(75)		22(72)	
Toilets	24(75)		22(72)	
Locker Rooms	26(78)		21(70)	
Electrical Closets	26(78)		13(55)	
Mechanical Spaces	35(95)		13(55)[8]	
Electrical Switchgear	35(95)		13(55)	
Elevator Mach. Room[10]	26(78)		13(55)	
Emergency Generator Room	40(104)		18(65)	
Transformer Vaults	40(104)			
Stairwells	none		18(65)	
Comm./Tel. Frame Room[7]	24(75)	45[4]	22(72)	30[12]
Storage Room	30(85)		18(65)	
Conference Room[11]	24(75)		22(72)	
Auditorium[10]	24(75)		22(72)	
Kitchen[10]	24(75)		22(72)	
Dining[10]	24(75)		22(72)	
Cafeteria[10]	24(75)		22(72)	
Courtrooms	24(75)		22(72)	454*

Figure 12.1 Indoor design considerations.
[1] Temperatures are degrees Celsius (Fahrenheit), to be maintained at +/−1°C (+/−2°F)
[2] Relative humidity is the minimum permissible, stated in percent. Maximum permissible relative humidity is 60% in conditioned areas.
[3] Dry bulb and relative humidity are to be maintained 150 mm (6 inches) to 1800 mm (6 feet) above the floor.
[4] Relative humidity should be maintained at +/−5% in automated data processing (ADP) spaces.
[5] Maximum temperature; space to be mechanically cooled if necessary.
[6] Room must not exceed temperature with generator running.
[7] Must comply with Energy Information Administration/Telecommunications Industry Association (EIA/TIA) Standard 569.
[8] Minimum temperature in the building must be 13°C (55°F) even when unoccupied.
[9] Confirm equipment manufacturer's requirements as more stringent. Provide in-room display and monitor device (such as wall-mounted temperature and humidity chart recorder).
[10] System should be designed for process cooling. Cooling system should be a dedicated independent system.
[11] Provide independent temperature control.
[12] Minimum relative humidity requirements may be omitted in moderate southern climate zones upon approval of local building code representatives.

Where occupancy requirements are likely to generate high levels of airborne particles, special air filtration should be provided on the return air system, or dedicated and localized exhaust systems should be used to contain airborne particulates.

12.3   AIR-HANDLING UNIT

Air-handling units (AHU) should be sized to not exceed 11,800 l/s (25,000 cfm). Smaller units are encouraged to facilitate flexible zone control, particularly for spaces that involve off-hour or high-load operating conditions.

- To the extent possible, *plug-n-play* AHU configurations should be considered, facilitating easy future adaptations to space-load changes.
- Psychrometric analyses (complete with chart diagrams) should be prepared for each air-handling unit.
- Air-handling unit/coil designs should assure that conditioned space temperatures and humidity levels are within an acceptable range, per programmed requirements, and ASHRAE Standards 55 and 62.
- Depending on sensible heat ratio characteristics, effective moisture control may require cooling coil air discharge dew point temperatures as low as 10°C (50°F).
- As required, provide face-by-pass or heat-recovery features to reheat cooling coil discharge temperatures for acceptable space entry.
- Provide a direct form of reheat and/or humidification only if space conditions require tight environmental control, or if recurring day-long periods of unacceptable humidity levels would otherwise result.

12.3.1   Supply, Return, and Relief Air Fans

Centrifugal double-width double-inlet forward curved and airfoil fans are preferable for VAV systems. All fans should bear the AMCA seal and performance should be based on tests made in accordance with AMCA Standard 210.

- Fans should be selected on the basis of required horsepower as well as sound power level ratings at full load and at part-load conditions.
- Fan motors should be sized so the motors do not run at overload anywhere on their operating curve.
- Fan operating characteristics must be checked for the entire range of flow conditions, particularly for forward curved fans.
- Fan drives should be selected for a 1.5 service factor and fan shafts should be selected to operate below the first critical speed.
- Thrust arresters should be designed for horizontal discharge fans operating at high static pressure.

12.3.2 Coils

- Individual finned tube coils should generally be between 6 and 8 rows with at least 2.1 mm between fins (12 fins per inch) to ensure that the coils can be effectively and efficiently cleaned.
- Dehumidifying coils should be selected for no more than negligible water droplet carryover beyond the drain pan at design conditions.
- All hot water heating and chilled water cooling coils should be copper tube and copper finned materials.
- Equipment and other obstructions in the air stream should be located sufficiently downstream of the coil so that the equipment/obstruction will not come in contact with the water droplet carryover.
- Cooling coils should be selected at or below 2.5 m/s face velocity (500 fpm) to minimize moisture carryover.
- Heating coils should be selected at or below 3.8 m/s face velocity (750 fpm).

12.3.3 Drains and Drain Pans

Drain pans should be made of stainless steel, insulated, and adequately sloped and trapped to assure drainage.

Drains in draw-through configurations should have traps with a depth and height differential between inlet and outlet equal to the design static pressure plus 2.54 mm (1 inch) minimum.

12.3.4 Filter Sections

Air filtration should be provided in every air-handling system.

- Air-handling units should have a disposable prefilter and a final filter.
- The filter media should be rated in accordance with ASHRAE Standard 52.
- Prefilters should be 30 to 35% efficient.
- Final filters should be 85% efficient and capable of filtering down to 3.0 microns. Filter racks should be designed to minimize the bypass of air around the filter media with a maximum bypass leakage of 0.5%.
- Filters should be sized at 2.5 m/s (500 FPM) maximum face velocity.
- Filter media should be fabricated so that fibrous shedding does not exceed levels prescribed by ASHRAE 52.
- The filter housing and all air-handling components downstream should not be internally lined with fibrous insulation. Double-wall construction or an externally insulated sheet metal housing is acceptable.

The filter change-out pressure drop, not the initial clean filter rating, must be used in determining fan pressure requirements. Differential pressure gauges and sensors should be placed across each filter bank to allow quick and accurate assessment of filter dust loading as reflected by air pressure loss through the filter and sensors should be connected to building automation system.

12.3.5 Access Doors

Access doors should be provided at AHUs downstream of each coil, upstream of each filter section, and adjacent to each drain pan and fan section. Access doors should be of sufficient size to allow personnel to enter the unit to inspect and service all portions of the equipment components.

12.3.6 Plenum Boxes

Air-handling units should be provided with plenum boxes where relief air is discharged from the air-handling unit. Plenum boxes may also be used on the return side of the unit in lieu of a mixing box.

Airflow control dampers should be mounted on the ductwork connecting to the plenum box.

12.3.7 Mixing Boxes

Air-handling units should be provided with mixing boxes where relief air is discharged from the air-handling unit.

Mixing boxes may also be used on the return side of the unit in lieu of a plenum box.

Airflow control dampers should be mounted within the mixing box or on the ductwork connecting to the mixing box.

12.3.8 Terminals

Variable air volume (VAV) terminals should be certified under the Air Conditioning and Refrigeration Institute (ARI) Standard 880 Certification Program and should carry the ARI Seal. Units should have BACnet or LonsWork self-contained controls. All terminals should be provided with factory-mounted direct digital controls compatible and suitable for operation with the building automation system (BAS).

If fan powered, the terminals should be designed, built, and tested as a single unit, including motor and fan assembly, primary air damper assembly, and any accessories. VAV terminals should be pressure-independent-type units.

Fan-powered terminals should utilize speed control to allow for continuous fan speed adjustment from maximum to minimum, as a means of setting the fan airflow. The speed control should incorporate a minimum voltage stop to ensure the motor cannot operate in the stall mode.

12.3.9 Air-Delivery Devices

Adequate ventilation requires that the selected diffusers effectively mix the total air in the room with the supplied conditioned air, which is assumed to contain adequate ventilation air. Booted plenum slots should not exceed 1.2 meters (4 feet) in length unless more than one source of supply is provided. "Dumping" action at reduced air volume and sound power levels at maximum m³/s (cfm) delivery should be minimized.

For VAV systems, the diffuser spacing selection should not be based on the maximum or design air volumes but rather on the air volume range where the system is expected to operate most of the time. The designer should consider the expected variation in range in the outlet air volume to ensure the air diffusion performance index (ADPI) values remain above a specified minimum. ADPI is assured with low temperature variation, good air mixing, and no objectionable drafts in the occupied space, typically 150 mm (6 inch) to 1830 mm (6 feet) above the floor. Terminal ceiling diffusers or booted-plenum slots should be specifically designed for VAV air distribution.

12.3.10 Motors

All motors should have premium efficiency as per ASHRAE 90.1.

- 1/2 HP and larger should be polyphase.
- Motors smaller than 1/2 HP should be single phase.

For motors operated with variable speed drives, provide insulation cooling characteristics as per the National Electrical Code (NEC) and National Fire Protection Association (NFPA).

12.4 MECHANICAL SYSTEM: SPACE REQUIREMENTS

A minimum of 4% of the typical floor's gross floor area should be provided on each floor for air-handling equipment. A minimum of 1% of the building's gross area should be provided for the central heating and cooling plant. Space requirements of mechanical and electrical equipment rooms should be based upon the layout of required equipment drawn to scale within each room.

12.4.1 Service Access

The HVAC design engineer should be cognizant of the necessity to provide for the replacement of major equipment over the life of the building, and should insure that provisions are made to remove and replace, without damage to the structure, the largest and heaviest component that cannot be further broken down.

Adequate methods of access should be included for items such as chillers, boilers, heat exchangers, cooling towers, reheat coils, VAV boxes, pumps, hot water heaters, and all devices that have maintenance service requirements. Space should be provided around all HVAC system equipment as recommended by the manufacturer and in compliance with local code requirements for routine maintenance.

- Access doors or panels should be provided in ventilation equipment, ductwork, and plenums as required for on-site inspection and cleaning.
- Equipment access doors or panels should be readily operable and sized to allow full access. Large central equipment should be situated to facilitate its replacement.

12.4.2 Vertical Clearances

Main mechanical equipment rooms generally should have clear ceiling heights of not less than 3.6 m (12 feet).

Catwalks should be provided for all equipment that cannot be maintained from floor level.

Where maintenance requires the lifting of heavy parts [45 kg (100 pounds) or more], hoists and hatchways should be installed.

12.4.3 Horizontal Clearances

Mechanical rooms should be configured with clear circulation aisles and adequate access to all equipment. The arrangement should consider the future removal and replacement of all equipment. The mechanical rooms should have adequate doorways or areaways and staging areas to permit the replacement and removal of equipment without the need to demolish walls or relocate other equipment.

Sufficient space areas (noted by outlining manufacturer's recommendations) for maintenance and removal of coils, filters, motors, and similar devices should be provided. Chillers should be placed to permit pulling of tubes from all units. The clearance should equal the length of the tubes plus 600 mm (2 feet).

Air-handling units require a minimum clearance of 750 mm (2 feet 6 inches) on all sides, except the side where filters and coils are accessed. The clearance on that side should equal the length of the coils plus 600 mm (2 feet).

12.4.4 Roof-Mounted Equipment

No mechanical equipment except for cooling towers, air-cooled chillers, evaporative condensers, and exhaust fans should be permitted on the roof of the building. Access to roof-mounted equipment should be by stairs, not by ship's ladders.

12.4.5 Housekeeping Pads

Housekeeping pads should be at least 150 mm (6 inches) wider on all sides than the equipment supported, and should be 150 mm (6 inches) thick. Mechanical equipment rooms must be designed in accordance with the requirements of ASHRAE Standard 15: Safety Code for Mechanical Refrigeration.

12.5   BASELINE SYSTEMS

Baseline systems illustrate a base reference for comparison when evaluating current systems. Separate systems should be provided for interior and perimeter zones where simultaneous heating and cooling operations may occur.

- Single AHUs should not serve multiple floors or scattered building loads.
- Multiple AHUs or floor-by-floor systems should be considered as baseline.
- Systems designed for federal courthouses should be limited to having no more than two courtrooms served by any single AHU, and that AHU should be dedicated to serving only those two courtrooms.

If a building program shows that an office building will have an open plan layout or if the program does not state a preference, assume that up to 40% of the floor plan will be occupied by closed offices at some point in the future.

12.5.1   Supply Zone

The supply of zone cooling and heating should be sequenced to prevent (or at the very least, minimize) the simultaneous operation of heating and cooling systems for the same zone. Supply air temperature reset control should be used to extend economizer operations and to reduce the magnitude of reheating, recooling, or mixing of supply air streams.

12.5.2   Perimeter Systems

Perimeter zones should be no more than 4.7 meters (15 feet) from an outside wall along a common exposure. Independent zones should be provided for spaces such as conference rooms, entrance lobbies, atriums, kitchen areas, dining areas, childcare centers, and physical fitness areas. Perimeter zones should not exceed 30m (300 sf).

12.5.3   Interior Zone

Interior zone(s) should have 100% outside air ventilation system(s) sized to meet the ventilation requirements of the interior zone. The ventilation system(s) may operate independently of any other air distribution system. However, these systems should connect to the return side of the VAV units serving interior zones. The interior zones should also have a baseline interior heating and cooling system such as:

- A ducted overhead VAV system with VAV boxes.
- A ducted overhead VAV system with fan-powered VAV boxes. Hot water heating coils in the fan-powered VAV boxes may be used on the top floor of a building for heating.
- An underfloor VAV air distribution system.

Enthalpy heat recovery should be used for interior zones and other special areas where outside air exceeds 30% of the total supply air quality.

Special areas such as auditoriums, atriums, and cafeterias should have an AHU with individual controls.

A dedicated AHU should be provided to maintain positive pressure in the main entry lobby.

AHUs with a capacity over 1416 LPS (3000 cfm) should have an enthalpy economizer cycle. A waterside economizer system should be employed where an airside enthalpy economizer is not practical or feasible. Systems dedicated to serving only unoccupied spaces with intermittent operation, such as elevator machine rooms, telephone equipment rooms, and similar spaces would be exempt from the requirements of having an economizer cycle.

Interior control zones must not exceed 180 m (1500 sf) per zone for open office areas or a maximum of three offices per zone for closed office areas. Corner offices should be designed as a dedicated zone.

12.6   PERIMETER OUTSIDE AIR VENTILATION SYSTEMS

Perimeter ventilation units should be self-contained direct expansion (DX) package units or air-handling units with a fan section having a variable speed drive, chilled water cooling coil, hot water heating coil, enthalpy heat recovery wheel, or desiccant wheel and supply air filtration.

- The perimeter ventilation units should provide 100% outside air.
- Reheat should be hot gas bypass, a heat pipe, or a run-around coil.
- Chilled water should be generated by an air-cooled chiller or a 24-hour chiller.
- If a desiccant wheel is used for controlling the specific humidity discharge at the wheel, condenser reheat should be used for regeneration of the desiccant, along with minimum electric backup.

Supply air dew point leaving the unit should be maintained at 10°C (50°F) and the supply air dry bulb temperature should be a minimum of 21.1°C (70° F) and not greater than 25.6°C (78° F).

- During occupied hours, this unit should operate to deliver conditioned ventilation air and maintain positive pressure in the perimeter zone with respect to outside air pressure.
- During unoccupied hours, the unit should run at 40% of its capacity to provide conditioned air at 10°C (50° F) dew point and at least 21.1°C (70°F) to help maintain positive pressure in the perimeter zone with respect to outside air.

In both the occupied and unoccupied modes, the system should operate to adjust the airflow as required to maintain a differential positive pressure in the perimeter zone relative to the prevailing pressure outside the building.

- When the outside air dew point drops below 2.8°C (37°F), the unit should have the capacity to maintain neutral pressure with respect to the outside by exhausting relief air from the return duct system.
- The ventilation unit should:
  - Have self-contained microprocessor controls capable of connecting to and interoperating with a BACnet or LONWORKS direct digital control (DDC) building automation system.
  - Be equipped with dampers to set the design airflow through the unit, and also an analog or digital display that measures and displays the amount of air flowing through the unit continuously.

12.7   VARIABLE AIR VOLUME

The VAV supply fan should be designed for the largest block load, not the sum of the individual peaks.

- The air distribution system up to the VAV boxes should be medium pressure and should be designed by using the static regain method.
- Downstream of the VAV boxes, the system should be low- and medium-pressure construction and should be designed using the equal friction method. Sound lining is not permitted.
- Double-wall ductwork with insulation in between is permitted in lieu of sound lining.
- All VAV boxes should be accessible for maintenance.
- Ducted return should be used at all locations.
- VAV fan-powered box supply and return ducts should have double-wall ductwork with insulation in between for a minimum distance of 5 feet.

12.7.1   Air Volume Control

Particular attention should be given to the volume control. VAV systems depend on air volume modulation to maintain the required ventilation rates and temperature set points. Terminal air volume control devices are critical to the successful operation of the system and should be provided.

Zone loads must be calculated accurately to avoid excessive throttling of air flow due to oversized fans and terminal units.

Diffusers should be high entrainment type (3:1 minimum) to maximize air velocity at low flow rates.

If ventilation air is delivered through the VAV box, the minimum volume setting of the VAV box should equal the larger of the following:

- 30% of the peak supply volume
- 0.002 m^3/s per m^2 (0.4 cfm/sf) of conditioned zone area
- Minimum m^3/s (cfm) to satisfy *ASHRAE Standard 62* ventilation requirements. VAV terminal units must never be shut down to zero when the system is operating.

Outside air requirements should be maintained in accordance with the Multiple Spaces Method, Equation 6-1 of *ASHRAE Standard 62* at all supply air flow conditions.

12.7.2   Airside Economizer Cycle

An airside enthalpy economizer cycle reduces cooling costs when outdoor air enthalpy is below a preset high temperature limit, usually 15 to 21°C (60 to 70°F), depending on the humidity of the outside air. Airside economizers should only be used when the economizers can deliver air conditions leaving the air-handling unit of a maximum of 10°C (50°F) dew point and a maximum of 70% relative humidity.

Enthalpy economizers should operate only when return air enthalpy is greater than the outside air enthalpy.

All air distribution systems with a capacity greater than 1,416 liters/second (LPS) (3,000 cfm) should have an airside economizer in accordance with *ASHRAE 90.1*, unless the design of the air-handling systems preclude the use of an airside economizer.

12.7.3   Variable Volume System with Shutoff Boxes

Systems with full shutoff VAV boxes should be used for perimeter zone applications only. VAV shutoff boxes should be used only with the perimeter air distribution systems in order to eliminate the need for reheat.

The air-handling unit and associated VAV boxes should have self-contained microprocessor controls capable of connecting to and interoperating with a DDC building automation system.

12.7.4   Variable Volume System with Fan-Powered Boxes

VAV systems with fan-powered VAV boxes may be used for both perimeter and interior zone applications.

The air-handling unit and associated VAV boxes should have self-contained microprocessor controls capable of connecting to and interoperating with a BACnet or LONWORKS DDC building automation system.

- Fan-powered boxes should be equipped with a ducted return, featuring a filter/filter rack assembly and covered on all external exposed sides with 2 inches of insulation. The return plenum box should be a minimum of 61 mm (24 inches) in length and should be double-wall construction with insulation in between or contain at least one elbow where space allows.
- Fan-powered boxes may have hot water heating coils used for maintaining temperature conditions in the space under partial-load conditions.
- Fan-powered boxes located on the perimeter zones and on the top floor of the building should contain hot water coils for heating.

12.7.5   Underfloor Air Distribution System

Underfloor air distribution systems should incorporate VAV units designed to distribute the supply air from under the floor using variable volume boxes or variable volume dampers running out from underfloor, ducted, or main trunk lines.

- Air should be distributed into the space through floor-mounted supply registers that should be factory fabricated with manual volume control dampers. Supply air temperature for underfloor systems should be between 10°C (50°F) dew point and 18°C (64°F) dry bulb.
- For perimeter underfloor systems, provide fan coil units or fin tube radiators located beneath the floor with supply air grilles or registers mounted in the floor.
- The air-handling unit, VAV boxes, and variable volume dampers should have self-contained microprocessor controls capable of connecting to and interoperating with a BACnet or LONWORKS DDC building automation system.
- The maximum zone size of an underfloor air distribution system should not exceed 2,360 l/s (5,000 cfm).
- Provide plenum zones both for perimeter and interior in order to control the underfloor variable volume dampers or boxes with separate plenum barriers between perimeter and interior zones. The underfloor plenum should be air tight and compartmentalized with baffles. Provisions should be provided for cleaning the plenum space. When underfloor supply air distribution is used, the ceiling plenum should be used for the distribution of the ducted return air.
- The perimeter and interior underfloor zones should be clearly separated in order to maintain proper pressurization, temperature, and humidity control. The perimeter wall below the raised flooring system should be provided with R-30 insulation and vapor barrier below the raised floor.
- All VAV boxes that are part of an underfloor air distribution system for both perimeter and interior systems should be located below the raised floor.
- The floor area used for an underfloor system should have the slab provided with a minimum of R-10 insulation and vapor barrier from below. This should incorporate the entire slab area used for the underfloor system.

12.7.6   Underfloor Air Displacement System

Underfloor air displacement systems should not reuse d incorporate VAV units designed to distribute supply air from under the floor or variable volume dampers from underfloor, ducted, or main trunk lines may be used.

- The VAV boxes or control dampers should be hard ducted or connected directly to the main trunk lines. Air should be distributed into the occupied space through floor-mounted, low-turbulence, displacement flow swirl diffusers, and should contain a dust collection basket situated below the floor.
- Supply air temperature for underfloor systems should be 10°C (50°F) dew point and 18°C (64°F) dry bulb.

- For perimeter underfloor systems, provide fan coil units or fin tube radiators located beneath the floor with supply air grilles or registers mounted in the floor.
- The air-handling unit, VAV boxes, and variable volume dampers should have self-contained microprocessor controls capable of connecting to and interoperating with a BACnet or LonsWork DDC building automation systems.
- The maximum capacity of an underfloor air distribution system should not exceed 2,360 l/s (5,000 cfm).

12.8   AIR DISTRIBUTION SYSTEMS

12.8.1   Supply, Return, and Exhaust Ductwork

Ductwork should be designed in accordance with *ASHRAE: Handbook of Fundamentals*, "Duct Design" chapter, and constructed in accordance with the *ASHRAE: HVAC Systems and Equipment Handbook,* "Duct Construction" chapter, and the *Sheet Metal and Air Conditioning Contractors' National Association* (SMACNA) design manuals.

- All ductwork joints and all connections to air-handling and air distribution devices should be sealed with mastic, including all supply and return ducts, any ceiling plenums used as ducts, and all exhaust ducts.
- Energy consumption, security, and sound attenuation should be major considerations in the routing, sizing, and material selection for the air distribution ductwork.
- Supply, return, and exhaust air ducts should be designed and constructed to allow no more than 3% leakage of total airflow in systems up to 750 Pa (3 inches WG).
- In systems from 751 Pa (3.1 inches WG) through 2,500 Pa (10.0 inches WG) ducts should be designed and constructed to limit leakage to 0.5% of the total air flow.
- Pressure loss in ductwork should be designed to comply with the criteria stated above. This can be accomplished by using smooth transitions and elbows with a radius of at least 1.5 times the radius of the duct.
- Where mitered elbows have to be used, double-foil sound-attenuating turning vanes should be provided. Mitered elbows are not permitted where duct velocity exceeds 10m/s (2000 FPM).

12.8.2   Ductwork Pressure

Figure 12.2 provides pressure classification and maximum air velocities for all ductwork.

Ductwork construction should be tested for leakage prior to installation. Each section tested must have a minimum of a 20-ft length straight run, a minimum of two elbows, and a connection to the terminal.

The stated static pressures represent the pressure exerted on the duct system and not the total static pressure developed by the supply fan.

Static Pressure	Air Velocity	Duct Class
250 Pa (1.0 WG)	>10 m/s DN (<2000 FPM DN)	Low Pressure
500 Pa (2.0 WG)	>10 m/s DN (<2000 FPM DN)	Low Pressure
750 Pa (3.0 WG)	>12.5 m/s DN (<2500 FPM DN)	Medium Pressure
1000 Pa (+4.0 WG)	<10m/s DN (<2000 FPM UP)	Medium Pressure
1500 Pa (+6.0 WG)	<10m/s DN (<2000 FPM UP)	Medium Pressure
2500 Pa (+10.0 WG)	>10 m/s DN (>2000 FPM DN)	High Pressure

Figure 12.2 Ductwork classification.

The actual design air velocity should consider the recommended duct velocities in Figure 12.3 when noise generation is a controlling factor.

- Primary air ductwork (fan connections, risers, main distribution ducts) should be medium pressure classification as a minimum.
- Secondary air ductwork (runouts/branches from main to terminal boxes, and distribution devices) should be low pressure classification as a minimum.

12.8.3 Ductwork Sizing

Supply and return ductwork should be sized using the equal friction method except for ductwork upstream of VAV boxes. Duct systems designed using the equal friction

Application	Controlling Factor Noise Generation (Main Duct Velocities)	
	m/s	fpm
Private Offices Conference Rooms Libraries	6	−1,200
Theaters Auditoriums	4	−800
General Offices	7.5	(1,500)
Cafeterias	9	(1,800)

Figure 12.3 Recommended duct velocities.

method place enough static pressure capacity in the supply and return fans to compensate for improper field installation and changes made to the system layout in the future.

In buildings with large areas of open plan space, the main duct size should be increased for revisions in the future.

Air flow diversity should also be a sizing criterion. Eighty percent diversity can be taken at the air-handling unit and decreased the farther the ductwork is from the source until air flow diversity is reduced to zero for the final portion of the system.

12.8.4 Ductwork Construction

Ductwork should be fabricated from galvanized steel, aluminum, or stainless steel sheet metal depending on applications.

Flex duct may be used for low-pressure ductwork downstream of the terminal box in office spaces. The length of the flex duct should not exceed the distance between the low-pressure supply air of diffusers in the future while minimizing replacement or duct, and the diffuser plus 20% to permit relocation modification of the hard ductwork distribution system. Generally, flex duct runs should not exceed 3 m (10 feet) nor contain more than two bends.

Joint sealing tape for all connections should be of reinforced fiberglass-backed material with field-applied mastic. Use of pressure-sensitive tape is not permitted.

12.8.5 Supply Plenums

12.8.5.1 Supply: Ceiling Plenum

Ceiling plenum supply does not permit adequate control of supply air and should not be used.

12.8.5.2 Supply: Raised-Floor Plenum

In computer rooms, underfloor plenum supplies are appropriate. As a general application in other areas (e.g., open offices), underfloor air distribution/displacement systems are appropriate. Where raised-floor plenums are used for supply air distribution, the plenums should be properly sealed to minimize leakage. R-30 insulation with vapor barrier should be provided for the perimeter of raised-floor walls.

12.8.6 Returns: Plenum and Ducted

12.8.6.1 Return Plenums

With a return plenum, care must be taken to ensure that the air drawn through the most remote register actually reaches the air-handling unit. The horizontal distance from the farthest point in the plenum to a return duct should not exceed 15 m (50 feet). No more than 0.8 m³/s (2,000 cfm) should be collected at any one return grille.

Return air plenums should be avoided. When deemed necessary for economic reasons, plenums should be sealed air-tight with respect to the exterior wall and roof slab or ceiling deck to avoid creating negative air pressure in exterior wall cavities that would allow intrusion of untreated outdoor air.

12.8.6.2 *Ducted Return Plenums*

All central multifloor-type return air risers must be ducted. Other less-flexible building spaces, such as permanent circulation, public spaces, and support spaces, should have ducted returns. Where fully ducted return systems are used, consider placing returns low in walls or on columns to complement ceiling supply air.

A return air duct in the ceiling plenum of the floor below the roof should be insulated.

Double-wall ductwork with insulation in between should be used in lieu of sound lining for a minimum of the last 5 feet before connecting to the AHU or a return air duct riser.

12.8.7 Kitchen Ventilation Systems

Products of combustion from kitchen cooking equipment and appliances should be delivered outside of the building through the use of kitchen ventilation systems involving exhaust hoods, grease ducts, and makeup air systems where required.

Commercial kitchen equipment applications should be served by a Type I hood constructed in compliance with Underwriters Laboratory (UL) 710 and designed in accordance with the code having jurisdiction.

Grease ducts should be constructed of black steel not less than 0.055 inch (1.4 mm) (No. 16 gauge) in thickness or stainless steel not less than 0.044 inch (1.1 mm) (No. 18 gauge in thickness).

Makeup air systems serving kitchen exhaust hoods should incorporate airside heat exchange to recover energy from the exhaust stream to be used for heating the supply air stream.

REFERENCES

Facilities Standards for the Public Building Service, U.S. General Services Administration, Public Buildings Service, Office of the Chief Architect, Washington, DC 20405, March 2003.

ADDITIONAL READING

Fundamentals of Protective Design for Conventional Weapons, TM 5-855-1. Washington, DC, Headquarters, U.S. Department of the Army, 1986.

Guidance for Protecting Building Environments from Airborne Chemical, Biological, or Radiological Attacks, National Institute of Occupational Safety and Health (NIOSH), U.S. Department of Health and Human Services (DHHS) (NIOSH) Pub No. 2002-139, May 2002.

Protective Construction Design Manual, ESL-TR-87-57. Prepared for Engineering and Services Laboratory, Tyndall Air Force Base, FL. 1989.

Security Engineering, TM 5-853 and Air Force AFMAN 32-1071, Volumes 1, 2, 3, and 4. Washington, DC, Departments of the Army and Air Force, 1994.

CHAPTER 13

Temperature Management Equipment

Martha Boss, Dennis Day, Charles Allen, and Randy Boss

CONTENTS

Although not often thought of as a vulnerability, the inability to manage and control temperature may, in fact, represent several opportunities for severe vulnerability. Certainly in a sheltering-in-place scenario, control of temperature is a key factor in keeping shelter users safe and comfortable. This temperature control must be accomplished without creating additional air pollutants in the shelter and without drawing attention to the shelter.

During normal building or facility usage, keeping temperature in control equates to using building systems correctly. Too much makeup air may undermine the temperature controls, and too little will lessen the interior air quality (assuming exterior contamination is not an issue). The very systems used may constitute a vulnerability if uncontrolled access can lead to pressure tank ruptures, cooling and humidification system contamination, or condensate cycles within the heating, ventilation, and air conditioning (HVAC) system that create biological (mold and bacteria) growth.

13.1 BOILERS

Boilers for hydronic hot-water heating applications should be low pressure, with a working pressure and maximum temperature limitation, and should be installed in a dedicated mechanical room with all provisions made for breeching, flue stack, and combustion air.

- For northern climates, a minimum of three equally sized units should be provided. Each of the three units should have equal capacities such that the combined capacity of the three boilers satisfies 120% of the total peak load of heating and humidification requirements.
- For southern climates, a minimum of two equally sized units at 67% of the peak capacity (each) should be provided. The units should be packaged, with all components and controls factory pre-assembled.

- Controls and relief valves to limit pressure and temperature must be specified separately.
- Burner control should be return water temperature actuated and control sequences, such as modulating burner control and outside air reset, should be used to maximum efficiency and performance.
- Multiple closet-type condensing boilers should be used, if possible.
- Boilers should have self-contained microprocessor controls capable of connecting to and interoperating with a BACnet or LonsWork (direct digital control) DDC building automation system.
- Boilers should have a minimum efficiency of 80% as per American Society of Heating Refrigeration, and Air Conditioning Engineers (ASHRAE) 90.1.
- Individual boilers with ratings higher than 29 MW (100 million Btu/hour) or boiler plants with ratings higher than 75 MW (250 million Btu/hour) are subject to review by the U.S. Environmental Protection Agency (EPA).
- Boilers should be piped to a common heating water header with provisions to sequence boilers online to match the load requirements.
- All units should have adequate valving to provide isolation of offline units without interruption of service.
- All required auxiliaries for the boiler systems should be provided with expansion tanks, heat exchangers, water treatment, and air separators, as required.

13.1.1 Gas Trains

Boiler gas trains should be in accordance with International Risk Insurance (IRI) standards.

13.1.2 Automatic Valve Actuators

Gas valve actuators should not contain NaK (sodium-potassium) elements since these pose a danger to maintenance personnel.

13.1.3 Venting

Products of combustion from fuel-fired appliances and equipment should be delivered outside of the building through the use of breeching, vent, stack, and chimney systems.

Breeching connecting fuel-fired equipment to vents, stacks, and chimneys should generally be horizontal and should comply with the National Fire Protection Association (NFPA) 54 standard.

Vents, stacks, and chimneys should generally be vertical and should comply with NFPA 54 and 211.

Breeching, vent, stack, and chimney systems may operate under negative, neutral, or positive pressure and should be designed relative to the flue–gas temperature and dew point, length and configuration of the system, and the value of the insulation techniques applied to the vent.

Venting materials may be factory fabricated and assembled in the field and may be double- or single-wall systems depending on the distance from adjacent combustible or noncombustible materials.

Material types, ratings, and distances to adjacent building materials should comply with NFPA 54 and 211.

13.2 STEAM HEATING SYSTEMS

District steam heating, if available, should be used if determined to be economical and reliable through a life cycle cost analysis. Steam furnished to the building under a district-heating plan should be converted to hot water with a heat exchanger in the mechanical room near the entrance into the building. If steam heating is used, the designer should investigate the use of district steam condensate for preheating of domestic hot water. Steam heating is not permitted inside the building other than conversion of steam to hot water in the mechanical room.

The use of steam for HVAC applications should be limited to the conversion of steam heat to hot water heat and for use in providing humidification. Steam should not be used as a heating medium for distribution throughout a building to terminal units, air-handling units (AHUs), perimeter heating units, coils, or any other form of heat transfer where steam is converted to a source of heat for use in space comfort control or environmental temperature control.

Only a clean steam generation system should be used to provide humidification. Steam delivered from a central plant, a district steam system, steam boilers, or any equipment where chemicals are delivered into the medium resulting in the final product of steam should not be used for the purpose of providing humidification to the HVAC system or occupied spaces.

13.3 HOT WATER HEATING SYSTEMS

The U.S. Government Services Administration (GSA) prefers low-temperature hot-water heating systems; 205 kPa (30 psi) working pressure and maximum temperature limitation of 93.3°C (200°F). The use of electric resistance and/or electric boilers as the primary heating source for the building is prohibited.

Design and layout of hydronic heating systems should follow the principles outlined in the latest edition of the *ASHRAE Systems and Equipment Handbook. Water Treatment*.

13.3.1 Temperature and Pressure Drop

Supply temperatures and the corresponding temperature drops for space heating hot water systems must be set to best suit the equipment being served.

- Total system temperature drop should not exceed 22°C (72°F).
- The temperature drop for terminal unit heating coils should be 11°C (52°F).

Design water velocity in piping should not exceed 2.5 meters per second (8 feet per second) or design pressure friction loss in piping systems should not exceed 0.4 kPa per meter (4 feet per 100 feet), whichever is larger, and not less than 1.3 meters per second (4 feet per second).

13.3.2 Freeze Protection

Propylene glycol manufactured specifically for HVAC systems should be used to protect hot water systems from freezing, where extensive runs of piping are exposed to weather, where heating operations are intermittent, or where coils are exposed to large volumes of outside air.

- A freeze protection circulation pump should be provided along with polypropylene glycol. Heat tracing systems are not acceptable for systems inside the building.
- Glycol solutions should not be used directly in boilers, because of corrosion caused by the chemical breakdown of the glycol.

The water makeup line for glycol systems should be provided with an inline water meter to monitor and maintain the proper percentage of glycol in the system. Provisions should be made for drain down, storage, and re-injection of the glycol into the system.

13.3.3 Radiant Heat

Radiant heating systems (hot water or gas fired) may be overhead or underfloor type.

- These systems should be considered in lieu of convective or all-air heating systems in areas that experience infiltration loads in excess of two air changes per hour at design heating conditions.
- Radiant heating systems may also be considered for high bay spaces and loading docks.

13.3.4 Instantaneous Hot Water

The use of instantaneous hot water generators is prohibited except for incidental use at terminal fixtures.

13.4   HEAT EXCHANGERS

Steam-to-water heat exchangers should be used in situations where district steam is supplied and a hot water space heating and domestic hot water heating system have been selected.

- Double-wall heat exchangers should be used in domestic hot water heating applications.
- Plate heat exchangers should be used for waterside economizer applications.

13.5   COOLING SYSTEMS

13.5.1   Chilled Water Systems

Chilled water systems include chillers, chilled water and condenser water pumps, cooling towers, piping, and piping specialties.

- The chilled water systems should have a 10°C (50°F) temperature differential in the central system, at the central plant, with a design supply water temperature between 4°C and 7°C (40°F and 45°F).
- In climates with low relative humidity, an 8°C (46°F) may be used.
- The chilled water system should have a 6°C (43°F) temperature differential in the secondary systems, at the terminal points of use, such as coils, with a design supply water temperature between 4°C and 7°C (40°F and 45°F).
- District chilled water, if available, should be used for cooling only if determined to be economical and reliable through a life cycle cost analysis.
- Mechanical equipment rooms must be designed in accordance with the requirements of ASHRAE Standard 15: Safety Code for Mechanical Refrigeration.
- Chiller leak detection and remote alarming should be connected to the BAS.
- Freeze protection: Propylene glycol manufactured specifically for HVAC Systems is used for freeze protection, primarily in low-temperature chilled water systems (less than 4°C, 40°F).
- The concentration of antifreeze should be kept to a practical minimum because of its adversary effect on heat exchange efficiency and pump life.
- The water makeup line for glycol systems should be provided with an inline water meter to monitor and maintain the proper percentage of glycol in the system.
- All coils exposed to outside airflow (at some time) should be provided with freeze protection thermostats and control cycles.
- Provisions should be made for drain down, storage, and re-injection of the glycol into the system.

13.5.2   Condenser Water

All water-cooled condensers must be connected to a recirculating heat-rejecting loop. The heat-rejection loop system should be designed for a 6°C (43°F) temperature differential and a minimum of 4°C (40°F) wet bulb approach between the outside air temperature and the temperature of the water leaving the heat-rejection equipment.

Heat tracing should be provided for piping exposed to weather and for piping down to 3 feet below grade.

13.5.3   Special Cooling Systems

13.5.3.1   Waterside Economizer Cycle

In certain climate conditions, cooling towers are capable of producing condenser water cold enough to cool the chilled water system without chiller operation.

This option should be considered in life cycle cost comparisons of water-cooled chillers.

- Waterside economizer cycles are particularly cost effective in the low humidity climates of the western United States. In the eastern United States, enthalpy airside economizer cycles tend to produce lower operating costs. However, where used, any airside economizer should be set so that no air with a dew point above 10°C (50°F) is allowed into the building.
- Waterside economizer systems should be used only in areas where the outside air temperature will be below 4.4°C (40°F) wet bulb.
- Waterside economizers should utilize a plate heat exchanger piped in parallel arrangement with its respective chiller.

13.5.3.2 Computer Room Air Conditioning Units

Mainframe computer rooms should be cooled by self-contained units for loads up to 280 kW (80 tons). These units should be specifically designed for this purpose and contain compressors, filters, humidifiers, and controls. The cooling units should be sized to allow for a minimum of 50% redundancy, either two units at 75% load or three units at 50%.

If the nature of the computer room is critical, three units sized at 50% of the design load should be used. Heat rejection from these self-contained units should be by air-cooled condensers or recirculating water-cooled condensers connected to a cooling tower or evaporative-cooled condenser. Waterside free cooling should be used when possible.

For cooling loads greater than 280 kW (80 tons), chilled water air-handling systems should be considered in a life cycle cost analysis. A dedicated chiller(s) is preferred, unless other parts of the building also require 24-hour cooling. The dedicated chiller plant should provide some means of redundant backup, by either multiple machines or a connection to the facility's larger chilled water plant.

In large computer installations (areas of 500 m^2 (5,000 ft^2)), segregate cooling of the sensible load (computer load) and control of the outside air ventilation and space relative humidity by using two separate air-handling systems.

- In this design, one unit recirculates and cools room air without dehumidification capability. This unit is regulated by a room thermostat.
- The second unit handles the outside air load, provides the required number of air changes, and humidifies or dehumidifies in response to a humidistat.

This scheme avoids the common problem of simultaneously humidifying and dehumidifying the air. The intent is to provide room temperature conditions that provide a higher available temperature for reduced fan power consumption and easier winter humidification. Verify with users to determine if the air conditioning system must be connected to the emergency power system. These systems should be provided with an alternative power source, connected to emergency generators, if the computer room houses critical components.

13.5.3.3 Desiccant Cooling Systems

For high-occupancy applications where moisture removal is required, solid desiccant with silica gel may be used in combination with mechanical cooling. Heat recovery wheels may be used prior to the mechanical cooling process.

Desiccant cooling units should be equipped with airflow-setting devices for both process and reactivation air flows, and should be equipped with gauges or digital displays to report those air flows continuously. The desiccant cooling system should have self-contained microprocessor controls capable of connecting to and interoperating with a direct digital control (DDC) building automation system.

Natural gas or condenser waste heat should be used as fuel for reactivation of the desiccant.

Lithium chloride liquid desiccants are not permitted.

13.5.3.4 Chillers

Chillers should be specified in accordance with the latest Air-conditioning and Refrigeration Institute (ARI) ratings procedures and latest edition of the Standard 90.1.

As a part of the life cycle cost analysis, the use of high-efficiency chillers with coefficient of performance (COP) and integrated part-load value (IPLV) ratings that exceed 6.4 (0.55 kW/ton) should be analyzed. Likewise, the feasibility of gas-engine driven chillers, ice storage chillers, and absorption chillers should be considered for demand shedding and thermal balancing of the total system.

BACnet or LonsWork microprocessor-based controls should be used. The control panel should have self-diagnostic capability; integral safety control and set point display, such as run time, operating parameters, electrical low voltage, and loss of phase protection; current and demand limiting, and output/input-COP [input/output (kW/ton)] information.

13.5.3.5 Chilled Water Machines

When the peak-cooling load is 1,760 kW (500 tons) or more, a minimum of three chilled water machines should be provided. The three units should have a combined capacity of 120% of the total peak-cooling load with load split percentages of 40-40-40 or 50-50-20. If the peak-cooling load is less than 1,760 kW (500 tons), a minimum of two equally sized machines at 67% of the peak capacity (each) should be provided.

- All units should have adequate valving to provide isolation of the offline unit without interruption of service.
- Cooling systems with a capacity of less than 50 tons should use air-cooled chillers.
- Chillers should be piped to a common chilled water header with provisions to sequence chillers online to match the load requirements.
- All required auxiliaries for the chiller systems should be provided with expansion tanks, heat exchangers, water treatment, and air separators, as required.

- If multiple chillers are used, automatic shutoff valves should be provided for each chiller.
- Chiller condenser bundles should be equipped with automatic reversing brush-type tube-cleaning systems.
- Chiller condenser piping should be equipped with recirculation/bypass control valves to maintain incoming condenser water temperature within the chiller manufacturer's minimum.
- Part-load efficiency must be specified in accordance with ARI Standard 550/590.

The design of refrigeration machines must comply with Clean Air Act amendment Title VI: Stratospheric Ozone Protection and Code of Federal Regulations (CFR) 40, Part Protection of Stratospheric Ozone. Chlorofluorocarbon (CFAN COIL) refrigerants are not permitted in new chillers. Acceptable non-chlorofluorocarbon refrigerants are listed in EPA regulations implementing Section 612 (Significant New Alternatives Policy (SNAP) of the Clean Air Act, Title Stratospheric Ozone Protection. (Note: Accept these criteria in documenting certification of Leadership in Energy and Environmental Design [LEED] ratings.)

Refrigeration machines must be equipped with isolation valves, fittings, and service apertures as appropriate for refrigerant recovery during servicing and repair, as required by Section 608 of the Clean Air Act, Title VI.

Chillers must also be easily accessible for internal inspections and cleaning.

13.5.3.6 Ice Storage Equipment

Ice-on-coil systems should be considered in locations where the demand costs of electricity are greater than $15.00 per kW (demand costs for peak generation, transmission, and delivery costs), including prefabricated tanks with glycol coils and water inside the tank.

The tank should be insulated. The tank's capacity and performance should be guaranteed by the vendor.

Self-contained, fabricated ice-storage systems should have self-contained BACnet LONWORKS microprocessor controls for charging and discharging the ice-storage system and should be capable of being connected to a central building automation system.

13.5.3.7 Cooling Towers

- Multiple cell towers and isolated basins are required to facilitate operations, maintenance, and redundancy. The number of cells should match the number of chillers. Multiple towers should have equalization piping between cell basins. Equalization piping should include isolation valves and automatic shutoff valves between cells.
- Supply piping should be connected to a manifold to allow for any combination of equipment use.
- Cooling towers should have ladders and platforms for ease of inspections and replacement of components.

- Variable-speed pumps for multiple cooling towers should not operate below 30% of rated capacity.
- Induced-draft cooling towers with multiple-speed or variable-speed condenser fan controls should be considered. Induced-draft towers should have a clear distance equal to the height of the tower on the air intake side(s) to keep the air velocity low.
- Consideration should be given to piping arrangement and strainer or filter placement such that accumulated solids are readily removed from the system. Clean-outs for sediment removal and flushing from basin and piping should be provided.
- Forced-draft towers should have inlet screens. Forced-draft towers should have directional discharge plenums where required for space or directional considerations. Consideration should be given to piping arrangement and strainer or filter placement such that accumulated solids are readily removed from the system. Clean-outs for sediment removal and flushing from basin and piping should be provided.
- The cooling tower's foundation, structural elements, and connections should be designed for a 44 m/s (100 mph) wind design load.
- Cooling tower basins and housings should be constructed of stainless steel.
- If the cooling tower is located on the building structure, vibration and sound isolation must be provided.
- Cooling towers should be elevated to maintain the required net positive suction head on condenser water pumps and to provide a 4-foot minimum clear space beneath the bottom of the lowest structural member, piping, or sump, to allow re-roofing beneath the tower.
- Special consideration should be given to de-icing cooling tower fills intended to operate in subfreezing weather, such as chilled water systems designed with a waterside economizer. A manual shutdown for the fan should be provided.
- If cooling towers operate intermittently during subfreezing weather, provisions should be made for draining all piping during periods of shutdown.
- For this purpose indoor drain down basins are preferred to heated wet basins at the cooling tower.
- Cooling towers with waterside economizers and designed for year-round operation should be equipped with basin heaters.
- Condenser water piping located above grade and down to 3 feet below grade should have heat tracing.
- Cooling towers should be provided with BACnet LonsWork microprocessor controls, capable of connecting to central building automation systems.

13.5.3.8 Chilled Water, Hot Water, and Condenser Water Pumps

- Each boiler cooling tower and chiller group pumps should be arranged with piping, valves, and controls to allow each chiller-tower group to operate independently of the other chiller and cooling tower groups.
- Pumps should be of a centrifugal type and should generally be selected to operate at 1,750 RPM. Both partial load and full load must fall on the pump curve.
- The number of primary chilled water and condenser water pumps should correspond to the number of chillers, and a separate pump should be designed for each condenser water circuit.

- Variable-volume pumping systems should be considered for all secondary piping systems with pump horsepower greater than 10 kW (15 horsepower [hp]). The specified pump motors should not overload throughout the entire range of the pump curve. Each pump system should have a standby capability for chilled, hot water, and condenser water pumps.

13.5.3.9 Water Treatment Humidifiers and Direct Evaporative Coolers

Makeup water for direct evaporation humidifiers and direct evaporative coolers, or other water spray systems should originate directly from a potable source that has equal or better water quality with respect to both chemical and microbial contaminants.

Humidifiers should be designed so that microbiocidal chemicals and water treatment additives are not emitted in ventilation air.

All components of humidification equipment should be stainless steel. Air washer systems are not permitted for cooling.

Humidification should be limited to building areas requiring special conditions.

- Courtrooms with wall coverings of wood should be provided with humidification.
- General office space should not be humidified unless severe winter conditions are likely to cause indoor relative humidity to fall below 30%.

Where humidification is necessary, atomized hot water, clean steam, or ultrasound may be used and should be generated by electronic or steam-to-steam generators.

- To avoid the potential for oversaturation and condensation at low load, the total humidification load should be divided between multiple, independently modulated units. Single-unit humidifiers are not acceptable.
- Humidifiers should be centered on the air stream to prevent stratification. All associated equipment should be constructed of stainless steel.
- Humidification systems should have microprocessor controls and the capability to connect to building automation systems.

When steam is required during summer seasons for humidification or sterilization, a separate clean-steam generator should be provided and sized for the seasonal load.

13.5.3.10 Water Treatment for Hydronic Systems

The water treatment for all hydronic systems, including humidification systems, should be designed to address the three aspects of water treatment: biological growth, dissolved solids and scaling, and corrosion protection.

- The performance of the water treatment systems should produce, as a minimum, the following characteristics; hardness: 0.00; iron content: 0.00; dissolved solids: 1,500 to 1,750 ppm; silica: 610 ppm or less; and a PH of 10.5 or above.

- The system should operate with an injection pump transferring chemicals from solution tank(s) as required to maintain the conditions described.
- The chemical feed system should have self-contained microprocessor controls capable of connecting to and interoperating with a DDC building automation system.

The methods used to treat the systems' makeup water should have prior success in existing facilities on the same municipal water supply and follow the guidelines outlined in the ASHRAE Applications Handbook.

13.6   HEAT RECOVERY SYSTEMS

Heat recovery systems house critical components. Heat recovery systems should be used in:

- All ventilation units (100% outside air units), and
- Where the temperature differentials between supply air and exhaust air are significant.

Heat recovery systems should operate at a minimum of 70% efficiency. The heat recovery systems must be capable of connecting to a microprocessor controller that can in turn be connected to a direct DDC building automation system.

Prefilters should be provided in all heat recovery systems before the heat recovery equipment.

13.6.1   Heat Pipe

For sensible heat recovery, a run-around type heat pipe should use refrigerant to absorb heat from the air stream at the air intake and reject the heat back into the air stream at the discharge of the air-handling unit. Systems should have solenoid valve controls to operate under partial-load conditions.

13.6.2   Run-around Coil

A glycol run-around coil could be used with control valves and a pump for part-load conditions. The run-around coils should be used at the exhaust discharge from the building and at the fresh air intake into the building.

The run-around coil system should be capable of connecting to a microprocessor controller that in turn can be connected to a DDC building automation system.

13.6.3   Enthalpy Wheel

A desiccant-impregnated enthalpy wheel with variable speed rotary wheel may be used in the supply and exhaust systems.

13.6.4 Sensible Heat Recovery

For sensible heat recovery, a cross-flow, air-to-air (z-duct) heat exchanger should recover the heat in the exhaust and supply air streams. Z-ducts should be constructed entirely of sheet metal. Heat-wheels may also be used for sensible heat recovery. The unit should have variable speed drive for controlling the temperature leaving the unit.

13.7 FAN COIL SYSTEM

For perimeter spaces, provide four-pipe fan coil units with a cooling coil, heating coil, 35% efficiency filters, internal condensate drain, and an overflow drain. The unit should have self-contained microprocessor controls and should be capable of connecting to and interoperating with a BACnet or LONWORKS DDC building automation system.

Fan coil units should be capable of operating with a unit-mounted or remote-mounted temperature sensor.

13.8 FIN-TUBE HEATING SYSTEMS

When fin-tube radiation is used, reheat should not be featured with perimeter air distribution systems.

Fin-tube radiation should have individual zone thermostatic control capable of connecting to a self-contained microprocessor that can interface with a BACnet or LONWORKS DDC building automation system.

13.9 HEAT PUMP SYSTEMS

Console perimeter heat pump system(s) may be considered for the perimeter zone.

For the interior zone, either a packaged heat pump variable-volume system or a central station AHU with a cooling-heating coil with variable air volume (VAV) boxes should be considered.

Condenser water loop temperatures should be maintained between 15°C (60°F) and 27°C (80°F) year round, either by injecting heat from a gas-fired, modular boiler if the temperature drops below 15°C (60°F) or by rejecting the heat through a cooling tower if the temperature of the loop rises above 35°C (95°F) dry bulb.

Outside air should be ducted to the return plenum section of the heat pump unit. Heat pumps should be provided with filter or filter rack assemblies upstream of the return plenum section of the air-handling unit.

13.10   HVAC INSULATION

All insulation materials should comply with the fire and smoke hazard ratings indicated by the American Society for Testing and Materials standard ASTM-E84, NFPA 255, and Underwriters Laboratory standard UL 723. Accessories such as adhesives, mastics, cements and tapes should have the same or better fire and smoke hazard ratings.

Insulation should be provided on all cold surface mechanical systems, such as ductwork and piping, where condensation has the potential of forming and in accordance with ASHRAE Standard 90.1. All equipment including air-handling units, chilled and hot water pumps, heat exchangers, converters, and pumps should be insulated as per ASHRAE Standard 90.1. All pumps should have jacketing.

Insulation that is subject to damage or reduction in thermal resistivity if wetted should be enclosed with a vapor seal (such as a vapor barrier jacket), and insulation should have zero permeability.

13.11   DUCT INSULATION

Materials used as internal insulation exposed to the air stream in ducts should be in accordance with UL 181 or ASTM C 1071 erosion tests, and should not promote or support the growth of fungi or bacteria, in accordance with UL 181 and ASTM G21 and G22.

- All exposed ductwork should have sealed canvas jacketing.
- All concealed ductwork should have foil-faced jacketing.

The insulation should comply with fire and smoke hazard ratings indicated by ASTM-E84, NFPA 255, and UL 723. Accessories (e.g., adhesives, mastics, cements, tapes) should have the same or better component ratings.

13.11.1   Return Air

The insulation of return air and exhaust air distribution systems needs to be evaluated for each project and for each system to guard against condensation formation and heat gain or loss on a recirculating or heat recovery system.

- Generally, return air and exhaust air distribution systems do not require insulation if located in a ceiling plenum or mechanical room used as a return air plenum.
- Return air and exhaust air distribution systems should be insulated in accordance with ASHRAE Standard 90.1.
- Ductwork with double-wall construction having insulation in between may be used for return air transfer grilles, but only if required for acoustic purposes.

13.11.2   Supply Air Duct Insulation

All supply air ducts must be insulated in accordance with ASHRAE Standard 90.1. Supply air duct insulation should have a vapor barrier jacket. The insulation should cover the duct system with a continuous, unbroken vapor seal.

13.11.3   Sanitary Sewer Vents

All sanitary sewer vents terminating through the roof should be insulated for a minimum of 1.83 meters (6 feet) below the roofline to prevent condensation from forming; a vapor barrier jacket should be included on this insulation.

13.11.4   Plenums

All piping exposed in plenums or above the ceiling should be insulated to prevent condensation.

REFERENCES

Facilities Standards for the Public Building Service, U.S. General Services Administration, Public Buildings Service, Office of the Chief Architect, Washington, DC 20405, March 2003.

ADDITIONAL READING

Fundamentals of Protective Design for Conventional Weapons, TM 5-855-1. Washington, DC, Headquarters, U.S. Department of the Army, 1986.

Guidance for Protecting Building Environments from Airborne Chemical, Biological, or Radiological Attacks, National Institute of Occupational Safety and Health (NIOSH), U.S. Department of Health and Human Services (DHHS) (NIOSH) Pub No. 2002-139, May 2002.

Protective Construction Design Manual, ESL-TR-87-57. Prepared for Engineering and Services Laboratory, Tyndall Air Force Base, FL. 1989.

Security Engineering, TM 5-853 and Air Force AFMAN 32-1071, Volumes 1, 2, 3, and 4. Washington, DC, Departments of the Army and Air Force, 1994.

CHAPTER **14**

Piping Systems

Martha Boss, Dennis Day, Charles Allen, and Randy Boss

CONTENTS

In order to understand the vulnerabilities of plumbing systems, the preferred design features must be understood. Even these preferred design systems have inherent vulnerabilities and may constitute preventative maintenance challenges. This chapter illustrates the basis of preferred pluming system design for most government buildings as a basis for understanding plumbing design principles for all buildings.

Modern systems require that power systems be available, even to control water movement. On September 6, 2005 this became very evident as fetid water covered 80% of New Orleans, Louisiana. The city's pumps were submerged and required drying prior to being put back in service; or were out of service due to lack of power.

Finally a substation in Jefferson Parish was identified as having power. Workers climbed into electrical towers, some reachable only by boat, to make repairs. These workers established a direct link from the substation to Pumping Station 6; and power was finally available to the pumps. An additional complication developed when the motors to the sluice gates did not work. Water being pumped out could only move through narrow openings around the gates. Workers created a tool to be used to move the sluice gates. (*New York Times*, September 8, 2005)

14.1 WATER SUPPLY

Water conservation should be a requirement of all plumbing systems. Use water-saving plumbing fixtures. All fixtures should have sensing devices for saving water.

14.1.1 Domestic Water Supply Systems, Cold Water Service

Cold water service should consist of a pressurized piping distribution system incorporating a separate supply line. This supply line should run from the tap in the existing outside water main to the equipment area inside the building. Incoming service should have double-check valves.

Water service should be metered inside the building using meters furnished by the local department of public works. Remote reading of meters will be accomplished by special equipment over telephone lines. Irrigation systems must be submetered for deduct billing of the sewer system.

Internal distribution should consist of a piping system that will supply domestic cold water to all necessary plumbing fixtures, water heaters, and all mechanical makeup water needs.

The water pressure at the fixture should be in accordance with the International Plumbing Code. The distribution system should include equipment that will maintain adequate pressure and flow in all parts of the system. If the water pressure is not adequate to provide sufficient pressure at the highest, most remote fixture, a triplex booster pumping system should be used.

14.1.2 Domestic Water Supply Systems, Hot Water Service

Hot water should be generated by heaters utilizing natural gas, electricity, or steam as an energy source. Selection should be supported by an economic evaluation incorporating first cost, operating costs, and life-cycle costs in conjunction with the heating, ventilation, and air conditioning (HVAC) energy provisions. Instantaneous hot water heaters are not permitted as a primary source.

Domestic hot water supply temperature should be generated at 60°C (140°F), and should be tempered to 49°C (120°F) using a three-way mixing valve, before supplying to all plumbing fixtures. Hot water supply to dishwashers should be at 82°C (180°F), and the temperature should be boosted from 60°C (140°F) to 82°C (180°F).

- Heat pump hot water heaters should be used where possible to save energy.
- Instantaneous hot water heaters are permitted for incidental use only.

The distribution system should consist of a piping system, which connects the water heater or heaters to all plumbing fixtures as required. Circulation systems or temperature maintenance systems should provide hot water at the furthest fixture from the heating source within 15 seconds of the time of operation.

14.1.3 Domestic Water Supply Equipment

Single water hammer arrestors should be provided at every branch to multiple fixtures and on every floor for both hot and cold water.

Domestic cold and hot water distribution systems should be insulated per the American Society of Heating, Refrigerating, and Air Conditioning Engineers standard ASHRAE 90.1 and all exposed piping should have polyvinyl chloride (PVC) jacketing.

14.2   SANITARY WASTE AND VENT SYSTEM

14.2.1   Waste or Vent Pipe and Fittings

A complete sanitary collection system should be provided for all plumbing fixtures, floor drains, and kitchen equipment. This system must be designed in compliance with applicable codes and standards.

- Piping should be cast iron soil pipe with hub and spigot joints and fittings.
- Aboveground piping may have no-hub joints and fittings.

14.2.2   Floor Drains

Floor drains should be provided in multi-toilet fixture restrooms, kitchen areas, mechanical equipment rooms, and parking garages and ramps. In general, floor drains should be of cast iron body type with 6-inch diameter nickel-bronze strainers for public toilets.

- Pressure booster systems will require large-diameter cast iron strainers, and parking garages will require large-diameter tractor grates.
- Drainage for ramps will require either trench drains or roadway inlets when exposed to rainfall.
- Isolation valves should be provided for all floor drains where drainage is not routinely expected from spillage, cleaning, or rainwater.

14.2.3   Sanitary Waste Equipment

Specific drains in kitchen areas should discharge into a grease interceptor before connecting into the sanitary sewer in accordance with the requirements of the state health department, and local authorities will determine which drains.

Floor drains and/or trench drains in garage locations are to discharge into sand/oil interceptors.

14.2.4   Automatic Sewage Ejectors

Ejectors should only be used where gravity drainage is not possible. Only the lowest floors of the building should be connected to the sewage ejector; fixtures on upper floors should use gravity flow to the public sewer. Sewage ejectors should be non-clog, screenless duplex pumps, with each discharge pipe not less than 100 mm (4 inches) in diameter. All ejectors should be connected to the emergency power system.

14.3 RAINWATER DRAINAGE EQUIPMENT

Roof drains should be of cast iron body type with high dome grates and membrane clamping rings, manufactured by any of the major foundries. Each roof drain should have a separate overflow drain located adjacent to the drain. Overflow drains will be the same drains as the roof drains except that a damming weir extension will be included. Piping systems should be in compliance with local codes and sized based upon local rainfall intensity.

A foundation drainage system should be provided, with perforated drain tile collecting into a sump containing a pumping system as required by the applicable codes.

14.4 POWER SUPPLY

14.4.1 Natural Gas Systems

Gas piping should not be placed in unventilated spaces, such as trenches or unventilated shafts, where leaking gas could accumulate and explode.

14.4.2 Service Entrance

Gas piping entering the building must be protected from accidental damage by vehicles, foundation settlement, or vibration. Where practical, the entrance should be above grade and provided with a self-tightening swing joint prior to entering the building.

14.4.3 Within Building Spaces and Confined Spaces

All spaces containing gas-fired equipment, such as boilers, chillers, and generators, should be mechanically ventilated. Vertical shafts carrying gas piping should be ventilated. Piping must not be placed through confined spaces that could potentially require a permit, such as interior trenches or unventilated shafts.

Gas meters should be located in a gas meter room, thus avoiding leakage concerns and providing direct access to the local gas utility. All gas piping inside ceiling spaces should have plenum-rated fittings. Diaphragms and regulators in gas piping must be vented to the outside.

14.4.4 Fuel Oil Systems

14.4.4.1 Fuel Oil Piping

The fuel oil piping system should use at least Schedule 40 black steel or black iron piping. Fittings should be of the same grade as the pipe material. Valves should be bronze, steel, or iron, and may be screwed, welded, flanged, or grooved. Double-wall piping with a leak-detection system should be used for buried fuel piping. Duplex fuel-oil pumps with basket strainers and exterior enclosures should be used for pumping the oil to the fuel-burning equipment.

14.4.4.2 Underground Fuel Oil Tanks

Underground fuel oil storage tanks should be of double-wall, nonmetallic construction or contained in lined vaults to prevent environmental contamination. Tanks should be sized for sufficient capacity to provide 48 hours of system operation under emergency conditions (72 hours for remote locations such as border stations). For underground tanks and piping, a leak detection system with monitors and alarms for both, is required. The installation must comply with local, state, and federal requirements, as well as U.S. Environmental Protection Agency (EPA) 40 CFR 280 and 281.

14.5 HVAC PIPING

All piping systems should be designed and sized in accordance with the *ASHRAE Fundamentals Handbook: HVAC Systems and Equipment Handbook*. Materials acceptable for piping systems are black steel and copper (Figure 14.1). No PVC or other types of plastic pipe are permitted.

14.5.1 Supports

Provide channel supports for multiple pipes and heavy-duty steel trapezes to support multiple pipes. The hanger and support schedule should have the manufacturer's number, type, and location. Comply with Manufacturers Standardization Society's MSS SP 2003. Pipe hangers and Supports–Selection and Application. Spring hangers and supports should be provided in all the mechanical rooms.

14.5.2 Flexible Pipe Connectors

Flexible pipe connectors should be fabricated from annular, close-pitched corrugated and braided stainless steel. All pumps, chillers, and cooling towers should have flexible connectors.

Standard Piping Material	Use	Comments
ASTM Schedule 40	Chilled water up to 300 mm (12 inch) dia. Condenser water up to 300 mm (12 in dia) Hot water Natural gas, fuel oil Steam 100 kPa (15 psig) to 1035 kPa (150 psi)	1035 kPa (150 psi) fittings standard weight pipe over 300 mm (12 in) diameter Test to 2100 kPA (300 psig) Weld and test to 2100 kPa (300 psig)
ASTM Schedule 30	Chilled water over 300 mm (4 in) dia condensor water over 300 mm (12 mi) dia.	1035 kPa (150 psi) fittings standard weight pipe over 300 mm (12 in) diameter
ASTM Schedule 80	Steam condensate	
Copper Tubing	Chilled water up to 102 mm (4 in) dia. Condensate water up to 102 mm (4 in) dia Domestic water Refrigeration Cast iron, sanitary, waste and vent storm	Builders option use type 1 below ground and type 1 above Lead free solder connections Type ACR.

Figure 14.1   Commercial standards for piping materials.

14.5.3   Cathodic Protection

The need for metal protection for underground piping must be evaluated by a soils resistivity test. The results of the soils resistivity test are provided in a geotechnical report. Cathodic protection and/or some other means of preventing pipe corrosion may be required, based on the geotechnical report's recommendations and conclusions.

14.5.4   Piping System and Equipment Identification

Valves and equipment in mechanical rooms, shafts, ceilings, and other spaces accessible to maintenance personnel must be identified with color-coded bands and permanent tags indicating the system type and direction of flow for piping systems or type and number for equipment. The identification system should also tag all valves and other operable fittings. Gas piping and sprinkler lines must be identified as prescribed by the National Fire Protection Association (NFPA).

14.6   PROVISIONS FOR PIPING IN EARTHQUAKE ZONES

In Seismic Zones 2, 3 and 4, sleeves for pipes should be at least 25 mm (1 in) larger than the pipe, to allow for movement. Flexible couplings should be provided at the bottom of pipe risers. Spreaders should be used to separate adjacent pipes, unless the distance is large enough to prevent contact in an earthquake (consult the Sheet

Metal and Air Conditioning Contractors' National Association [SMACNA] Seismic Restraint Manual and ASHRAE Application Handbook).

14.7   PARALLEL PIPING SYSTEMS

The system should utilize parallel piping systems with a two-pipe main distribution system arranged in a reverse return configuration. Reverse return is preferred because it provides the best overall control and maintenance of a balanced system as the system is modified.

Series loop piping for terminal or branch circuits should be equipped with automatic flow-control valves at terminal units (all types of heat transfer units).

14.8   BOILERS: CONTROL AND PIPING

Each boiler should be provided with a control and piping arrangement that protects the boiler from thermal shock. A primary–secondary piping arrangement with a modulating mixing control valve and higher primary flow rate will assure that the boiler return water temperature does not drop too low, as commonly occurs with night setback.

14.9   PUMPS

14.9.1   Hydronic Hot Water Space Heating Pumps

Hydronic hot water space heating pumps should generally be selected to operate at 1750 RPM.

14.9.2   Pumping Systems

Each pumping system should be provided with two pumps, one operating while the other is in standby mode. These pumps should be configured for automatic lead/lag operation.

14.9.3   Variable Volume Pumping Systems

Variable volume pumping systems should be provided for all secondary-piping systems with pump horsepower greater than 10 kW (15 HP).

14.10   EXPANSION TANKS

Pressurized diaphragm expansion tanks should be used when available in appropriately sized manufactured products.

14.11   AIR SEPARATORS, VENTS, VALVES, DRAINS

Isolation valves, shutoff valves, bypass circuits, flanges, and unions should be provided as necessary for piping at equipment to facilitate equipment repair and replacement. Equipment requiring isolation includes boilers, chillers, pumps, coils, terminal units, and heat exchangers.

14.11.1   Isolation Valves

Each terminal unit or coil should be provided with isolation valves on both the supply and return, and a flow-indicating balance valve on the return line. Isolation valves should be provided on all major pipe branches, such as at each floor level, building wing, or mechanical room.

Air separators and vents must be provided on hot water systems to remove accumulated air within the system.

- Automatic bleed valves should only be used in accessible spaces in mechanical rooms where these valves can be observed by maintenance personnel and must be piped directly to open drains.
- Manual bleed valves should be used at terminal units and other less accessible high points in the system.

Valves should also be provided for zones off vertical risers.

14.11.2   Vents

Air vents should be provided at all localized high points of the piping systems and at each heating coil.

14.11.3   Drains

Likewise, system drains should be provided at all localized low points of the heating system and at each heating coil.

14.12   HYDRONIC, CLOSED-LOOP SYSTEMS

Closed piping systems are unaffected by static pressure; therefore, pumping is required only to overcome the dynamic friction losses. Pumps used in closed-loop hydronic piping should be designed to operate to the left of the peak efficiency point on their curves (higher head, less flow). This compensates for variances in pressure drop between calculated and actual values without causing pump overloading. Pumps with steep curves should not be used, as these pumps tend to limit system flow rates.

14.13   VARIABLE-FLOW PUMPING

Variable flows occur when two-way control valves are used to modulate heat transfer. The components of a variable volume pumping system include pumps, distribution piping, control valves, and terminal units, and will also include boilers and chillers unless a primary–secondary arrangement is used.

All components of the system are subject to variable flow rates. Sufficient pressure differential must be provided across every circuit to allow design flow capacity at all times. Flow may be varied by variable speed pumps or staged multiple pumps. Pumps should operate at no less than 75% efficiency on their performance curve.

Variable flow pumping must be designed carefully. Package systems should be used, complete with pumps and controls that are factory-tested prior to shipment. Chillers and most boilers may experience flow-related heat exchange problems if flow is not maintained above a minimum rate. For this reason, separate, constant flow primary water pumps are recommended for variable volume pumping systems.

14.14   PRIMARY AND SECONDARY PUMPING

In this application, primary and secondary circuits are separate, with neither having an effect on the pumping head of the other. The primary circuit serves source equipment (chiller or boiler), while the secondary circuit serves the load.

Primary–secondary pumping arrangements allow increased system temperature design drops, decreased pumping horsepower, and increased system control. The primary loop and pumps are dedicated and sized to serve the flow and temperature differential requirements of the primary source equipment. This permits the secondary pump and loop to be sized and controlled to provide the design flow rate and temperature differential required to satisfy the heating or cooling loads.

Primary–secondary systems are recommended for larger buildings (circulation of more than 76 L/s [1,000 gpm]) and campus facilities.

14.15   CHILLED WATER AND CONDENSER WATER PIPING

In general, HVAC systems should utilize parallel piping systems with a two-pipe main distribution system arranged in a reverse return configuration. If applied, series loop piping for terminal or branch circuits should be equipped with automatic flow-control valves at terminal units (all types of heat transfer units).

Each terminal unit or coil should be provided with isolation valves on both the supply and return and a flow-indicating balance valve on the return line. Isolation valves should be provided on all major branches, such as at each floor level, building wing, or mechanical room.

For new chilled water HVAC distribution, a pumping and piping arrangement is generally appropriate, with constant volume primary pumping and variable volume secondary pumping. The primary and secondary circuits should be separate, with neither having an effect on the pumping head of the other. The primary circuit serves the source equipment (chillers), while the secondary circuit serves the load.

14.16 PIPING INSULATION

Piping systems conveying fluids having design temperatures less than 18°C (65°F) or greater than 40°C (105°F) should be insulated.

All piping systems with surface temperatures below the average dew point temperature of the indoor ambient air and where condensate drip will cause damage or create a hazard should be insulated with a vapor barrier to prevent condensation formation regardless of whether piping is concealed or exposed. Chilled water piping systems should be insulated with nonpermeable insulation (of perm rating 0.00) such as cellular glass.

All exposed and concealed piping should have PVC jacketing.

REFERENCES

ASTM Schedule 40 PVC Pipe, ASTM D 2949 Schedule 30 Manufacturers Standardization Society MSS SP2003 Pipehangers and Supports—Selection and Application, June 2002.

Behind the first roar of machinery to drain the city, A tale of pluck and luck, John Schwartz, *New York Times*, September 8, 2005.

Facilities Standards for the Public Building Service, U.S. General Services Administration, Public Buildings Service, Office of the Chief Architect, Washington, DC 20405, March 2003.

ADDITIONAL READING

Fundamentals of Protective Design for Conventional Weapons, TM 5-855-1. Washington, DC, Headquarters, U.S. Department of the Army, 1986.

Guidance for Protecting Building Environments from Airborne Chemical, Biological, or Radiological Attacks, National Institute of Occupational Safety and Health (NIOSH), Department of Health and Human Services (DHHS) (NIOSH) Pub. No. 2002-139, May 2002.

Protective Construction Design Manual, ESL-TR-87-57. Prepared for Engineering and Services Laboratory, Tyndall Air Force Base, FL. 1989.

Security Engineering, TM 5-853 and Air Force AFMAN 32-1071, Volumes 1, 2, 3, and 4. Washington, DC, Departments of the Army and Air Force, 1994.

CHAPTER 15

Decontamination Methods, Biocides, and Personal Protective Equipment

Martha Boss, Dennis Day, and Jon A. Cummins

CONTENTS

Chapters 15 through 17 discuss decontamination from the perspective of water intrusion and mold amplification site decontamination. Mold amplification sites will also be sites of bacterial growth. Compromised heating, ventilation, and air conditioning (HVAC) systems will not transfer air correctly, and the latent spores or active mold growth within an HVAC system may exacerbate air quality decline during a chemical, biological, or radiological (CBR) contaminant influx.

A mold amplification scenario was chosen to illustrate decontamination needs. The structural, physical, and contaminant loading circumstances associated with mold growth should be addressed to reduce building vulnerability prior to any potential CBR event. In addition, if a threat proceeds to requiring sheltering in place, the most likely contamination within a sustained shelter area is biological, caused by human habitation and restricted use of interior spaces. Mold was also chosen as the biological to exemplify other biologicals since the vegetative state of molds has parallel features to desiccant-prone bacterial cells. And the spore stage of mold reproduction has parallel features to the nondesiccant-prone encysted bacteria and viruses.

Biological contamination associated with pathogenic bacteria or viruses may require more extensive decontamination. The biocide choice should be based on the level of risk given the biological agent of concern and the (building) materials contaminated. The decontamination procedures discussed herein relative to mold decontamination can, in most cases, be used for Risk Group 2 bacteria. Viruses, prions, encysted bacteria, and eggs from pathogens (worms, flatworms) are resistant to biocides that function by causing desiccation. Some viruses and encysted bacteria can be destroyed using oxidative chemicals—especially quick-acting hydrogen peroxide.

If the ventilation system is contaminated with an airborne pathogen, a knowledge of the HVAC system as discussed in Chapters 8 through 13 will be needed (in addition to concepts presented in Chapters 16 through 18). Likewise if the plumbing system (including the plumbing system associated with the HVAC system) is compromised, the information presented in Chapter 14 must be considered.

Some decontamination methods may be very similar to those used to abate asbestos or lead. The main difference is the use of biocides. Biocides are chemicals designed to kill life. Thus at concentrated levels, biocides are a danger to workers. Manufacturer's instructions must be carefully followed. These instructions and a stocking of appropriate biocides are a required planning element.

In some cases, only partial decontamination will result and sufficient pathogenic loading may still be present to cause illnesses. The vulnerability assessments must, therefore, address when the building and facility are anticipated to be unusable and the recourse to inhabitants or users of those facilities. An example of this type of vulnerability is when epidemics or pandemics render use of public places far too vulnerable, despite the best protective measures available, including sequential decontamination.

15.1 DECONTAMINATION (DECON) TERMS

15.1.1 Amplification

Amplification is the process whereby development and growth of bacteria or mold is enhanced due to conditions that increase the potential for these organisms to survive and thrive.

15.1.2 Bags

Bags may be used to carry materials for use or disposal. The bags must be leak resistant with strong seams and sufficient thickness to resist breakage. A minimum of 4-millimeter thickness is required, with 6 millimeters preferred. Hefty Steel Sacks® are reinforced with steel fibers and may also be used.

15.1.3   Desiccants

Various desiccants (drying agents) are marketed that chemically react with water and thus *pull* the water chemically away from materials such as drywall. Most of these chemicals are also called *hydrophilic*, which means *water loving*. Unfortunately, these products are very dangerous because skin contains water and the removal of this water on skin causes burns. Less caustic desiccants usually contain filler materials so that the concentration of desiccant is less per unit area, and this fact must be considered when calculated the amount of desiccant needed. Decisions as to desiccant use must be made on a site-specific basis.

- The most common solid desiccants used are a form of lye (sodium hydroxide). Products containing sodium hydroxide are usually applied by dispersing dry granules over a surface area and allowing several days for the desiccation process to proceed. This treatment is common in crawl space locations where soils need to be dried. The area where the lye is used must be restricted for entry only by trained personnel. Training must include hazard communication information as required by 29 CFR 1910.1200. Various formulations of silica dioxide, lava rocks, and clay compounds are also used. These products are usually hung in bags or dispersed over limited surfaces.
- Alcohol can also be considered a desiccant in that the alcohol causes drying through evaporation to proceed more quickly. Alcohol is present in many decontamination solutions. These solutions may contain sufficient alcohol to render the immediate atmosphere where use is occurring flammable. As such, alcohol-based solutions if used must be applied taking into account the flammability hazard as well as the toxic hazard to humans. These products, if used as directed, provide the desiccant properties of alcohol without the flammability hazard.

15.1.4   Dehumidification

Air-movement devices use air movement without assist, refrigeration dehumidification, and chemical hygroscopic dehumidification. Air movement without assist essentially removes moisture by displacing the moisture-bearing air with drier air. Fans are an example of air-movement devices. Dehumidification devices may be used with or without fans.

- *Refrigeration-based dehumidifiers* create a cold surface onto which excess moisture condenses. Refrigeration dehumidifiers are used to control humidity down to 40% Rh and at temperatures down to 3°C. These dehumidifiers may also be used for room air conditioning, however their use for mold avoidance or abatement situations is limited if their primary usage is for room air conditioning. If these dehumidifiers are already available within a building area, ramping up their air intake can help in the dehumidification of room areas. Portable units may be needed in addition to any other air conditioning system elements already installed.
- *Adsorption/desiccant dehumidifiers* use hygroscopic media to adsorb moisture. Adsorption dehumidifiers control humidity levels below 40% and are ideal for

low-temperature environments. The dehumidifier manufacturer must be consulted as to the hygroscopic media required and the replenish rate required for effective use. These dehumidifiers are usually not readily available for maintenance use aimed at air drying structural materials in large air spaces.

Both dehumidification types add heat to the air being dried.

15.1.5   Gloves

Gloves used to protect skin from decontamination solutions should be nitrile and at least 4 millimeters in thickness.

15.1.6   Goggles

Goggles are used with half-face respirators and also at any time when splash protection is needed around the eyes. The goggles used should be plastic with a flexible elastic headband. Vented goggles may be used when biocide is not being sprayed. When biocide is being sprayed, only nonvented goggles should be used. Note: Safety glasses with side shields are not a substitute for goggles when biocide is being sprayed.

15.1.7   HVAC

Heating, ventilation, and air conditioning system.

15.1.8   Mold Staining

This involves stains caused by mold presence or die back. Die back occurs when mold colonies die. The cellular degradation products then ooze from the surface where contamination is present. This or other cellular degradation products may produce a stain that is either close to the mold growth site or leaks away from this site. The most common stains are black or brown.

15.1.9   Material Safety Data Sheets (MSDS)

MSDS contain the chemical information required by the 29 CFR 1910.1200 Hazard Communication standard.

15.1.10   Pails

Pails are either soft or hard plastic round containers. These containers should be checked for leaks prior to usage.

15.1.11   Respirator

Respirators are barrier filter or cartridge devices that are used to prevent inhalation of contaminants and certain chemicals. The respirators described in these standard operating procedures (SOPs) are air-purifying respirators (APRs) and protect by filtering the air prior to inhalation by workers.

- For work situations where particulate including mold spores are of concern, the National Institute for Occupational Safety and Health (NIOSH) has certified N95 respirators as acceptable. This certification is for the type of filter used. Depending on the concentration of contaminants, either half- or full-face respirators may be required.
- Some biocide and other chemical use may require additional respiratory protection or different types of respirator cartridges or filters. MSDS should be consulted to determine respirator use when chemicals are used.
- Respirators must be clean and must be stored in clean areas.
- Under no circumstances should the respirators be worn around the neck for sequential reuse. If reuse is anticipated, the respirator must remain on the face until the next reuse cycle is complete.

15.1.12   Scrub

To scrub is to apply decontamination solution with pressure while wet washing in both a vertical and horizontal movement of the scrubbing media (rags, mops, etc.).

15.1.13   Swab

To swab is to apply decontamination solution with only a light amount of vertical pressure and no washing with pressure or horizontal scrubbing. Note: Swab or swabbing may also refer to a method for taking biological samples. Essentially this method uses light pressure and a swab that resembles a Q-tip brand swab to take samples. Using a swab to take a sample must not be confused with using the swabbing technique for applying decontamination solution.

15.1.14   Vacuums: Wet and Dry

Standard wet and dry vacuums can be used. Preference should be given for hard plastic collection containers rather than metal. Alto®, Craftsman®, Euroclean®, MI-T-M®, Nilfisk®, Porter Cable®, or Shop-Vac® are examples of acceptable product lines. Consideration must be given both to the original equipment and the availability of service and parts in each geographic area. Replacement parts include regularly replaced items such as filters, belts, and ring seals.

15.1.15   Visible Mold Growth

Mold that can be seen during a visual examination of a surface is visible mold growth. Not all mold growth is readily visible, in particular *Penicillium* may appear as a dustlike gray coating, and may thus be missed as visible mold growth.

15.1.16   Wash

To wash is to apply decontamination solution using a combination of initial swabbing and scrubbing techniques. Swab first and then scrub.

15.1.17   Wiping

Wiping is a limited form of scrubbing. Wiping down a surface means to apply the decontamination solution with a single downward movement of a rag or mop.

15.2   EQUIPMENT DECONTAMINATION OR DISPOSAL

Personal protective equipment (PPE) and other equipment used on-site must be either decontaminated or properly disposed. Since biocides can destroy some equipment, more disposal may be needed than that estimated for asbestos or lead-abatement projects.

Respirators can be decontaminated using chlorine solution and sequential rinses. Respirator filters and cartridges cannot be decontaminated and should not be used for more than one workday. Storage of biologically contaminated respirator filters and cartridges may cause residual biological contaminants to amplify through reproduction. Thus, if respirator filters and cartridges are worn on successive days, workers will be exposed over and over again to the growing biological contamination in their filters and cartridges.

Amplification is particularly troublesome in HEPA filters, whether used on respirators or to filter interior air streams. When used in respirators, these filters, if contaminated, may continually expose workers within the very small breathing space provided by the interior of respirators. If filtration units are contaminated, air containing unacceptable levels of biological contaminants may be vented from the decontamination area.

All area air filtration units must be checked to assure filtration continues to be adequate for occupied spaces. When at all possible, venting should be outdoors where human receptors are not present. Filtration may be checked using the same air-monitoring techniques used to gauge biological contamination in workplace air. If air is vented when biocides are in use, the level of biocides present in vented air may also need to be checked.

Vessels used to store biocides may not be usable again. The manufacturer's recommendations should be consulted and documented as to reuse of pails, buckets, mops, handheld tools, polyethylene sheeting, and other equipment exposed to biocides.

Porous materials used, such as fibrous booms and spill mats, cannot be decontaminated and should be disposed.

Vacuums whether dry or wet, and negative air machines should be decontaminated in accordance with the manufacturer's recommendations.

15.3   DECONTAMINATION SOLUTIONS

In all cases, material safety data sheets (MSDS) must be consulted to determine the hazards associated with using decontamination solution.

15.3.1   Decontamination Solution Types

Decontamination solutions are chemicals that when applied assist in the removal of contaminants from surfaces.

15.3.1.1   Surfactant Decontamination Solution

An example of a surfactant decontamination solution is soap, which aids in washing the surfaces clean by increasing the potential for material to slide off the surface. All consumer products labeled as soap are in fact this type of surfactant decontamination solution. The differences in the soap products are primarily associated with added fragrance or color. Usually the soap with the least amount of additives is preferred.

15.3.1.2   Detergent Decontamination Solution

A detergent decontamination solution such as trisodium phosphate (TSP), chemically bonds contaminants. Liquid TSP is also available in a variety of decontamination products sold for janitorial use.

15.3.1.3   Universal Solvent Decontamination Solution

Universal solvent decontamination solutions are solutions such as quaternary ammonia that dissolve both ionic and organic constituents of biologicals. Quaternary ammonia is available both as a powder and as a powder packed in alcohol-based wrappings. The alcohol-based wrapping contains the quaternary ammonia until the packet is placed in water, at which time the wrapping dissolves releasing the ammonia. This type of packaging is often used for equipment decontamination since a premeasured amount of powder is delivered below the surface of the wash water. Quaternary ammonia is also available as a liquid cleaner.

Ammonia that is not quaternary ammonia is also available in various ammonia–water solution formulations. Ammonia in this form emits a strong odor and must be handled with care. The higher the percentage of ammonia in water, the greater the hazards. Concentrated ammonia solutions are very dangerous and must not be used.

Ammonia in any form must not be used when chlorine solutions are used. Special biocide products may include both chlorine and ammonia as a chloramine. These chloramine products require special preparation and stabilizers and cannot be safely made by simply mixing chlorine and ammonia products.

15.3.1.4 Desiccant Decontamination Solution

Desiccant decontamination solutions contain alcohol, which destroys bacteria and mold cells by drying them out. These solutions evaporate quickly and have little residual property. The higher the percentage of alcohol in the solution, the greater the flammability risks during use.

15.3.1.5 Biocide Decontamination Solution

Biocide decontamination solutions include a form of chlorine (chlorine dioxide, calcium hypochlorite) that reduces the development of bacteria and mold cells. Note: Decontamination solutions with biocide properties that contain alcohols and phenols are not recommended for large areas due to the flammable properties of alcohols and toxic properties of phenols.

15.3.1.6 Hydrogen Peroxide

Various forms of hydrogen peroxide provide an effective and quick decontamination solution. The hydrogen peroxide is an oxidizer, which in higher concentrations is toxic to vegetative and cellular structures. In addition, these solutions lyse the outer membranes of both cellular and noncellular structures. These products do not have residual properties unless combined with residual biocide decontamination solutions.

15.3.1.7 Residual Biocide Decontamination Solution

Residual biocide decontamination solutions have biocide properties and also leave a film of residual chemical on surfaces to kill newly developing bacteria and mold colonies.

15.3.2 Decision Logic for Biocide Usage

- Biocide decontamination solution must be used if mold contamination has been identified.
- Interiors of HVAC units and associated ductwork may require the use of biocide decontamination solution, even if visible mold is not observed on these units. The decision to use biocides must be based on site-specific conditions and the presence of mold. An example of a situation in which such decontamination might be required is when the areas serviced by these HVAC units are contaminated with airborne mold spores.

Detergent decontamination solution may be used if mold contamination has not been identified either on the surfaces being cleaned or within the areas being serviced by air-treating devices.

15.4 SELECTING BIOCIDES

In researching which biocides to use and the concentrations required, the following should be considered:

- Type of biological contamination. Does the manufacturer have data showing that their product is effective against the biological contaminants and at their current contamination level? Dwell time required after application of the biocide should also be calculated in terms of biocide effectiveness.
- Presence of electrical wiring or ventilation equipment that could be corroded by biocides.
- Flammability hazards posed by the biocides and their application methods
- General building ventilation, workplace zoning, and level of occupancy of the building
- Can the heating ventilation and air conditioning (HVAC) system can be isolated or shut down?
- Usage in or on:
  - False plenums, ductwork, flexible duct work, horizontal plenums, and ventilation hoods
  - Furnace, boiler, or other combustion-chamber areas
  - Condensers, face and bypass systems, and evaporative coils
  - Cooling towers, sumps, liquid-filled plumbing lines and vessels, misters
  - InHumidifiers, swamp coolers, and sprinkler systems
- Slip, trip, and fall potential when used on flooring or polyethylene sheeting
- Dermal hazard potential if PPE is breached
- Respiratory hazards given chosen application methods
- Waste disposal requirements

These and other biocide and decontamination design questions must be approached on a site-specific basis. Questions as to potential effects on carpets, porous materials, heirlooms, antique finishes, and other real property not slated for disposal must also be answered.

15.4.1 Types of Biocides

Biocides come in several forms, the most common being:

- Acids or bases that are pH altering
- Chlorines, bromines, or iodines
- Chlorine dioxide
- Hydrogen peroxide
- Hypochlorite granules
- Alcohols
- Phenols like the active ingredients in Lysol
- Sulfur compounds
- Quaternary ammonia
- Stabilized chemical mixes
- Ozone gases

Soaps have limited biocide properties and mainly function to remove biological contamination from surfaces. Some liquid soaps have added biocides to enhance their bacterial biocide effect. Bar soaps should not be used on decontamination sites as without sufficient drying, the bar and surrounding liquids may harbor biological contaminants.

Ozone gas treatments used alone or in combination with ultraviolet light treatment have shown some success in eliminating the viability of airborne spores. These treatments must, however, be repeated several times to cover all spore release cycles following initial decontamination events.

15.4.2 Biocides and MSDS

MSDS for biocides may inappropriately discuss hazards on the assumption that the biocides will be used for surface applications only. Manufacturers should be consulted whenever fogging, concentrated soaking applications, high-pressure delivery systems, or usage within confined areas is anticipated.

15.4.3 Biocide and Biocide-Treated Material Disposal

Spent biocides and materials treated with biocides may be considered a hazardous waste. During the development of MSDS, manufacturers are required to determine disposal options for spent chemicals. The requirement does not extend to materials that contain biocides after a decontamination event.

In determining appropriate disposal and labeling requirements, the following general regulatory requirements should be considered: OSHA per 29 CFR 1910.1200 Hazard Communication and equivalent requirements in 29 CFR 1926; Resource Conservation and Recovery Act (RCRA) as regards storage, treatment, and disposal of wastes; and Department of Transportation requirements as to labeling, marking, placarding, and transportation protocols on public byways.

15.5   STEAM CLEANING

Steam cleaning without the use of biocide washes or rinses is usually not effective, and may make the situation worse. Steam cleaning applies water with the core of the stream being very hot and perhaps producing steam. The peripheries of this core or the applied waters often do not retain sufficient heat to kill biological contaminants.

15.6   VACUUMING

Wet and dry vacuuming, if used in a cleaning cycle, may aerosolize additional biological fragments, spores, and particulates to which biological contaminants are attached. Thus vacuuming and sweeping may require additional personal and area protection be used to protect workers and building inhabitants. In general, dry removal without prior treatment of areas with biocides should be carefully evaluated in terms of increased hazards.

15.7   PERSONAL PROTECTIVE EQUIPMENT (PPE) AND PERSONNEL DECONTAMINATION

Biological contaminants, like chemical contaminants, may require the use of PPE. Engineering controls are, of course, the first choice. Examples of engineering controls are reducing ventilation rates, overwrapping contaminated materials prior to removal, and other controls that lessen the potential for contaminants to migrate into the air.

The main difference between biological and chemical contamination is that biological contaminants can reproduce. Thus, improper methods, use of PPE, and decontamination of PPE can lead to worker exposures in the future and, thus, after the initial decontamination attempt.

The OSHA Bloodborne Pathogen Standard, 29 CFR 1910.1030, covers workplace exposures to pathogenic biologicals that may be bloodborne or are carried in blood-derived tissue fluids. This standard requires the use of barrier methods and medical consultation for workers.

Barrier methods for all mold or fungi and yeasts follow similar conventions as those contained in 29 CFR 1910.1030. If airborne levels exceed or may exceed those judged to be healthy, respirators must be worn. Usually anyone involved in a biological decontamination event should assume the need to wear respirators during some part of the decontamination work efforts.

Other barrier methods such as splash shields, gloves, protective coveralls, boot covers, and hoods may also be needed. These barrier methods are designed to prevent skin exposure through broken skin or mucous membranes (nose, mouth, and genitals). The added benefit is that these barrier methods, when used correctly, eliminate most of the potential for workers to *carry contamination home on their own clothing.*

Workers must always keep in mind that PPE when used for any contamination, whether chemical or biological is usually not totally protective. Exposures are reduced but not eliminated. Thus the choice of initial protective equipment should be the responsibility of a competent person. If visible mold is present, respirators and goggles must be worn as the contaminated material is removed.

15.7.1 PPE List

- Goggles: Eye protection provided will be in accordance with American National Standards Institute standard ANSI Z87.1 and goggles used must be without side vents.
- Protective headgear (hard hats) should be provided as required when overhead hazard is present.
- Washable uniforms (shirts, pants) and cloth socks
- Plastic-coated disposable paper coveralls
- Footwear
  - Steel-toed leather boots are required where heavy equipment could fall or pinch the toes.
  - Deconnable boots (Neoprene without lining) are required where overspray or floor application of biocide will impact the boot surface. An alternative is to wear impermeable latex or other plastic booties over leather boots or shoes.
  - In other workplace situations where physical and chemical hazards are not present, regular footwear is acceptable.
- Nitrile 4-mil thickness gloves
- Disposable respirators; N95 respirators (for work where dusts of visible mold particulate are of concern). Additional respiratory protection may be required based on the MSDS for the chemical used.

15.7.2 Respirators

For work situations where particulates including mold spores are of concern, the NIOSH has certified N95 respirators as acceptable. This certification is for the type of filter used. Depending on the concentration of contaminants either half- or full-face respirators may be required. For these SOPs the assumption is that half-face respirators are sufficient.

- Some biocide and other chemical use may require additional respiratory protection or different types of respirator cartridges or filters. MSDS should be consulted to determine respirator use when chemicals are used.
- Respirators must be clean and must be stored in clean areas.
- Under no circumstances should the respirators be worn around the neck for sequential reuse. If reuse is anticipated, the respirator must remain on the face until the next reuse cycle is complete.

15.7.3 Barrier Clothing

Barrier clothing is clothing that prevents biocide solution from coating the skin. Barrier clothing includes:

- Plastic-coated disposable paper coveralls. These coveralls should be used when normal work clothes would become wet due to decontamination solution usage.
- Gloves.
- Deconnable boots or boot covers.

Gloves will be worn whenever cleaning includes use of any decontamination solution. The only exception to this requirement is when soap is being used to clean the hands. Goggles will be worn whenever biocides are being sprayed. Safety glasses or goggles will be worn throughout cleaning events.

If biocide use has the potential to cause worker's feet to be placed in biocide solutions (such as in floor cleaning), workers must don either deconnable boots or moisture-resistant boot covers.

15.7.4 Personnel Decontamination

If mold decontamination has occurred, PPE decontamination will consist of:

1. Wet wiping of goggle exterior with prepackaged wet wipes or paper towels wetted with isopropyl alcohol. Bagging and later washing of goggles with soap and water to clean the goggle interior.
2. Wet wiping of hardhats (if used) with biocide decontamination solution.
3. Wet wiping of boot exteriors with biocide decontamination solution.
4. Removal and bagging of disposable respirators for later cleaning.
5. Removal and bagging of gloves for disposal.

Uniforms and cotton socks should be laundered at the end of the workday. If visible contamination is transferred to the uniforms during the workday, the uniforms should be doffed and replaced with clean uniforms.

15.7.5 Disposal

Equipment and disposal bags will be decontaminated prior to removal from a mold abatement area. Decontamination should be by wet wiping with biocide decontamination solution.

Materials used for wet wiping will be disposed of as sanitary waste.

All equipment, disposal equipment, and transfer bags will also be visibly cleaned prior to removal from the regulated areas.

All biologically contaminated waste, including contaminated wastewater filters, scrap, debris, bags, containers, equipment, and contaminated clothing, will be collected and placed in leak-tight containers such as double plastic bags, 6-mil disposal bags, sealed double-wrapped polyethylene sheets, or other approved containers.

For temporary storage, sealed impermeable containers will be stored in a waste load-out unit or in a storage or transportation conveyance (i.e., dumpster/roll-off).

Procedures for hauling and disposal will comply with state, regional, and local standards.

15.8   VACUUM USAGE

15.8.1   Quick Facts

Vacuuming is the process of creating a vacuum within a machine interior and then drawing air toward that vacuum. The concept assumes that dust and debris will also be drawn into the machine interior. Sometimes additional abrasion (e.g., beater-bar method) is used to ensure that the dust and debris is moved into the air stream.

Consideration must be given both to the original equipment and the availability of service and parts in each geographic area. Replacement parts include regularly replaced items such as filters, belts, and ring seals.

15.8.2   Filters

The air stream is exhausted through a filter. The filter may be a simple flat paper filter or a pleated paper filter. The highest level of paper filtration is obtained by using a HEPA filter that is made of special material and is very tightly pleated. Unfortunately these filters can easily clog and be rendered unusable; consequently prefilters often must be used to remove the larger debris.

HEPA vacuums, when used for mold abatement, may serve to clear the air of mold spores and vegetative material. These vacuums may also be used for surface cleanup. Such HEPA vacuum usage must be with carefully controlled site-specific plans. These plans will ensure that the vacuums and the containments around these vacuums are correctly positioned. This level of mold abatement is beyond the scope of this SOP.

Mesh plastic filters are also used. These plastic filters are used in wet vacuuming and may also be used to prevent water from entering the vacuum motor housing portal, which extends into the debris and water collection container.

15.8.3   Collection Bags and Containers

If a bag-equipped vacuum is used, a certain portion of the smaller dust and debris exits through the bag itself.

Vacuums pick up debris through nozzles that lead to circular piping to either a collection bag or hard-sided container. The nozzle collection efficiency is dependent upon how efficiently the nozzle moves debris up the nozzle to the collection container. If the efficiency is low, debris moves around the nozzle area rather than moving the debris toward the collection container. Even the most efficient vacuum will fail to pick up some of the debris.

15.8.4 Wet and Dry Vacuums

Wet and dry vacuums move debris toward a collection container that may also receive water. These vacuums have protected motor housings and wiring to ensure (if intact) that moisture does not contact electrical supply lines. In addition, these vacuums have three-prong grounded plugs. All such vacuums must be checked for electrical safety according to the manufacturer's instructions prior to use.

Decontamination solution can be removed from surfaces by using wet and dry vacuums. Since these vacuums exhaust air from the collection container through an air outlet exhaust portal, the collection container may be filled with decontamination solution prior to vacuum usage. Filling the collection container with the decontamination solution will aid in ensuring that the air exhausted has some contact with aerosolized decontamination solution prior to exhausting through the outlet exhaust portal. The vacuum container (that received the vacuumed debris and liquids) should be filled to a depth of 2 inches with the decontamination solution. The solution type should be the same as the decontamination solution chosen for surface application on the surfaces being vacuumed. *Alcohol- and phenol-containing solutions must, however, not be used. No flammable solutions can be used as wet vacuuming decontamination solutions.*

15.8.5 Vacuum Use

In situations where no mold is present, vacuums can be used according to manufacturer's directions. If standing water is present, only a wet and dry vacuum can be used.

15.8.5.1 *Vacuum Use with Mold Present*

Wet and dry vacuum usage in areas where visible mold has been determined to either be currently present or present prior to recent cleaning events (within the past 36 hours) requires that precautions be taken.

If visible mold is being cleaned up, always mist the general area with biocide decontamination solution prior to any vacuum usage. Once this is done follow the procedures below.

15.8.6 Decontamination Solution Misting

Mist the wet dry vacuum exhaust portal area (not directly into the portal, just the air around this vacuum air outlet) with biocide decontamination solution prior to turning the vacuum on, and at regular intervals during vacuum use:

- Every 5 minutes if areas with visible mold are being vacuumed. Note: These areas should also be misted as the vacuuming is occurring.
- Every 15 minutes if visible mold has been removed through immediately prior cleaning efforts (within the past 24 hours).

15.8.7 Alternate Procedures: When Wet Cleaning Is Not Possible

The vacuum is *not* to be used to remove mold in lieu of washing down the surfaces or misting these surfaces and removing the mold-contaminated materials. In some cases, however, an alternate procedure may be needed.

If for some reason the vacuum must be used in areas where either misting and/ or wet cleaning is not possible, the vacuum must be used in the wet mode with 1.5 inches of biocide decontamination solution placed in the collection container prior to vacuum usage. The biocide decontamination solution in the base of the vacuum container will serve to capture any mold that is being vacuumed up and reduce the viability of the mold spores as these spores enter the solution.

The areas where alternate procedures are required and the reasons why these alternate cleaning methods were used must be stated in the maintenance documentation.

15.8.8 Vacuum Clean-Out after Use

Clean out the vacuum after each use if visible mold or former visible mold areas have been vacuumed in the following manner.

- Dump any waters or dry debris collected into either waste containers or into the sanitary sewer.
- Wipe out the interior of the vacuum with biocide decontamination solution. Assure that all debris has been removed from the vacuum interior.
- Submerse the vacuum plastic mesh filter(s), hoses, and the attachments to the hoses in the biocide decontamination solution. The vacuum canister may be used as a container in which to wash these attachments and hoses.
- Air-dry the vacuum and vacuum accessories. Alternately, the vacuum may be used to essentially dry itself by running it in dry mode.

15.8.9 Vacuum Filters: Mesh and Paper

Two types of filters are standard with wet and dry vacuums: a plastic mesh filter that is used for wet vacuuming and a dry pleated filter used for dry vacuuming.

- All vacuum mesh filters can be cleaned with the biocide decontamination solution during the cleaning event described above.
- If the dry filters are used, these should be disposed of after any use where visible mold was (unexpectedly) vacuumed.

15.8.10 Vacuum Storage

Note: If the vacuum is obtained from storage and has not been cleaned as described below, cleaning will be required prior to use of the vacuum. The same decontamination solutions used during the cleaning events should be used prior to vacuum storage. If detergents were used, use detergents; if biocides were used, use biocides.

Prior to any extended storage of vacuums that have been used in mold decontamination (extended meaning for longer than 1 month) the vacuums should be:

• Emptied of loose debris contained within the vacuum housing (barrel or cyclone assembly)
• Thoroughly cleaned with decontamination solution

The vacuum paper filters can be used until:

• Debris covers the filter to the extent that filter usage is no longer efficient
• Paper becomes wet

Prior to extended storage, used filters should be removed for disposal. Mesh plastic filters can be cleaned with decontamination solution, dried, and retained.

The decontamination solution used prior to extended storage will be:

• Detergent if the vacuums were not used in mold-contaminated areas.
• A biocide if vacuums were used in mold-contaminated areas.

If no visible mold was *ever* observed, the vacuum can be used over and over again, and be cleaned using normal cleaning procedures.

ADDITIONAL READING

Centers for Disease Control/National Institutes of Health. (2006). *Biosafety in Microbiological and Biomedical Laboratories*. 5th ed, Richmond, J.Y. and R. W. Mckinney, eds. Washington, DC: U.S. Government Printing Office.

University of Toronto Biosafety Committee. (1979), *Guidelines for the Handling of Recombinant DNA Molecules, Animal Viruses and Cells, Microorganisms and Parasites*. Ottawa: Medical Research Council of Canada.

University of Toronto Biosafety Committee. (1996). *Laboratory Biosafety Guidelines*, 2nd edition. (UTBSC).

University of Toronto Biosafety Committee. (2000). *Biosafety Policies and Procedures Manual 2000*, (UTBSC).

World Health Organization. (1999). *WHO Infection Control Guidelines for Transmissible Spongiform Encephalopathies. Report of a WHO Consultation*. Geneva, Switzerland, 23-26, March 23–26.

CHAPTER 16

Building Materials and Decontamination
Discussion of General Decontamination Terms and Method for Specific Building Materials

Martha Boss, Dennis Day, and Jon A. Cummins

CONTENTS

16.1   METALS

16.1.1   Quick Facts

Metal studs, flooring material, ceiling material, wall paneling, and furnishings are usually not subject to mold growth. Mold growth may, however, be associated with coating (paint or other surfacing materials) applied to the metal.

If these metals are located next to a visible mold growth area, but do not have visible mold on their surfaces, maintenance procedures may be used for decontamination.

Maintenance procedures include the use of cleaning products that do not oxidize metal to clean uncoated metal. Chemicals that will oxidize (rust) uncoated metal include chlorines. Material safety data sheets (MSDS) should be consulted to determine which chemicals are acceptable for metal cleaning, including the cleaning of coated metal.

If chlorine is a component of the decontamination solution (whether biocide or added to detergent), additional treatment of the metal surfaces may be required after cleaning to assure that further oxidization (rusting) of the metal does not occur. This treatment is usually by application of an oil-based sealant. The manufacturer of the metal part or equipment piece may have special cleaning instructions, and these instructions must take precedence over this standard operating procedure (SOP).

To lessen the formation of rust after treatment with decontamination solution, metal should be coated with biocide encapsulants. If previously unsealed ferrous (iron-containing) metal is coated, repainting with anticorrosive paint will be required to control future rusting. Aluminum does not need to be treated for rust, however the bindings, bolts, joist hangers, and other nonaluminum attachments associated with the aluminum may need to be treated if rust-colored breakthrough is observed. The breakthrough will look like an orange-brown stain that shows through the encapsulant coating.

16.1.2 Coated Metal Surfaces

This cleaning sequence is for metal that has a coating.

- Mist the suspect area with detergent decontamination solution.
- Sequentially wash the surface with detergent decontamination solution.
- Wash (swab or wet wipe) any newly exposed surfaces with detergent decontamination solution.

16.1.3 Coated Metal Surfaces: Mold Present

The cleaning method discussed above should be used, however the decontamination solution must contain a biocide.

Apply shellac-based sealant or equivalent to the exposed surfaces after cleaning. Allow the surfaces and coatings to dry. Wipe down the metal with clean dry rags as needed prior to air-drying.

16.1.4 Uncoated Surfaces and Metal Wiring

Uncoated metal surfaces must be evaluated on a case-by-case basis to determine the cleaning methods that will cause the least damage to the metal surfaces while assuring that mold contamination is removed.

Metal wiring is considered an uncoated metal surface.

- Metal wiring and metal-wiring housings must not be wet cleaned.
- Prior to any cleaning event, the electrical energy source must be deenergized and locked out. Questions as to electrical safety should always be brought to the attention of licensed electricians. *Electrical safety always takes precedence over the risk associated with any surface contamination.*
  - Dry vacuuming should be used to remove any debris from surfaces.
  - Misting of detergent decontamination solution in the general vicinity of unenergized metal wiring may be an acceptable procedure. The misting must not render the wiring wet (coated with moisture). Care must be taken to protect wiring.

16.1.5 Uncoated Surfaces and Metal Wiring: Mold Present

Cleaning should proceed as described above.

- For any wet cleaning, the decontamination solution must include a biocide.
- Cleaning may include dry vacuuming with biocide contained in the vacuum canister that receives the dust and debris from these surfaces.

16.2 PAINTED SURFACES

16.2.1 Quick Facts

Paint is either water based (latex or acrylic), oil based (oil), polyurethane, or epoxy. Pigments are suspended or dissolved in the paint bases to provide paint colors.

Material safety data sheets and design specifications offer valuable information as to the types of paints used within a facility.

16.2.2 Lead and Cadmium

Some paints may contain lead or cadmium pigments. Cadmium is usually associated with bright yellow pigmented paints. Both of these heavy metals, if present, only become a danger if either inhaled or ingested. Consequently, cleaning methods that do not generate dusts or paint chips can be used on lead- and/or cadmium-containing paints. Cleaning methods that remove these paints would be considered abatement.

Lead abatement is regulated in some states, and state regulations need to be consulted as to when lead-based paint removal becomes a state-regulated process. If lead-based paint is suspected and if the paint must be removed, consult an environmental expert as to removal techniques. The removal of lead-based paint is beyond the scope of this book and will not be discussed further herein.

16.2.3 Urethanes

Urethane paints may be either water based or oil based. The term *polyurethane* is used to mean a mixture of urethanes. Such paints may also be described as stain varnishes and varnishes.

16.2.4 Isocyanates

Isocyanate-bearing paints and sealants are special types of paint and in general are much more toxic than other paints. Isocyanate paints are used to coat metal surfaces and if suspected, cleaning and maintenance of the surfaces coated with these paints requires special procedures and is beyond the scope of this book.

16.2.5 Exterior versus Interior

Paints are also identified as either exterior or interior. This definition relates to whether the paints will adequately withstand the demands of either an exterior or interior application. In some cases, the paints may also have varying levels of toxicity upon application. This toxicity, even after curing, may determine whether the paint can be used in an interior location.

16.2.6 Curing

Curing is the process of changing the paint from the initial liquid coating to a solid paint surface. For water-based paints, the water essentially evaporates and leaves behind the latex or acrylic paint pigments. For oil-based paints, the oily solvent, which is a petroleum-derived chemical, volatilizes and leaves behind the paint pigments. Epoxy paints cure in a very different fashion; essentially the two components of the epoxy are mixed and thermal curing takes place. Thermal curing forms strong chemical bonds between the epoxy and the substrate on which the paint is applied.

All curing processes emit chemicals. While water-based paints are generally described as less toxic, the additives used (especially for exterior water-based paints) may render them toxic. These additives often serve to reduce the potential for the water-based paints to be *food* for molds. The terms *mold inhibitor* and *fungicide* are used to describe these paint additives.

After paint is cured, the majority of the paint coating is comprised of the cured pigments that adhere to the surface. This book will use the term *paint* to mean this paint coating.

16.2.7 Gloss, Semigloss, and Flat

Paint is formulated to cure with a variety of finishes. In addition, texture can also be added by placing sand, drywall joint compounds, cellulose, and other such materials in paints. Depending upon the curing process and the desired finish, paints may be more or less able to withstand abrasion and aging. Some paints are formulated to age. For example, flat paint is made to continually slough off thus maintaining a flat finish. This process also means that dirt affixed to the paint also sloughs off. In contrast semigloss and gloss finishes are intended to form a more solid and unchanging surface than flat paint.

16.2.8 Substrates with Metal Flaking

Peeling paint may also pick up the substrate material. For example, paint on metal may contain metal flakes, particularly if the metal has rusted. These metal flakes are usually mostly iron and do not significantly increase the toxicity levels. These flakes may, however, constitute a cutting hazard and if large enough may pierce the skin surface.

16.2.9 Paint

The decontamination solution must be chosen so that contaminants can be removed without removing the paint. Cleaning of painted surfaces is really the process of selecting a chemical for cleaning that will not resuspend or dissolve the pigment. The phrase "like dissolves like" should be remembered. Essentially a decontamination solution that is too like the original paint solution (prior to curing) will be more likely to remove the paint as well as any contamination on the paint.

In addition to concerns as to paint solubility, temperature and abrasion must be evaluated in painted surface cleaning methods.

16.2.10   Paint: Mold Present

Because paint contains many different chemicals and is applied in layers, some paints may provide food or a hiding place for mold. Painted surfaces may have visible mold in one area or throughout the paint surface.

If the painted surface has a dispersed *measles* pattern, air dispersal of mold spores should be suspected. Usually a measles pattern is associated with widespread dispersal of water or oil within a given area. The measles pattern looks like dots over a wide surface area. The dots are often black or brown, although other colors may be present. If this type of pattern and air dispersal is discovered, a professional mold expert should be consulted.

- Painted concrete surfaces with a pattern of mold growth that looks like dots spread over a large surface indicates that mold is moving through the air in a viable state.
- If mold appears to be imbedded in the paint surface or spread throughout a surface area from baseboard to ceiling level, mold abatement may be beyond that discussed in this book.
- If these mold-contaminated surfaces are in excess of 30 square feet, the concrete cleaning must occur using a site-specific plan.

For other more limited mold growth on painted surfaces, decontamination includes the following:

- Mist the suspect area with biocide decontamination solution.
- Sequentially wash the surface with biocide decontamination solution.
- Wash (swab or wet wipe) any newly exposed mold growth surfaces with biocide decontamination solution. Note: A mold growth surface is one that has visibly been determined to be a site where mold is growing.
- Assess whether paint delamination or peeling has occurred. Determine if this delamination was caused by condensate layer formation. Determine if moisture can be controlled and institute those measures.
- Remove delaminated or peeling paint using dust control methods. Dust control can usually be achieved by removing the paint using wet methods such as scrubbing with a wet rag.
- Apply shellac-based sealant or equivalent to the exposed surfaces. Allow the surfaces and coating to air-dry. Wipe down the metal with clean dry rags as needed.

16.2.11   Painted Concrete

If the concrete was covered with a paint sealer, this paint sealer surface may be compromised by the presence of contamination. Painted concrete surfaces will actually be addressed more as paint than as concrete. However, the concrete substrate can continually provide moisture, and thus the drying status of the concrete continues to be an issue.

Etching chemicals (muriatic acid) are not considered sealants.

Painted concrete is a special circumstance because of the porosity and small cavities present on the concrete surfaces. Painted concrete surfaces often have cracks, fissures, and small circular cavities. These areas can be very difficult to clean if contamination has occurred.

If the concrete surface is smooth, detergent decontamination solution can be applied as either a liquid or a paste. Wet wiping or mopping of this solution can be accomplished with concurrent application of detergent decontamination solution applied using a low-pressure misting device.

If the painted surface begins to delaminate or dissolve in the decontamination solution, less abrasive methods may be needed and more sequential washings.

16.2.12   Painted Concrete: Mold Present

If the concrete was covered with a paint sealer, this paint sealer surface may be compromised by the presence of mold.

If the concrete surface is smooth, biocide decontamination solution can be applied as either a liquid or a paste. Wet wiping or mopping of this solution can be accomplished with concurrent application of biocide decontamination solution applied using a low-pressure misting device.

Painted concrete surfaces often have cracks, fissures, and small circular cavities. These areas can be very difficult to clean if mold contamination has occurred. In order to increase the dwell time for biocides, a paste application must often be used. The biocide paste should be formed by combining Clorox 2® and water. After applying the paste, cheesecloth can be taped over the area.

- If fissures are deep, a liquid solution of Clorox 2 may be used to penetrate these fissures.
- After the decontamination solution or paste has remained in place for one hour, the area can be mechanically washed. All visible mold should be removed or the process must be repeated.

If the painted surface begins to delaminate or dissolve in the decontamination solution, less abrasive methods may be needed and more sequential washings.

16.3   WOOD

16.3.1   Quick Facts

Wood includes hardwoods, softwoods, and compressed woods. Compressed woods include woods made from:

- Chips: oriented strand board (OSB), oaktag
- Sheets: plywood

Wood may also be present as laminates or veneers. Laminates have a sealed surface affixed to a compressed wood surface. Veneers are very thin woods glued to other wood surfaces.

Wood studs, flooring material, ceiling material, wall paneling, and furnishings contaminated with mold growth should be evaluated by an expert prior to attempting abatement of the mold.

16.3.2 Efflorescence

Wood, upon water intrusion, sometimes exhibits efflorescence. Efflorescence looks like white or grayish crystalline material oozing from wood. Efflorescence is caused by minerals leaching through the wood and may be mistaken for mold because of its color and association with known water intrusion. On the other hand, efflorescence may in fact be associated with mold growth.

16.3.3 Intrusion

Certain molds may intrude (grow into) past the wood surface. Despite repeated cleaning efforts, these molds may grow back. These molds may appear to be gone and may no longer be emitting spores to the air, however their vegetative structures remain viable and the mold will often come back.

Unsealed wood surfaces near mold growth are suspect for hidden mold growth and must be evaluated on a case-by-case basis. This evaluation is beyond the scope of this book.

16.3.4 Sealants

Wood surfaces may be sealed with stain, varnishes, paints, and other chemicals that make the wood surface less porous. Cleaning of painted wood surfaces is essentially cleaning of the sealant layer.

16.3.5 Cleaning

Wood cleaning is conducted on two types of wood surfaces:

- Sealed: covered with varnish, stain and sealer, paint
- Unsealed: unfinished wood

Cleaning of sealed wood surfaces usually involves a combination of waxing and polishing. The MSDS for the various products must be consulted as to personal protective equipment use. Resurface the areas where sealants have been removed with new sealants.

Cleaning of unsealed wood is more difficult since the unsealed wood will retain the decontamination solution; thus excess solution if used may cause the wood to warp. Cleaning should proceed with a minimum of wetting, followed by air-drying.

16.3.6 Cleaning: Wood near Mold-Contaminated Area

- This cleaning sequence is for wood that is near (juxtapositioned to or within 6 inches of) mold growth areas, but does not show signs of mold growth on its surface.
- Mist the suspect area with biocide decontamination solution.
- Sequentially wash the wood surface with biocide decontamination solution. Wash only the area where the wood has been exposed by the removal of other building materials (exposed remaining surfaces).
- Apply shellac-based sealant or equivalent to the exposed surfaces. Allow the surfaces and coating to air-dry.

16.3.7 Cleaning Prior to Removal: Mold Present

This removal sequence is required if the wood shows signs of visible mold growth.

- Mist the suspect wood area with biocide decontamination solution.
- Wash (swab or wet wipe) any newly exposed mold growth surfaces with biocide decontamination solution.
- If washing does not remove the mold growth surface:
  - Sequentially remove the wood by cutting above and to the side of any mold or former mold growth areas. Removal must occur with repetitive misting of the surrounding nearby environs with biocide decontamination solution.
  - If removal is not an option, some cleaning organizations recommend sanding of wood surfaces to remove contamination. This type of cleaning (mechanical abrasion) must only be done within negative air pressure containments and is beyond the scope of this book.
- Apply shellac-based sealant or equivalent to the exposed surfaces. Allow the surfaces and coating to air-dry.

16.4 CONCRETE

16.4.1 Quick Facts

Concrete is a mixture of cement and various additives. In some cases, aggregate is also added to change the surface features of the concrete. Even cured concrete remains porous. Surface coatings and treatments alter just the surface of the concrete.

Concrete grout used between ceramic tiles is also included as "concrete" in the following discussion. The ceramic tile surface itself is glazed and is an impermeable surface similar to marble and terrazzo. These impermeable surfaces do not readily absorb water and unless cracked do not constitute potential mold growth surfaces. If water intrusion, cracking, or mold growth is present due to structural damage or overlays of food materials (grease, oils), the formerly impermeable surfaces can be considered to have been rendered permeable. The concrete cleaning methods discussed herein can be used for these surfaces.

One of the difficulties in cleaning concrete is the potential to change the concrete moisture level or in some other way damage the concrete surface.

Since concrete can transmit and hold water, it can be a building material where mold grows. Usually, however, a food source is also needed, such as mineral oil from refrigerant systems sprayed on the concrete or paint. Yes, paint is a food source for molds.

16.4.2 Asbestos Concrete

One of the additives to concrete in the past was asbestos. The resulting concrete is often called *transite*. If transite is mechanically abraded (scratched) through aggressive washing or fracturing, the transite may become friable asbestos. Friable asbestos and the abatement of this asbestos must be handled by trained asbestos workers and is beyond the scope of this standard operating procedure.

Consequently, one of the first decisions that must be made is whether concrete surfaces have been adequately characterized as to asbestos content. If the concrete is suspected of containing asbestos, asbestos sampling and asbestos abatement measures may be required.

16.4.3 Concrete and Liquid Intrusion

Moisture vapor emission at each temperature and pressure gradient (various levels) from concrete slabs and coverings is measured in *pounds per square inch*. Depending on the porosity of the concrete and the type of covering that has been placed over the concrete, a certain amount of moisture will pass through the concrete and be trapped either in the concrete layer itself or directly beneath covering materials.

If volatile organic compounds (VOCs) are present in the concrete substrate layers, these chemicals will also move upward, especially if the concrete covering is at a higher temperature than the substrate. Examples of these VOCs are uncured mastics or any oil-like material that has moved into the concrete layers.

16.4.4 Concrete Permeability

Permeability of the concrete is a direct function of porosity, which is a direct linear function of the water-to-cement ratio in the original concrete mix. With increases in the water-to-cement ratio (meaning the water component increases), porosity and, thus, permeability increase exponentially.

16.4.5 Concrete Curing

As the concrete is mixed and readied for use, drying is used to remove excess water not needed to hydrate the cement mix into a *glue*. During curing as the concrete is *aged* in place after pouring, the concrete's gluelike cement mix undergoes a chemical reaction that turns the concrete and any aggregate into an agglomerate rock. A moist curing process is often used.

If the concrete is not adequately cured prior to the installation of other coatings or adhesives, these coatings will also move into the concrete surfaces and will be subject to the concrete permeability factors noted above.

The internal alkaline state of the concrete is also a factor. When the concrete surface has an alkalinity over 9 on a pH scale, adhesive and bonding systems are compromised. If high-alkaline concrete is used and moisture intrusion or buildup occurs, the resulting basic concrete surface will cause additional adhesion problems. This type of event occurs in laundry rooms where basic detergents and liquid spillage may occur.

16.4.6 Unsealed Concrete

Concrete that has been subject to water intrusion must first be assessed as to other soilage that may have occurred during the water intrusion. Unsealed concrete will absorb liquids, and these liquids may transfer across the concrete's thickness.

For concrete drying, the water conductivity of the concrete must be less than the surrounding surfaces. If this is not the case, the concrete will pull the water toward itself and will not adequately dry, even though the surface coatings of the concrete may look dry. Establishing that the concrete is dry may require repeated observations of the concrete and delamination or removal of any surface coatings may be required.

Products can be used on unsealed concrete to prevent future moisture intrusion across a concrete surface. A preferable product sinks into the concrete and forms a crystalline matrix with the concrete, thereby sealing the concrete's pores. However, these products will not penetrate throughout the concrete's thickness in most cases, and only the surface can be considered sealed. Water that intrudes toward the sealed surface will not penetrate; however the water may build up to unacceptable levels behind the sealed surface (and be invisible to the observer).

Clean the exposed concrete by:

- Wet wiping with detergent decontamination solution, or
- Low-pressure spraying of detergent decontamination solution.

Apply the solution so as to wet the concrete surface and all concrete gaps, fissures, grooves, and holes.

- Remove all visible debris and soilage through the washing event.
- Dry the concrete.

16.4.7 Unsealed Concrete: Mold Present

If the concrete surface is smooth, biocide decontamination solution can be applied as either a liquid or a paste. Wet wiping or mopping of this solution can be accomplished with concurrent application of biocide decontamination solution applied using a low pressure-misting device.

Concrete surfaces often have cracks, fissures, and small circular cavities. These areas can be very difficult to clean if mold contamination has occurred. In order to increase the dwell time for biocides, a paste application must often be used. The biocide paste should be formed by combining Clorox 2 and water. After applying the paste, cheesecloth can be taped over the area.

- If fissures are deep, a liquid solution of Clorox 2 may be used to penetrate these fissures.
- After the decontamination solution or paste has remained in place for one hour, the area can be mechanically washed. All visible mold should be removed, or the process must be repeated.

If the surface begins to delaminate or dissolve in the decontamination solution, less abrasive methods may be needed and more sequential washings.

16.4.8 Painted or Sealed Concrete

If the concrete was covered with a paint sealer, this paint sealer surface may be compromised by the presence of contamination. Painted concrete surfaces will actually be addressed more as paint than as concrete. However, the concrete substrate can continually provide moisture, and thus the drying status of the concrete continues to be an issue.

Etching chemicals (muriatic acid) are not considered sealants.

Painted concrete is a special circumstance because of the porosity and small cavities present on the concrete surfaces. Painted concrete surfaces often have cracks, fissures, and small circular cavities. These areas can be very difficult to clean if contamination has occurred.

If the concrete surface is smooth, detergent decontamination solution can be applied as either a liquid or a paste. Wet wiping or mopping of this solution can be accomplished with concurrent application of detergent decontamination solution applied using a low pressure misting device.

If the painted surface begins to delaminate or dissolve in the decontamination solution, less abrasive methods may be needed and more sequential washings.

16.4.9 Painted or Sealed Concrete: Mold Present

If the concrete was covered with a paint sealer, this paint sealer surface may be compromised by the presence of mold.

If the concrete surface is smooth, biocide decontamination solution can be applied as either a liquid or a paste. Wet wiping or mopping of this solution can be accomplished with concurrent application of biocide decontamination solution applied using a low pressure-misting device.

Painted concrete surfaces often have cracks, fissures, and small circular cavities. These areas can be very difficult to clean if mold contamination has occurred. In order to increase the dwell time for biocides, a paste application must often be used. The biocide paste should be formed by combining Clorox 2 and water. After applying the paste, cheesecloth can be taped over the area.

- If fissures are deep, a liquid solution of Clorox 2 may be used to penetrate these fissures.
- After the decontamination solution or paste has remained in place for one hour, the area can be mechanically washed. All visible mold should be removed, or the process must be repeated.

If the painted surface begins to delaminate or dissolve in the decontamination solution, less abrasive methods may be needed and more sequential washings.

16.4.10   Follow-up Concrete Sealing

All concrete where visible mold was present will be sealed with shellac-based sealant or equivalent. Surrounding surfaces may also be sealed with the shellac-based sealant or equivalent. All surfaces should be allowed to air-dry.

16.5   FLOOR TILE AND SHEET

16.5.1   Quick Facts

This chapter discusses floor tiles and sheet vinyl made of semiporous, flexible vinyl or linoleum.

Vinyl asbestos tile (VAT), asbestos-lined sheet surfacing (linoleum or vinyl), and flooring secured with asbestos-laden mastic constitute asbestos-containing materials. Removal and maintenance of these materials may require asbestos protocols and is beyond the scope of this chapter.

The condition of wood and concrete underlayment of the subfloor must be evaluated. The subfloor if wet, must be dried. If visible mold is discovered, the subfloor surface must be decontaminated.

16.5.2   Floor Tiles

Floor tiles are placed on their substrate surfaces with cracks between each tile. These tiles have a hard top surface and this top surface contains the largest percentage of vinyl. The lower layers contain lesser amounts of vinyl and increasingly more filler/backing material. Floor tile thickness varies, as does the resistance of the tile layers to delamination.

16.5.3 Sheet Surfacing

Sheet surfacing materials contain layers similar to those described for floor tiles. In addition, some older surfacing materials contain linoleum rather than vinyl or mixtures of both. Linoleum is a petroleum-based product like vinyl, however linoleum does not contain plastics. As such linoleum is generally less pliable than sheet vinyl.

Both sheet vinyl and linoleum have backing materials. The newer products have backing material that does not readily retain moisture. Older products, however, may be backed with felt, papers, horsehair, or mixtures of these porous materials.

If the underlayment consists of horsehair padding, the horsehair padding cannot be saved. The subsequent removal of the horsehair pad and the overlaying carpet must be handled using a site-specific plan rather than the guidance in this chapter. (Horsehair pad is a known repository for pathogenic encysted bacteria.)

If the underlayment is felt or felt and paper, the sheet surfacing material and this underlayment must be evaluated to determine if asbestos is present. These materials were generally produced in a time when asbestos fibers were used to stabilize such materials. In addition, these materials would have absorbed any asbestos-containing mastic used in their installation.

16.5.4 Cleaning

Tiles and floor surfacing is cleaned using detergent decontamination solution. Follow-on waxing or acrylic coating may be required. Some newer products combine a detergent with acrylic coating and can be used in a one-step cleaning process.

16.5.5 Removal: Water Intrusion

Flooring materials that do not currently have mold growth, but that have been subject to water intrusion may be removed using normal maintenance procedures. These procedures include:

- Removal of surfacing materials
- Bagging (using standard garbage bags) of the tiles for disposal

If the flooring material is backed with porous backing (felt, paper) and if this paper has been wetted, drying must occur within a 24-hour period. Usually air-drying with assist (fans, dehumidifiers) is used. If the flooring remains wet after this 24-hour period, removal will be required.

16.5.6   Cleaning and Removal: Mold Present

If the contaminated area exceeds 30 square feet, the removal must occur using a site-specific plan. This type of removal is beyond the scope of this chapter and will not be discussed further herein.

If visible mold staining or growth is observed and the area contaminated with mold is less than 30 contiguous square feet, the following protocols can be used:

- Mist the suspect area with biocide decontamination solution.
- Sequentially remove any delaminated flooring material. Removal must occur with repetitive misting of the surrounding nearby environs with biocide decontamination solution.
- Wash (swab and wet wipe) any newly exposed mold growth surfaces with biocide decontamination solution. Note: A mold growth surface is one that has visibly been determined to be a site where mold is growing.
- Apply shellac-based sealant or equivalent to the exposed remaining surfaces if removal of visible mold was required on these surfaces. Allow surfaces to air-dry.

16.6   CARPET

16.6.1   Quick Facts

Steam cleaning should only be conducted with the concurrent application of carpet cleaning chemicals with biocide properties. Steam cleaning does not generate sufficient heat at all application areas to be an effective biocide. Carpets must be totally dried as quickly as possible.

The process of steam cleaning reaerosolizes materials previously trapped in carpets. While these aerosolized materials remain in the air, individuals may be able to *breathe them in*. Consequently, occupying (for habitation purposes) of the carpeted areas should be delayed for 4 to 6 hours after the carpet has been determined to be dry by touch. Small children in particular must not be allowed to crawl or walk across wet or newly dried carpet.

Carpet wetted with unsanitary waters through sewer backup, floods, or spillage may be very difficult to clean. Mold spores and bacterial endospores will be activated by moisture and fed by the deposition of organic residues.

- Cleaning of substrate and padding may be even more difficult. Steam from steam cleaning of these carpets is more dangerous than steam cleaning of carpets soiled through intended usage.
- The usual recommendation is to dispose of all carpet that has remained wet for more than 24 hours or has been soiled with sewage or other biocontaminated materials. This disposal must be done in a controlled manner so as not to spread the biological contamination into the airstream.
- Site-specific planning is required when areas greater than 2 square feet are contaminated with unsanitary waters. This type of carpet removal is beyond the scope of this chapter and will not be discussed further herein.

The installation or reinstallation of carpet into areas where biocontamination is likely to reoccur is *not* advisable. This essentially means that carpet should not be reinstalled in bathroom areas near toilets; an alternate nonporous flooring should be used.

16.6.2 Drying after Water Intrusion: No Mold and No Sewage Present

Drying of carpet, porous underlayment, and subfloor and flooring materials must occur within 24 hours of the wetting event. Drying sequences will vary with the different carpets and carpet underlayment used. In general the drying sequence will include using wet and dry vacuums followed by or concurrent with air exchanges. Air exchanges include replacing wet moist air with heated air or regular (cooling) fans to move the air across surfaces. Dehumidifiers may also be used.

- Moisture removal should proceed until the carpet is only moist to the touch.
- Biocide decontamination solution should be applied to all surfaces using misting or low-pressure spraying techniques. Note: This application is as a preventative measure (not an abatement measure) to lessen the potential for mold growth as drying proceeds.
- Moisture removal may be complicated by carpet padding and other porous underlayment materials. If the porous pad or underlayment are wet, the carpet may need to be folded back to allow access to the underlayment for wet and dry vacuuming.
- If the pad or underlayment consists of horsehair padding, the horsehair padding cannot be saved. The subsequent removal of the horsehair pad and the overlaying carpet must be handled using a site-specific plan. (Horsehair pad is a known repository for pathogenic encysted bacteria.)

The condition of wood and concrete underlayment of the subfloor must be evaluated. The subfloor if wet, must be dried. This drying can be accomplished by the use of wet and dry vacuums, wiping with absorbent materials, and air exchanges over the floor area. Air exchanges may include the use of fans or dehumidifiers. If visible mold is discovered, the subfloor surface must be decontaminated.

16.6.3 Treatment and Removal: Mold Present

This cleaning sequence is for carpet that has been compromised by mold growth either by being in the area where mold is growing (within 2 feet) or by actual mold growth on the carpet surface. Carpet by definition includes both the carpet and any porous underlayment (pad).

- Soak the carpet boundary (six inches around the mold contamination) with biocide decontamination solution.
- Mist the suspect carpet area with biocide decontamination solution.

- Sequentially apply (using low-pressure spraying or pouring) biocide decontamination solution into the carpet fibers. Soak the carpet and the carpet pad with biocide decontamination solution.
- Do not make cuts into the contaminated areas of the carpet. Cut in "clean" carpet areas where visible mold is not evident.
- Remove all the wet carpet and pad by sequentially rolling the carpet and pad toward the disposal bag. The carpet must be removed wet. Note: The carpet may have been made wet by water intrusion or the application of biocide.
- Dry the remaining carpet and substrate as described above (See Section 16.6.2, "Drying after Water Intrusion: No Mold and No Sewage Present").

If cutting must occur *within* the contaminated carpet surface boundaries and these boundaries cannot be soaked with biocide, or if the carpet area exceeds 30 square feet, the carpet removal must occur using a site-specific plan. This type of carpet removal is beyond the scope of this SOP and will not be discussed further herein.

16.7   DRYWALL: SHEETROCK, WALLBOARD, GYPBOARD

16.7.1   Quick Facts

Drywall is formed from gypsum and paper. Gypsum is a rock that varies in color due to impurities. In its pure form or after processing, gypsum is white. Chemically, gypsum is hydrous calcium sulfate ($CaSO_4.2H_2O$). Note that even naturally occurring gypsum contains water (H_2O).

Gypsum boards are formed by sandwiching a core of wet gypsum plaster between two sheets of heavy paper. When the gypsum core sets and is dried, the sandwich becomes a strong, rigid, fire-resistant building material. The gypsum boards are known by various names in the construction industry including:

- Sheetrock
- Drywall
- Wallboard
- Gypboard
- Gypsum wallboard and lath
- Prefinished wallboard
- Gypsum sheathing

16.7.2   Green Board

Green board is designed as an indoor tile backer board, and can be used as sheathing. Green board has a tapered edge while sheathing is a square-edged product. Green board has a moisture-resistant paper and a treated gypsum core. The core is treated with moisture repellents, such as the asphalt and wax emulsions.

These moisture repellants are lubricants that cause the gypsum crystals to slip easily. This slippage may result in sagging of green board applied to a ceiling.

- Due to this sagging potential, ceiling applications of green board need 12 in on center (o.c.) support. Note: *On center* means that the distance from the center of one supporting board to the next supporting board's center does not exceed 12 inches.
- If supports are 16 in o.c., 5/8 in green board should be used. Note: *On center* means that the distance from the center of one supporting board to the next supporting board's center does not exceed 16 inches.

If improper installation has occurred, green board applied to a ceiling can sag, even if additional water sources have not intruded into the ceiling area. Thus, just the appearance of sagging ceiling drywall may not mean that water intrusion has occurred.

Regular green board or Moisture-Guard® (green board) may have been used in high-humidity areas. In some cases, however, regular drywall may also be found in these areas.

For any new installation of drywall in high-humidity areas (either as the result of renovations or mold decontamination), cement board must be used. Such high-humidity areas include: toilet stalls, sink backsplash areas, tub and shower surrounds, laundry room walls behind and immediately to the side of washers, walls constructed around sumps, walls constructed around garbage compactors or chutes, and vertical chutes that enclose water-carrying piping (either potable, gray water, or sanitary carrier waters).

16.7.3   Fire-Resistant Drywall

Fire-resistant drywall is fire resistant because in its natural state, gypsum contains water, and when exposed to heat or flame, this water is released as steam, retarding heat transfer. This water can also become available when, in addition to other water sources (water intrusion from foundations, pipe leakage, etc.), the gypsum board becomes wetted.

The American Society for Testing and Materials (ASTM) standard ASTM C 36 designates two types of gypsum wallboard:

1. Regular gypsum wallboard
2. Type X wallboard, which is typically required to achieve fire resistance ratings, is formulated by adding noncombustible fibers to the gypsum.

These fibers help maintain the integrity of the core as shrinkage occurs providing greater resistance to heat transfer during fire exposure.

By ASTM definition, type X gypsum wallboard must provide:

- Not less than a one-hour fire resistance rating for 5/8-in board, or
- A ¾-hour fire resistance rating for ½-in board applied in a single layer, nailed on each face of load-bearing wood framing members, when tested in accordance with the requirements of ASTM E 119, Methods of Fire Test of Building Constructions and Materials.

Additionally, the Gypsum Association requires ½-in type X gypsum board to achieve a one-hour fire resistance rating when applied to a floor–ceiling system, as described by GA File Number FAN COIL 5410, in GA 600, the *Gypsum Association Fire Resistance Design Manual*.

16.7.4   High-Impact Drywall and Wallboard

High-impact wallboard is a specially designed product consisting of 5/8-in fire-shield type X wallboard with a stronger core and face paper backed with Lexan polycarbonate film manufactured by GE Plastics.

16.7.5   Drywall and Moisture Intrusion

If drywall becomes wet either before or after installation, the drywall may become a growth area for mold. The moisture source may be a leak or the formation of a condensate layer. Condensate layers form when temperature differences cause water to "rain" out onto surfaces. Dew on grass after a cool night is an example of a condensate layer.

- When drywall retains additional moisture over time, the drywall may begin to become structurally compromised and increasingly pliable. Drywall surfaces will appear to bulge as the drywall swells.
- Hidden surfaces of drywall may be subject to moisture intrusion and may become mold growth sites. This situation may occur long before visual evidence on the drywall surfaces that face occupied areas is evident.

Any time moisture intrusion or mold growth on drywall is discovered, the first question must be: Where is the moisture source and can that source be eliminated? If the moisture source cannot be eliminated, drywall may need to be removed and replaced with a more appropriate building material.

The practice of drilling holes in drywall to dry the wall has not been shown to be effective in most cases. This procedure may be a temporary method to limit mold growth, however *in all cases where drywall has retained moisture for more than 24 hours, the drywall should be removed*.

16.7.6   Drywall: Water Intrusion

Visibly examine the drywall, if water intrusion has caused the drywall to be damaged (but no mold growth is visible), remove all the water-damaged drywall if not dried within 24 hours.

- Removal the drywall by cutting around the water intrusion area. Continue examining the newly exposed areas for mold, and if mold is discovered, move to Section 16.7.9, "Drywall Removal: Mold Present."

- Use dust-suppression methods (i.e., misting of the air with water, vacuuming, or slow removal) to control dust during the drywall removal.
- Bag waste (using standard garbage bags) for disposal.

Removal of wet drywall must occur within 24 hours; alternately, drywall may be dried. Any drywall that remains in place after a drying action has occurred must be monitored to ensure that mold does not develop. This monitoring must occur for at least 3 weeks and should consist of visual examination of the drywall. The use of borescopes to look behind the drywall face may be necessary in some cases (determined on a site-specific basis).

Drywall may have water intrusion on just one plane. This means that the drywall is not soaked with all surfaces from front to back of the drywall sheet wetted. Soaking implies that the drywall has lost structural integrity and is extremely pliable to the touch. Drywall that is soaked to this extent cannot be cost-effectively saved.

If the drywall remains intact and rigid, and only one surface has been wetted (i.e., the front of the drywall is wet and not the back, or vice versa), the drywall can often be dried. Drying requires either the use of desiccants or air movement.

16.7.7   Chemical Desiccants

Various desiccants (drying agents) are marketed that essentially chemically react with water and thus *pull* the water chemically away from the drywall. Most of these chemicals are also called *hydrophilic* which means *water loving*. Unfortunately, these products are very dangerous because skin contains water and the removal of this water on skin causes burns. Less caustic desiccants usually contain filler materials so that the concentration of desiccant is less per unit areas where application occurs. Decisions as to desiccant use must be made on a site-specific basis.

- The most common solid desiccants used are a form of lye (sodium hydroxide). This product is usually applied by dispersing the lye over a surface area and allowing the lye to remain in place for several days. This treatment is common in crawl space locations where soils need to be dried. The area where the lye is used must be restricted for entry only by trained personnel. Training must include hazard communication information as required by 29 CFR 1910.1200. Examples of solid desiccant products include sodium hydroxide– and potassium hydroxide–based product lines. Various formulations of silica dioxide, lava rock, and clay compounds are also used. These products are usually hung in bags or dispersed over limited surfaces.
- Alcohol can also be considered a desiccant in that the alcohol causes drying through evaporation to proceed more quickly. Alcohol is present in many decontamination solutions. These solutions may contain sufficient alcohol to render the immediate atmosphere where use is occurring flammable. As such, alcohol-based solutions if used must be applied taking into account the flammability hazard as well as the toxic hazard to humans. Examples of alcohol-based products include Alconox™ and LiquiNox™. These products, if used as directed, provide the desiccant properties of alcohol without the flammability hazard. Dawn® dishwashing liquid also provides alcohol in solution.

16.7.8 Dehumidification

Air-movement devices use air movement without assist, refrigeration dehumidification, and chemical hygroscopic dehumidification. Air movement without assist essentially removes moisture by displacing the moisture-bearing air with drier air. Fans are an example of an air-movement device. dehumidification devices may be used with or without fans.

- *Refrigeration-based dehumidifiers* create a cold surface onto which excess moisture condenses out. Refrigeration dehumidifiers are used to control humidity down to 40% Rh and at temperatures down to 3°C. These dehumidifiers may also be used for room air conditioning, however their use for mold avoidance or abatement situations is limited if their primary usage is for room air conditioning. If these dehumidifiers are already available within a building area, ramping up their air intake can help in the dehumidification of room areas.
- Portable units may be needed in addition to any other air conditioning system elements already installed. The product lines chosen should cover areas up to 800 to 1,100 square feet. However, their efficiency may be less when the larger surface areas are water impacted and require treatment. Industrial dehumidifiers that cover 500 to 1,500 square feet may also be used.
- *Adsorption/Desiccant dehumidifiers* use hygroscopic media to adsorb moisture. Adsorption dehumidifiers control humidity levels below 40% and are ideal for low-temperature environments. The dehumidifier manufacturer must be consulted as to the type or hygroscopic media required and the replenish rate required for effective use. These types of dehumidifiers are usually not readily available for maintenance use aimed at air-drying structural materials in large air spaces.

16.7.9 Drywall Removal: Mold Present

Drywall removal proceeds with these primary goals:

- Limiting the viable mold spores that get into the air
- Removing the drywall and all other readily removed building materials that are next to the contaminated drywall (i.e., insulation, wallpaper)
- Treating or removing stud work that is next to the drywall

Drywall removal requires that only the boundary around the contaminated drywall be the area where any cutting occurs. If cutting must occur *within* the contaminated drywall surface boundaries, and if these contaminated surfaces where such cutting must occur are in excess of 30 square feet, the drywall removal must occur using a site-specific plan. This type of drywall removal is beyond the scope of these standard operating procedures and will not be discussed further herein.

If visible mold growth is observed on the drywall:

- Mist or swab the drywall surfaces with biocide decontamination solution.
- Cut around the contaminated drywall while misting the area with biocide decontamination solution.

- Carefully and slowly remove the drywall to control dust generation. If dust is observed, further misting or covering of the drywall removal area with plastic sheeting may be needed.
- Remove any insulation found behind the drywall.
- Inspect the remaining drywall surfaces, including the drywall above, below, and to the side of the drywall removal area. Also inspect the drywall that is now exposed and constitutes the *other* side of the wall. Depending on the penetration of the mold layer the following drywall will be removed:
  - Drywall that faces the occupied space (room)
  - Drywall that faces the occupied space (room) and drywall on the other side of the space
- All porous insulation, debris, and other materials contained within the mold-contaminated drywall space will also be misted with biocide decontamination solution and removed. Immediately bag the material removed for disposal.
- Apply shellac-based sealant or equivalent to the remaining exposed vertical face of the drywalls. This is the drywall that is not removed due to lack of either water damage or visible mold growth, and is directly in back of or above the drywall that has been removed.

16.7.10   Wood or Metal Studs

Refer to the sections on *Wood* SOPs for treatment of wood studwork. Refer to the section on *Metal* SOPs for treatment of metal studwork.

16.8   WALL COVERINGS

16.8.1   Quick Facts

Wall coverings can be made of a variety of materials in addition to paper. Paper is a cellulose product and as such is food for mold. If wallpaper is made by covering the paper with plastic (vinyl wallpaper), metallic coverings, or flocking, additional protocols may be required. These protocols include:

Removal of wallpaper from the substrate and cleaning of any back surfaces; vinyl- and plastic-coated wallpaper cannot be cleaned from the front surface inward.

Special products may be required when metallic or flocked wallpaper is being treated. The wallpaper manufacturer should be consulted if the wallpaper is to be saved, as the decontamination solution will be specific to the types of wallpaper being treated.

Other wall coverings may be generically called *wallpaper* and include fabrics or in some cases vinyl without a paper backing.

Depending on the glues and overcoats used (paints, shellacs, polyurethanes), the wall coverings may be quite different from the original products purchased as wallpaper once they are installed. The substrate (drywall, plaster, wood paneling) also interacts with the wall covering and may alter decisions that can be made if the wall covering becomes wet.

In general, the more impermeable the wall covering is to water on its surface, the more unlikely a drying process is to remediate the backing to that wall covering. As an example, vinyl wallpaper can be wiped down on its surface with wet rags and efficiently cleaned. However, this same vinyl wallpaper, if made wet behind the vinyl surface coating, will retain water. This retained water in combination with the wallpaper paste often makes an excellent amplification site for molds.

Conversely, paper wall coverings are often destroyed by water intrusion and may crinkle, delaminate, and even fall off the walls. However, if the paper remains intact, this type of wallpaper may be dried. Of course the drying process must also dry the substrate and further water intrusion must be controlled.

16.8.2 Cleaning

If water intrusion causes bulging or delamination of wallpaper, these areas must be opened to expose the substrate. The wallpaper that remains intact may be dried from the backside out. If the wallpaper looses its structural integrity and cannot be reapplied, the wallpaper and the wallpaper paste must be removed from the substrate.

- If the wallpaper appears to be wet, but is not bulging or delaminating, decisions as to salvaging the wallpaper depend on the ability to dry not only the surface, but also the layers beneath the surface. In some cases, the moisture source is actually formation of a condensate layer rather than water leaks or cascading behind the wall covering.
- Some wall covering is of sufficient thickness to be delaminated from the substrate and still be retained for reinstallation. In these circumstances, cleaning with detergent decontamination solution will occur on the water-damaged surfaces and a 6-inch boundary around these surfaces.

After attempting to dry the wall coverings, an area (or several areas) should be examined to determine if the drying effectively dried all layers beneath the wallpaper. This examination should occur 36 hours after the drying event and 2 weeks after the drying event. If this examination reveals either retained water or visible mold, the wall covering should be either removed or cleaned as described below.

16.8.3 Cleaning Prior to Removal: Mold or Retained Water Present

This sequence is required if the wall coverings show signs of *visible* mold growth.

- Mist or swab the suspect wall covering area with biocide decontamination solution.
- Wash (swab or wet wipe) any newly exposed mold growth surfaces beneath the wall covering with biocide decontamination solution (refer to the substrate-specific SOPs).

- If washing removes the mold contamination, the wall covering may be:
  - Sufficiently thick to be delaminated from the substrate and still be retained for reinstallation. In these circumstances, cleaning with biocide decontamination solution will occur on the water-damaged surfaces and a 6-inch boundary around these surfaces.
  - Destroyed by efforts required to clean and/or dry the wall covering. In this case, the wall covering will need to be removed and the substrate dried. The wall covering adhesive should also be removed at this time.
  - Cleaned on one side only. If the other side is clean (based on visual assessment) and the wall covering is relatively impermeable, the wall covering can be assumed to be clean. This assumption does not take the place of testing the wall covering to determine whether residual spores are present. Consequently, in this case some risk is inherent in assuming that the wall covering is now clean without testing to prove that case.
- If washing does not remove the mold growth surface:
  - Sequentially remove the wall covering by cutting above and to the side of any mold or former mold growth areas. Removal must occur with repetitive misting of the surrounding nearby environs with biocide decontamination solution.
  - If cleaning destroys the wall covering, remove the wall covering with misting and washing of the area with biocide decontamination solution.
  - Apply shellac-based sealant or equivalent to the exposed substrate surfaces. Allow the surfaces and coating to air-dry.

16.8.4 Reapplication

Decisions as to reapplication of the wall covering will depend on other factors, including the appearance of the cleaned wall covering. If the moisture source is a condensate layer, the current wall covering system and the wall substrate should be carefully reevaluated as to adequacy.

16.9 INSULATION

16.9.1 Quick Facts

Insulation is most often a porous material. The intent of insulation may be thermal (heat and cold) or noise control. In some cases, insulation is also used to control condensate from piping; this use is a form of thermal control.

The insulation material may be backed with paper, plastic, metallic foil, canvas, hardened glues, acrylic coatings, and concrete-type coatings. These coatings should always be evaluated for asbestos content prior to proceeding with cleaning or removal. If the insulation is suspected of containing asbestos, asbestos sampling and asbestos abatement measures may be required.

In addition to the original insulation material, air movement or construction practices may have coated the insulation with paints, wallpaper paste, glues, oils, debris, and soils. This insulation is often a better place for mold amplification if any of the above have been added.

16.9.2 Cleaning

If insulation appears clean (no additional material as described above) and free of mold contamination, drying may be possible. This drying is seldom effective unless the insulation is completely exposed. Complete exposure means opening the insulated cavity so that air movement occurs on all sides of the insulation. Often this type of exposure is not possible because the insulation is affixed to a substrate. In those cases, the substrate may be of more concern than the insulation, and removal of the insulation may be required.

If the insulation can be dried within 24 hours and the insulation is penetrated only by clean water (no soils, gray water, or sewage), the insulation may be reused. However, the reuse will always bear some risk, as the relative dryness of insulation is very difficult to measure. Seemingly dry insulation if reapplied in areas where condensate layers can form, or air movement is restricted so as to increase humidification within a space, may not be dry enough.

In most cases, wet insulation should simply be removed and replaced. In other circumstances, an environmental professional should be consulted if insulation salvage is being considered.

Apply shellac-based sealant or equivalent to the remaining surfaces that are next to the insulation. Allow the surfaces and coating to air-dry.

16.9.3 Cleaning Prior to Removal: Mold or Retained Water Present

This removal sequence is required if the insulation shows signs of visible mold growth or has retained water for more than 24 hours.

- Mist the suspect insulation area with biocide decontamination solution.
- Wash (swab or wet wipe) any newly exposed mold growth surfaces beneath the insulation with biocide decontamination solution.
- Sequentially remove the insulation by cutting above and to the side of any mold or former mold growth areas. Removal must occur with repetitive misting of the surrounding nearby environs with biocide decontamination solution. These mistings must be sufficient to render the insulation dripping wet.
- Apply shellac-based sealant or equivalent to the remaining surfaces that are next to the insulation. Allow the surfaces and coating to air-dry.

16.10   CEILING TILES AND GRID

16.10.1 Quick Facts

Ceiling tiles may be directly affixed to concrete or wood ceiling structures. Often the ceiling tiles in commercial buildings are instead suspended by placement on a metal grid. This metal grid has an upturned T formation, with the two legs of the T supporting the tile.

The ceiling tiles may be:

- Semicompacted porous compressed fiber (acoustic tile):
  - With a coated colored side on one side (facing the occupied space) and a brown unsealed open side facing the false plenum (created between the ceiling tile and the upper floor decking)
  - With a coated colored side on one side (facing the occupied space) and a brown unsealed open side that is covered with fiberglass facing the false plenum (created between the ceiling tile and the upper floor decking)
- Hard compacted compressed fiber (nonacoustic tile)

16.10.2 False Plenums

These tiles if placed on a grid may form a space above the tiles where air is moved throughout the building. Often this air is called *return air* because the air is pulled into return air ducts. These return air ducts return the air to the HVAC system. The spaces are called *false plenums.*

When false plenums are involved in any type of cleanup effort, the effect of the return air stream on the integrity of the HVAC system must be considered. In particular, if biological or particulate (lead dust) or asbestos contamination is suspected, the abatement measures must consider whether the past or present contamination has already spread to the HVAC system.

16.10.3 Tile Aging, Contamination, and Discoloration

Ceiling tiles may discolor with age and exposure to air contaminants such as tobacco smoke. If the backside of the tile is brown originally, a gray discoloration may indicate that mold growth is present, or may just indicate that the tiles are dusty. If however, this gray discoloration is accompanied by moisture intrusion, mold contamination should be suspected.

- Because *Penicillium* is a common mold that grows on ceiling tiles, and due to the fact that the *Penicillium* growth will look like a light coating of gray dust, visual comparison of suspect ceiling tiles to new ceiling tiles may be required.
- Other types of discoloration and tile staining should also be noted, and may indicate moisture intrusion or mold growth.

16.10.4 Tile Grid

The tile grid may also be subject to aging and material breakdown. Metal grid work may rust, and may generate rust flakes or particulates that move into the general air space. Metal grid when cleaned must be protected from follow-on rusting.

Placement of tiles on the tile grid must also be evaluated. Sagging ceiling tiles may indicate water intrusion has occurred or that the humidity level in the false plenum is accelerating the uptake of moisture on some surfaces of the ceiling tiles.

16.10.5 Suspended Ceiling Tile

16.10.5.1 Removal after Water Intrusion

Tiles that have been subject to water intrusion that do not currently have mold growth may be removed using normal maintenance procedures. These procedures include:

- Removal of ceiling tiles
- Bagging (using standard garbage bags) of the tiles for disposal

16.10.5.2 Removal: Mold Present

Compressed fiber ceiling tile cleaning is not possible. If these tiles are contaminated with mold growth, they will be removed for disposal.

If the contaminated area exceeds 30 square feet, the removal must occur using a site-specific plan. This type of removal is beyond the scope of this SOP and will not be discussed further herein.

If visible mold staining or growth is observed and the area contaminated with mold is less than 30 contiguous square feet, the following protocols can be used:

- Mist the ceiling area with biocide decontamination solution.
- Sequentially remove the ceiling tiles and all backing layers (insulation). Removal must occur with repetitive misting of the surrounding nearby environs with biocide decontamination solution.
- Wash (swab or wet wipe) the ceiling tile grid with biocide decontamination solution. Dry the grid and evaluate whether further sealing (painting) will be required to prevent rust formation.

16.10.6 Fixed Ceiling Tile, Cleaning and Removal: Mold Present

- Mist the ceiling area with biocide decontamination solution.
- Sequentially remove the ceiling material that has visible mold contamination. Remove a boundary around these areas. The boundary area should be at least 6 inches in radius from the mold contamination boundary. Removal must occur with repetitive misting of the surrounding nearby environs with biocide decontamination solution.
- Wash (swab or wet wipe) any newly exposed mold growth surfaces with biocide decontamination solution. Note: A mold growth surface is one that has visibly been determined to be a site where mold is growing.
- Apply shellac-based sealant or equivalent to the exposed remaining surfaces if removal of visible mold was required on these surfaces. Allow surfaces to air-dry.

ADDITIONAL READING

Centers for Disease Control/National Institutes of Health. (2006). *Biosafety in Microbiological and Biomedical Laboratories*, 5th ed., Richmond, J. Y. and R. W. Mckinney, eds. Washington, DC: U.S. Government Printing Office.

World Health Organization. (1999). *WHO Infection Control Guidelines for Transmissible Spongiform Encephalopathies. Report of a WHO Consultation*, Geneva, Switzerland, March 23–26.

CHAPTER 17

Decontamination of Ventilation Systems
Discussion of General Decontamination Terms and Methods for Ventilation Systems

Martha Boss, Dennis Day, and Jon A. Cummins

CONTENTS

17.1   FAN COIL UNITS

17.1.1   Quick Facts

The Quick Facts section must be specific to each type of fan coil unit. Attach the manufacturer's specifications, equipment list, and standard operating procedures to this standard operating procedure (SOP). If conflicts are noted between this SOP and the manufacturer's instruction, these conflicts must be resolved prior to proceeding with cleaning events.

All electrified systems must be locked out and tagged out (LOTO). The LOTO must remain in place until all wet cleaned surfaces have dried.

Examine the fan coil units with the unit faceplate or other coverings removed.

17.1.2   Wet Cleaning

All vacuuming will be done with a wet and dry vacuum. Clean all surfaces using a combination of vacuuming and cleaning (swabbing or wiping) of surfaces with detergent decontamination solution. Assuming that no mold is discovered (by visible evidence of mold growth), the decontamination solution may be a detergent for all nonporous surfaces.

- Clean the sheet metal interiors and their associated wall mounting surfaces by vacuuming as detergent decontamination solution is applied (misted) to the components, or
- Carefully wet wipe detergent decontamination solution (so as to minimize dripping).
- Use washing techniques as needed to remove stubborn deposits within the unit. Washing will include using a metal brush on the metal grids and metal coils if laden with debris (hair, lint).

17.1.3 Foam Liners

Compressed or uncompressed foam padding and liners should be sprayed with decontamination solution. If the foam is uncompressed (soft foam with no plastic overlay), the decontamination solution used must be a biocide. This biocide must have residual properties to ensure that the uncompressed foam remains free of mold growth between cleaning events.

- After spraying this foam (where accessible), blot the foam with a disposable paper towel or rag to remove excess water.
- Remove any debris that is visible within the foam surfaces (hair, lint) and bag for disposal.
- Check the foam for delamination from the affixing surface. If delamination has occurred, determine if the foam should be reglued or replaced. If replacement is required, replace the foam with a compressed foam product.

17.2 CONDENSATE DRAIN PANS AND TUBING

Inspect any drip or condensate trays associated with fan coil units. Inspect all tubing leading to or from these trays. Examine the conduit in which the tubing may be directed below the fan coil unit.

- Apply biocide decontamination solution to all of these surfaces and wash the surfaces.
- If the interior of the drainage tubing can be examined (i.e., the tube is clear) and the tubing contains visible debris, remove and replace the tubing. Treat the tubing entrance and exit piping or trays with biocide decontamination solution by misting and washing, prior to replacing the tubing.

If the drainage tubing is not clear, consider replacing with clear tubing. If replacement is not feasible, a representative number (10%) of opaque tubing locations should be examined during each maintenance cycle. This examination will require removing the tubing and visually examining the tubing bore for signs of debris.

During the cooling season, the condensate drip pans will require frequent cleaning to eliminate the accumulation of bioslime within the standing waters. This cleaning will include the use of biocide decontamination solution.

17.3 FANS: BLOWER CHAMBER

The blower chamber is the area where air is received. The fan moves the air from the blower chamber toward the thermal treatment areas (heating or cooling) and then into the ductwork or the room.

- The fans in the blower chamber will also require cleaning. Motor housings must not be wet cleaned and must be protected from decontamination solution overspray.

- Most surfaces can be dry cleaned using a wet and dry vacuum operated in the dry mode.
- Fins on fans will require individual fan blade cleaning to remove debris. This cleaning must not remove the lubricant needed to ensure efficient fan turning. If removal is required, consult the manufacturer's information to apply needed lubricant.
- Fan belts should be examined; if cracked or deformed these belts must be replaced. Fan belts can be dry cleaned only. Wet cleaning should not be used.

17.4   HEATING UNITS (IF PRESENT)

Accessory heating units also require cleaning. These units are usually self-contained and manufacturer's instructions must be followed for any cleaning. Use only cleaning materials that will not cause rusting within the combustion chambers. Rust within the combustion chambers can cause gaps in the chamber and potentially leakage of combustion gases (including carbon monoxide) from the combustion chamber. Any rusting or gaps noted within the combustion chamber must be immediately reported for follow-on maintenance or replacement.

The manufacturer's instructions may not be available for some older units. In these cases, only dry cleaning should be used.

17.5   FILTER MEDIA

Removal and replacement of filter media must be accompanied by cleaning of the filter housing. Cleaning will include washing the filter housing, frame, and grid with biocide decontamination solution.

17.5.1   Dry Cleaning

If the fan coil unit is dry, the fan coil unit interiors may be vacuumed using a wet and dry vacuum in the dry mode.

17.5.2   Cleaning: Mold Present

Cleaning if mold has been discovered, proceed as above. However, in all cases where mold has been identified in any part of the fan coil unit, the decontamination solution used must be a biocide.

- After all visible mold has been removed, bag all solid waste and transfer the solution within the wet and dry vacuum to a disposal pail. Disposal bags will be 6-mil polyethylene (i.e., Hefty Steel Sack or equivalent). Disposal pails will be 5-gallon plastic buckets with tight sealing lids.
- Decontaminate the interior of the wet and dry vacuum.

17.6 HVAC COMPONENTS: DECONTAMINATION

17.6.1 Quick Facts

The Quick Facts section must be specific to each type of heating, ventilation, and air conditioning (HVAC) unit. Attach the manufacturer's specifications, equipment list, and standard operating procedures to this SOP. If conflicts are noted between this SOP and the manufacturer's instructions, these conflicts must be resolved prior to proceeding with cleaning events.

All electrified systems must be locked out and tagged out (LOTO). The LOTO must remain in place until all wet cleaned surfaces have dried.

17.6.2 General Cleaning

If visible mold is not observed and has not been reported in any of the areas that receive air from the HVAC units, then the HVAC units will be cleaned using normal maintenance cleaning procedures. The decontamination solution used is either a detergent or a universal solvent. The most common universal solvent is ammonia, which will dissolve both ionic and organic molecules. Ammonia must never be used where chlorine products are being used, unless these two chemicals are present in stabilized solutions. Consult the material safety data sheets (MSDS) to determine which solutions can be used together or sequentially.

All vacuuming will be done with a wet and dry vacuum. Clean all surfaces using a combination of vacuuming and cleaning (swabbing or wiping) of surfaces with decontamination solution. Assuming that no mold is discovered (by visible evidence of mold growth), the decontamination solution may be a detergent for all nonporous surfaces.

- Clean the sheet metal interiors and their associated wall mounting surfaces by vacuuming as detergent decontamination solution is applied (misted) to the components, or
- Carefully wet wipe detergent decontamination solution (so as to minimize dripping).
- Use washing techniques as needed to remove stubborn deposits within the unit. Washing will include using a metal brush on the metal grids and metal coils if laden with debris (hair, lint).

17.6.3 Foam Liners

Compressed or uncompressed foam padding and liners should be sprayed with detergent or biocide decontamination solution. If the foam is uncompressed (soft foam with no plastic overlay), the decontamination solution used must be a biocide. This biocide must have residual properties to ensure that the uncompressed foam remains free of mold growth between cleaning events.

- After spraying this foam (where accessible), blot the foam with a disposable paper towel or rag to remove excess water.

- Remove any debris that is visible within the foam surfaces (hair, lint) and bag for disposal.
- Check the foam for delamination from the affixing surface. If delamination has occurred, determine if the foam should be reglued or replaced. If replacement is required, replace the foam with a compressed foam product.

17.6.4 Condensate Drain Pans and Tubing

Inspect any drip or condensate trays. Inspect all tubing leading to or from these trays. Examine the conduit in which the tubing may be directed below the unit.

- Apply biocide decontamination solution to all of these surfaces and wash the surfaces.
- If the interior of the drainage tubing can be examined (i.e., the tube is clear) and the tubing contains visible debris, remove and replace the tubing.
  - Treat the tubing entrance and exit piping or trays with biocide decontamination solution by misting and washing, prior to replacing the tubing.
  - Tube replacement must be noted on maintenance logs. If debris continues to be a problem within these condensate return lines, the HVAC unit should be marked as requiring mold inspection.

If the drainage tubing is not clear, consider replacing with clear tubing. If replacement is not feasible, a representative number (10%) of opaque tubing locations should be examined during each maintenance cycle. This examination will require removing the tubing and visually examining the tubing bore for signs of debris.

During the cooling season, the condensate drip pans will require frequent cleaning to eliminate the accumulation of bioslime within the standing waters. This cleaning will include the use of biocide decontamination solution.

17.6.5 Fans: Blower Chamber

- The fans in the blower chamber will also require cleaning. Motor housings must not be wet cleaned and must be protected from decontamination solution overspray.
- Most surfaces can be dry cleaned using a wet and dry vacuum operated in the dry mode.
- Fins on fans will require individual fan blade cleaning to remove debris. This cleaning must not remove the lubricant needed to ensure efficient fan turning. If removal is required, consult the manufacturer's information to apply needed lubricant.
- Fan belts should be examined; if cracked or deformed these belts must be replaced. Fan belts can be dry cleaned only. Wet cleaning should not be used.

17.6.6 Heating Units

Accessory heating units also require cleaning. These units are usually self-contained and manufacturer's instructions must be followed for any cleaning. Use only cleaning materials that will not cause rusting within the combustion chambers. Rust within the combustion chambers can cause gaps in the chamber and potentially leakage of combustion gases (included carbon monoxide) from the combustion chamber. Any rusting or gaps noted within the combustion chamber must be immediately reported for follow-on maintenance or replacement.

The manufacturer's instructions may not be available for some older units. In these cases, only dry cleaning should be used.

17.6.7   Filter Media

Removal and replacement of filter media must be accompanied by cleaning of the filter housing. Cleaning will include washing the filter housing, frame, and grid with biocide decontamination solution.

17.6.8   HVAC Canvas Duct Riser (If Applicable)

If visible mold growth is observed on the canvas flexible junctures on the HVAC duct risers:

- Mist the canvas with biocide decontamination solution and allow to air-dry.
- Clean the metal air duct juncture by wet wiping with the biocide decontamination solution and allow to dry.
- Inspect the interior of the canvas liner:
  - If visible mold is present within the interior, remove the canvas and replace.
  - If visible mold is not present, do not remove or alter the existing canvas liner.

17.6.9   Ductwork Interiors

Ductwork interiors if contaminated are beyond the scope of this SOP. Interior duct-work contamination as evidenced by visible mold must be immediately reported to management.

17.6.10   HVAC Closet

Inspect the closet, alcove, or other area immediately around the HVAC unit. Dust and debris may be removed by vacuuming or wet wiping. Visible dirt and debris not removed by vacuuming will be removed by washing the area with detergent decon-tamination solution.

Clean the door of the HVAC closet, including the grate behind the door filter assembly and the top of the door proper.

- Mist the door filter with biocide decontamination solution saturating the filter grid. Allow to air-dry prior to replacement.

- Wipe down the door grate with detergent decontamination solution. Wipe dry.
- Clean HVAC closet surfaces by vacuuming as detergent decontamination solution is applied (misted) to the components, or carefully wet wipe detergent decontamination solution (so as to minimize dripping).

17.6.11   Dry Cleaning

If the HVAC unit and closet is dry, the unit interiors may be vacuumed using a wet and dry vacuum in the dry mode.

17.6.12   Cleaning: Mold Present

Cleaning if mold has been discovered will proceed as above. However, in all cases where mold has been identified in any part of the HVAC unit, the decontamination solution used must be a biocide.

Uncompressed foam, plasticized open-mesh web liners, and fiberglass liners if contaminated with mold must be removed. Removal will include initial misting with biocide decontamination solution and wetting with biocide decontamination solution during the delamination of these materials. All such materials should be immediately bagged for disposal.

After all visible mold has been removed, bag all remaining solid waste and transfer the solution within the wet and dry vacuum to a disposal pail.

Decontaminate the interior of the wet and dry vacuum.

17.6.13   Decontamination Systems Identification

Provide a coordinated system of identification that includes:

- Proposed access portal locations into ductwork and air-handling units (AHUs)
- Exhaust portal locations for all vacuum units to include those used for the establishment of negative pressure enclosure (NPE)

Affix this identification system to the ductwork and to the exhaust portal locations during the mobilization phase.

17.6.13.1   Diagrams

Provide diagrams both for planning purposes and to assess the decontamination process. These drawings should include mechanical drawings with associated calculations attached. Control diagrams must be provided for the pressure and mechanical cleaning delivery systems. Design analysis and calculations for decontamination equipment should include:

- Flow rates for biocides and pressure differentials needed to maintain these flow rates

- Pressure differentials for vacuuming and the creating of negative air pressure within ductwork and AHU interiors
- Pressure differentials for creating NPEs
- Mechanical cleaning system access portals
- Mechanical cleaning system rotation rate and anticipated vibration against sheet metal housings

Submit schematics to illustrate access and project phasing for each HVAC duct system to include dampers, regulating devices, terminal units, supply outlets, return and exhaust inlets, and the location of required cleaning access points. Drawings must include AHU and ductwork access points and all other schematics needed to evaluate the planned decontamination strategy.

17.6.13.2 Service Labeling

- Piping, including that concealed in accessible spaces, exposed, bare, painted, or insulated, must be labeled to designate when lockout or tagout (LOTO) is required.
- Electrical system LOTO tags must be affixed to the system lockout site (circuit panel or main). Labels denoting that LOTO has occurred must be affixed to the HVAC units where LOTO is controlling the electrical energy flow.
- Scaffolding and power lifts used must be labeled to indicate maximum loading. Additional information needed to determine if scaffolds and power lifts as assembled meet manufacturer's instructions should be provided.

17.6.14 Equipment and Materials

Equipment and materials stored at the site must be fully protected from damage, dirt, debris, and weather. All equipment should be secured from the general public. All equipment must be thoroughly cleaned before going into storage on-site.

17.6.14.1 Pressure Delivery, Cutting, Power Lifts

Pressure delivery devices and cutting devices used should be controlled by fail-safe mechanical controls.

Power lifts must be controlled as specified by the manufacturer.

17.6.14.2 Products

Mechanical materials and equipment to be provided or used should be current standard catalog products from manufacturers regularly engaged in the manufacture of the products. Biocides used must have been proven effective for decontamination of mold-contaminated surfaces. Manufacturer's information as to the biocide's proven effectiveness should be considered as defining whether the biocide is a standard product that can be used.

Mechanical material and equipment should meet the specified requirements and be suitable for the chosen decontamination method. Materials and equipment must be new and free from defects. Biocides should be new and in previously unopened containers. All biocide mixing must occur in equipment that is leak free and does not react with the biocide.

Provide standard manufacturer's identification plates for each individual piece of equipment. This includes, but is not limited to, scaffolds, power lifts, power washers, power mixers, handheld tools, and powered mechanical washing devices.

17.6.15   Decontamination

All methods used must conform to National Fire Protection Association (NFPA) and National Electrical Code (NEC) requirements to assure that electrical safety is maintained.

Materials and equipment must be used in accordance with the requirements of the contract drawings and approved recommendations of the manufacturers.

Workers skilled in this type of work with documented training in the hazards associated with exposure to molds are required to perform this work. Decontamination must be accomplished in a manner such that no degradation of the designed fire ratings of walls, partitions, ceilings, and floors occurs in any areas where HVAC equipment remains operational. This may require a reinstallation of needed building materials.

17.6.16   Cleaning Exposed Surfaces

Thoroughly clean exposed surfaces of piping and equipment that has become covered with dirt, plaster, or other material using a detergent solution. If visible mold is present, biocide must be added to the detergent solution in accordance with the manufacturer's directions.

Mechanical equipment, including piping, ducting, and fixtures, must be rendered clean and free from dirt, grease, and any visible mold.

17.6.16.1   *Insulation*

Insulation and system components within ductwork and AHU interiors must be removed. If visible mold is present, the insulation must be treated with biocide solution before removal. Biocide solutions must be allowed to thoroughly soak into the insulation so that the insulation is removed wet.

17.6.16.2   *Ductwork*

Ductwork exteriors and interiors must be accessed for cleaning. All ductwork must be sequentially vacuumed and washed with biocide. The most effective sequence for all ductwork cleaning should be chosen. All ductwork to be retained must be treated with biocide and air-dried.

- Assembled ducting must be dried by subjecting main and branch interior surfaces to air streams moving at velocities two times the specified working velocities, at static pressures within maximum ratings. This may done with filter-equipped, wheel-mounted, portable blowers; compressed air–operated perimeter lances that direct the compressed air and that are pulled in the direction of normal airflow. Compressed air used for cleaning ducting must be water and oil free. All air streams must remain within the confines of the ductwork and must exhaust either to an exterior portal (to the outside air) or to an area designated within the negative pressure enclosure (NPE).
- Flexible ductwork must be removed. Removal will require that the attachment locations for the flexible ductwork are misted with biocide as the flexible ductwork is disassembled. The ductwork must be immediately bagged or overwrapped. A combination of biocide misting and overwrapping should be used to prevent the contents of the flexible ductwork from spilling during the ductwork removal.
- Interior insulated ductwork must be removed. Removal requires that the attachment locations for the ductwork are misted with biocide as the ductwork is disassembled. The ductwork must be immediately bagged or overwrapped. A combination of biocide misting and overwrapping should be used to prevent the contents of the ductwork from spilling during the ductwork removal.

17.6.16.3   Filters

Filters must be removed. Filters with visible mold growth must be treated with a biocide mist before removal.

Holding frames must be cleaned with biocide solution and air-dried.

17.7   DECONTAMINATION: LAUNDRY ROOM AREAS

The washer area, the back of the dryers, and the dryer exhaust vents require repetitive preventative maintenance (PM). All electrified systems must be locked out and tagged out (LOTO). The LOTO must remain in place until all wet-cleaned surfaces have dried. Areas under washers must be regularly examined for moisture intrusion. Any floor conduits (open holes, fractures in concrete decking) near the washers must be closed (plugged).

Piping associated with the washers may be both a source of moisture from leaks and a source of condensate. The condensate may be forming due to improperly insulated pipes and/or cold and hot water piping being position too closely to one another. Water intrusion, no matter what the source, must be stopped. All piping insulation must be evaluated to determine the potential for asbestos contamination. Fiberglass insulation may have been placed over the former asbestos insulation, so all areas must be examined. Black uncompressed foam insulation may be backed with asbestos papers or may contain imbedded asbestos.

If the piping insulation is contaminated with mold, the insulation must be removed using misting with biocide decontamination solution as the insulation is removed. If this piping is contaminated with asbestos or lead paint (some insulation may have been painted on the exterior wrapping), removal is beyond the scope of this SOP.

Vacuum and clean all dryer vents, including the vents that conduct air from the dryers to any outside exhaust and the vents associated with lint filters. If mold is not present, dry vacuuming can be used. If mold is present:

- Remove and replace the vents, or
- Clean the vents using wet vacuuming methods while misting and washing with biocide decontamination solution.

Some dryer casements also contain slotted openings in their casements. These slots should be wet wiped with detergent decontamination solution. If mold is observed, additional cleaning may be required within the casements. Opening the casements and additional cleaning of the dryer is beyond the scope of this SOP.

ADDITIONAL READING

Centers for Disease Control/National Institutes of Health. (2006). *Biosafety in Microbiological and Biomedical Laboratories*, 5th ed., Richmond, J. Y. and R. W. Mckinney, eds. Washington, DC: U.S. Government Printing Office.

World Health Organization. (1999) *WHO Infection Control Guidelines for Transmissible Spongiform Encephalopathies. Report of a WHO Consultation*, Geneva, Switzerland, March 23–26.

CHAPTER 18

Emergency Plans

Martha Boss and Randy Boss

CONTENTS

18.1   INTRODUCTION

As the vulnerability of a facility or site is assessed, one of the key components should be the status of emergency planning. When appropriately designed, these plans, policies, and procedures can have a major impact upon occupant survivability. This chapter presents emergency planning elements that are often overlooked, but it is not an exhaustive emergency planning template.

All buildings should have current emergency plans to address fire, weather, and other types of emergencies. These plans should be updated to consider chemical, biological, and radiological (CBR) attack scenarios, procedures for communicating instructions to building occupants, identifying suitable shelter-in-place areas, identifying appropriate use and selection of personal protective equipment (i.e., clothing, gloves, and respirators) and directing emergency evacuations.

Individuals developing emergency plans and procedures should recognize that fundamental differences exist among chemical, biological, and radiological agents. In general, chemical agents will show a rapid onset of symptoms, while the response to biological and radiological agents will be delayed. Whether assumed to be possible due to intentional CBR release or natural events, both chemical and biological security must be discussed during planning events.

- Chemical security may include the use of building materials and ventilation system components that do not impart physical, chemical, or biological risk. The interaction of these materials with the general building environs must be considered.
- Concurrent with chemical security, biological security is also a concern in all facilities. The amplification biologicals (molds, bacteria, viruses) within building envelopes may be spread via ventilation systems and magnified by chemical contamination including contamination associated with deteriorating building materials. When and if such spread occurs, areas where further amplification can occur are areas where the building is potentially not secure.

Issues such as limited- and controlled-access designated areas and procedures for chemical storage, heating, ventilation, and air conditioning (HVAC) control or shutdown, and communication with building occupants and emergency responders should all be addressed.

18.1.1   Communication Systems

Communication systems must always be evaluated in terms of actual efficacy in an emergency. Wireless communication systems, including cell phone transmission systems, may be overloaded and as a result rendered ineffective. Reliance on external providers to guarantee emergency communications is not acceptable. Some form of site controlled alarm and communication system (e.g., radio frequency devices, walkie-talkies) must be available for use.

18.1.2 Drills

Staff training, particularly for those with specific responsibilities during an event, is essential and should cover both internal and external events. Holding regularly scheduled practice drills, similar to the common fire drill, provides plan-testing events and an opportunity for occupants and key staff to rehearse the plan. These drills increase the likelihood for success in an actual event. For protection systems in which HVAC control is done via the energy management and control system, emergency procedures should be exercised periodically to ascertain that the various control options work (and continue to work) as planned.

18.2 HAZARDOUS MATERIALS TRANSPORTATION AND CBR

Hazardous materials in transportation are particularly vulnerable to sabotage or misuse. Security of hazardous materials in the transportation environment poses unique challenges as compared to security at fixed facilities. Hazardous materials are frequently transported in substantial quantities. Such materials are already mobile and are frequently transported in proximity to large population centers. Further, hazardous materials in transportation are often clearly identified to ensure safe and appropriate handling during transportation and to facilitate effective emergency response in the event of an accidental release. While the hazardous materials regulations (HMRs) provide for a high degree of safety with respect to avoiding and mitigating unintentional releases of hazardous materials during transportation, the HMRs do not specifically address security threats.

If a component of emergency response planning includes the need to secure or transport hazardous materials, the requirements defined in the *The Hazardous Materials; Security Requirements for Offerors and Transporters of Hazardous Materials* (68 FR 14510) must be met. The U.S. Department of Transportation (DoT) issued this rule to enhance the security of hazardous materials (hazmat) transported in commerce. This rule requires certain shippers and carriers to develop and implement security plans. In addition, all shippers and carriers of hazardous materials must assure that their employee training includes a security component.

18.2.1 Security Plans

All people engaged in offering a specific hazmat for transportation in commerce or transporting hazmat in commerce must have written security plans. In addition, the U.S. Department of Defense (DoD) requires compliance with applicable DoT requirements in DoD Regulation 4500.9-R, Part 2, Chapter 204, para. A.1, dated May 2003. Thus, DOD security plans must also be in compliance with 49 CFR 172.800 and a portion of those compliance requirements are quoted here:

Each person who offers for transportation or transports one or more of the following hazardous materials (hazmat) must develop and adhere to a security plan for hazardous materials:

1. A highway route-controlled quantity of a Class 7 (radioactive) material in a motor vehicle, rail car, or freight container;
2. More than 25 kilograms (kg) (55 pounds) of a Division 1.1, 1.2, or 1.3 (explosive) material in a motor vehicle, rail car, or freight container;
3. More than one L (1.06 quart) per package of a material poisonous by inhalation that meets the criteria for Hazard Zone A;
4. A shipment of a quantity of hazardous materials in a bulk packaging having a capacity equal to or greater than 13,248 Liter (L) (3,500 gallons) for liquids or gases or more than 13.24 cubic meters (468 cubic feet) for solids;
5. A shipment in other than a bulk packaging of 2,268 kg (5,000 pounds) gross weight or more of one class of hazardous materials for which placarding of a vehicle, rail car, or freight container is required;
6. A select agent or toxin regulated by the Centers for Disease Control and Prevention under 42 CFR 73; or
7. A quantity of hazardous material that requires placarding under the provisions of subpart F Part 172.

The security plan must include an assessment of possible transportation security risks for shipments of the hazardous materials and appropriate measures to address the risks. Under 49 CFR 172.802, the components of a security plan must include:

- Personnel security: measures to confirm information provided by job applicants hired for positions that involve access to and handling of the hazardous materials covered by the security plan.
- Unauthorized access: measures to address the assessed risk that unauthorized persons may gain access to the hazmat covered by the security plan or transport conveyances being prepared for transportation of the hazmat covered by the security plan.
- En route security: measures to address the assessed security risks of shipments of hazmat covered by the security plan en route from origin to destination, including shipments stored incidental to movement.

The security plan must be in writing, must be specific to the hazmat to be transported, and must be retained for as long as the security plan remains in effect. Copies of the security plan, or portions thereof, must be available to the employees who are responsible for implementing the plan, consistent with personnel security clearance or background investigation restrictions and a demonstrated need to know. The security plan must be revised and updated as necessary to reflect changing circumstances. When the security plan is updated or revised, all copies of the plan must be maintained as of the date of the most recent revision.

18.2.2 Training

49 CFR 172 Subpart H (final rule) provides training requirements applicable to all persons considered hazmat employees. Hazmat employees are those employees who during the course of employment load, unload, or handle hazmat; those that prepare hazmat for transportation (prepare shipping papers, label, mark, placard, package); those responsible for the safe transportation of hazmat; or those operating a vehicle to transport hazmat. All DoT/DoD hazmat employees must be trained in accordance with the requirements of 49 CFR 172 Subpart H.

Two separate requirements are extant for security training:

1. **In-Depth Training:** If a security plan is required, then affected employees are required to be trained on the specifics of that plan. Each hazmat employee must be trained concerning the security plan and its implementation. Security training must include facility or site security objectives, specific security procedures, employee responsibilities, actions to take in the event of a security breach, and the organization security structure.
2. **Security Awareness Training:** Hazmat employees, as defined in 49 CFR 171.8 and trained under 49 CFR 172.704, must receive training that provides an awareness of the security issues associated with hazardous materials transportation and possible methods to enhance transportation security.

No later than March 24, 2006, each hazmat employee was to receive training that provides an awareness of security risks associated with hazardous materials transportation and methods designed to enhance transportation security. This training must include a component covering how to recognize and respond to possible security threats. After March 25, 2003, new hazmat employees were required to receive the security awareness training within 90 days after employment.

18.3 BIOLOGICAL RISK AND PHARMACEUTICAL SUPPORT

Pharmaceutical support planning is an increasingly important component of emergency planning. Such support has proven invaluable in radiological emergencies. To deal with biological threats, additional planning to meet supply needs should occur. Since a real-time, biological threat–detection system is not currently widely available, emergency planners may be forced to assume the worst and provide prophylactic medical support.

Planning conducted in February 2006 illustrated the Health and Human Services Department's Project Bioshield countermeasures. A five-year, $21.9 million contract was awarded to deliver 390,000 doses of Ca-DTPA (Pentetate Calcium Trisodium Injection Sterile Solution) and 60,000 doses of its Zn-DTPA (Pentetate Zinc Trisodium Injection Sterile Solution). These drugs combat internal exposure to plutonium, americium, and curium, according to an HHS statement. The number of doses requested by HHS was based upon a threat assessment of the medical affects of a nuclear or radiological attack. Note: This assessment was conducted by the Homeland Security Department and the interagency Weapons of Mass Destruction Medical Countermeasures Subcommittee.

The pandemic planning initiatives are an example of biological risk evaluations that will require pharmaceutical support elements.

18.4   RADIATION EMERGENCIES

Needed information may be provided by others since the potential for monitoring outside the shelter by the shelter's inhabitants will be limited. After a release of radioactive materials, local authorities will monitor the levels of radiation and determine what protective actions to take. The most appropriate action will depend on the situation. The local emergency response network or news station should provide information and instructions during any emergency. If a radiation emergency involves the release of large amounts of radioactive materials, sheltering in place may be required or evacuations may be ordered.

If you are advised to shelter in place, you should do the following:

- Close and lock all doors and windows.
- Turn off fans, air conditioners, and forced-air heating units that bring in fresh air from the outside. Only use units to recirculate air that is already in the building.
- Close fireplace dampers.
- Move to an inner room or basement.

Obviously a knowledge of the building's shielding capacity and ventilation system options will be crucial in determining if shelter can be provided in that building. Following are some steps recommended by the World Health Organization (WHO) if a nuclear blast occurs.

18.4.1   Near the Blast

1. Turn away.
2. Close and cover your eyes.
3. Drop to the ground face down.
4. Place your hands under your body.
5. Remain flat until the heat and two shock waves have passed.

18.4.2   Outside: When the Blast Occurs

1. Cover your mouth and nose (i.e., scarf, handkerchief, or other cloth).
2. Remove any dust from your clothes by brushing, shaking, and wiping in a ventilated area.

3. Proceed to a shelter, basement, or other underground area, preferably located away from the direction that the wind is blowing.
4. Prior to entering the shelter, remove all clothing. If possible, take a shower, wash your hair, and then enter the shelter.

18.4.3 Inside Shelter: When the Blast Occurs

1. Cover your mouth and nose (i.e., scarf, handkerchief, or other cloth).
2. Clean and cover any open wounds on your body.
3. Shut off ventilation systems and seal doors or windows until the fallout cloud has passed. However, after the fallout cloud has passed, unseal the doors and windows to allow some air circulation.
4. Stay inside until authorities have alerted people in your area that it is safe to leave their shelters. When in doubt, stay inside.
5. Listen to the local radio or television for information and advice. Authorities should provide advice on shelter or evacuation options.
6. Cover your mouth and nose with a damp towel when and if you leave the shelter.
7. Use stored food and drinking water. Do not eat local fresh food or drink water from open water supplies.

18.4.4 Evacuation

1. Listen to the radio or television for information about evacuation routes, temporary shelters, and procedures to follow.
2. Before leaving a shelter, close and lock windows and doors and turn off air conditioning, vents, fans, and furnace. Close fireplace dampers.
3. Take disaster supplies (i.e., flashlight and extra batteries, battery-operated radio, first aid kit and manual, emergency food and water, nonelectric can opener, essential medicines, cash and credit cards, and sturdy shoes).

18.4.5 Potassium Iodide

Potassium iodide (KI) should only be taken in a radiation emergency that involves the release of radioactive iodine, such as an accident at a nuclear power plant or the explosion of a nuclear bomb. A "dirty bomb" most likely will not contain radioactive iodine.

A person who is internally contaminated with radioactive iodine may experience thyroid disease later in life. The thyroid gland will absorb radioactive iodine and may develop cancer or abnormal growths later on. KI will saturate the thyroid gland with iodine, decreasing the amount of harmful radioactive iodine that can be absorbed. KI only protects the thyroid gland and does not provide protection from any other radiation exposure.

Note: Some people are allergic to iodine and should not take KI. Check with your doctor about any concerns you have about potassium iodide.

18.5 FIRE PROTECTION SYSTEMS

Fire protection assets are always critical assets. All the infrastructure planning will be for naught if the building catches fire! The fire protection system inside the building should maintain life safety protection after an incident and allow for safe evacuation of the building when appropriate.

While fire protection systems are designed to perform well during fires, these systems are not traditionally designed to survive a bomb blast. The three components of the fire protection system are:

- Active features, including fire alarms, sprinklers, stand pipes, fire hydrants, smoke control;
- Passive features, including fire resistant barriers; and
- Operational features, including system maintenance, testing intervals, and employee training.

18.5.1 Smoke Removal Systems

In the event of a blast, the available smoke removal system may be essential to smoke removal, particularly in large, open spaces. This equipment should be located away from high-risk areas such as loading docks and garages. The system controls and power wiring to the equipment should be protected. The system should be connected to emergency power to provide smoke removal.

Smoke removal equipment should be provided with stand-alone local control panels located in the fire command center that can continue to individually function in the event the control wiring is severed from the main control system.

During an interior bombing event, smoke removal and control is of paramount importance. If window glazing is hardened, a blast may not blow out windows, and smoke may be trapped in the building.

18.5.2 Water Supply and Secondary Water Supply

A dependable public or private water supply capable of supplying the required flow for fire protection must be provided in accordance with the requirements of NFPA 24. The local fire department must be consulted as to the required fire hydrant locations and thread types for hydrants. The supply to the fire pump should include an auxiliary bypass (normally closed) from the municipal water supply. To increase the reliability of the fire protection system in strategic locations, a dual pump arrangement could be considered, with one electric pump and one diesel pump.

The fire protection water system should be protected from single point failure in case of a blast event. The incoming line should be encased, buried, or located 50 feet away from high-threat areas. The interior mains should be looped and sectionalized where provided. The interior standpipes should be cross-connected on each floor.

The secondary water supply fire suppression systems should have a 30-minute duration capacity in accordance with NFPA requirements.

- For buildings located in rural areas where established water supply systems for fire fighting are not available, the water supply should be obtained from a tank, reservoir, or other source that can supply a minimum of 10,000 gallons.
- A secondary water supply for high-rise buildings should be provided. In seismic zones 2, 3, and 4 the on-site reservoir supplying fire pumps must be installed in accordance with NFPA 20.

18.5.3 Fire Department Access

At least one side of all buildings should be accessible to fire apparatus. Fire department vehicle access should be provided and maintained in accordance with the requirements of the National Model Fire Code that is used, NFPA 241 and NFPA 1141. The local fire department should be consulted as to their specific requirements regarding the surface material of the access roadway(s), minimum width of fire lane(s), minimum turning radius for the largest fire department apparatus, weight of the largest fire department apparatus, and minimum vertical clearance of the largest fire department apparatus. At least one access road having a minimum unobstructed width of 26 feet should be located within a minimum of 15 feet and a maximum of 30 feet from the building.

18.5.4 Window Glazing

During emergency planning and if glazing has been used on windows, firefighters must be consulted to determine if their normal tools (i.e., pick-head axe, halligan tool) can readily overcome the glazing barriers. If the use of more specialized tools, such as a rabbit tool, a k-tool, circular saws, rams, or similar devices is necessary to break through the glazing barrier or if the glazing itself is hardened so that a blast may not blow out the windows, alternative methods or systems must be designed to ensure smoke from the incident is not trapped inside the building.

18.5.5 Aerial Apparatus

Buildings or portions of buildings exceeding 30 feet in height from the lowest point of fire department vehicle access should be provided with aerial access. Pavements and curbed roads capable of accommodating fire department aerial apparatus must be installed. Overhead utility and power lines should not be within the aerial access roadway.

18.6 REPLACEMENT EQUIPMENT AND RELIEF SUPPLIES

The potential for replacement of critical equipment is an important consideration. If conduits for replacement equipment will not be available, then critical equipment and replacement parts for that equipment may need to be stored on-site.

Consider the events in New Orleans Louisiana after hurricane Katrina:

> Existing pumps had to be dried prior to pump repair, which involved replacing motors and other parts. The extra time to rehabilitate these pumps delayed the pumping of water back into Lake Pontchartrain. So the draining of waters from New Orleans, which was 80% submerged, was prolonged.

Similarly, supply chains must be analyzed. If the supplies cannot be guaranteed, planning documents must analyze this risk and make decisions about stockpile efficacy. Examples of supply chain failures include:

> After Hurricane Wilma, which occurred in 2005, residents in South Florida waited for ice and water that in some cases didn't arrive. Many of the residents were elderly, lack transportation and/or lived 25 blocks from the closest county-run distribution site.
>
> Drivers were given the wrong directions, drivers didn't gas up before the hurricane, and drivers were given cell phones that didn't work—all leading to supply chain disruptions. Private industry owned the ice, water, packaged foods, and delivery trucks. When the requirement went from 0 to 600 trucks in a day, private industry could not supply trucks quickly enough. (*Miami Herald*, October 28, 2005)

18.7 DEBRIS MANAGEMENT

Debris management includes both debris from the initial emergency and that generated during an emergency response. Even after the debris is disposed of, issues may arise as hidden contamination is assessed. Common contaminants include asbestos, lead, mercury, and Freon.

> Representatives of the U.S. Army Corps of Engineers discussed the debris left after hurricanes Katrina and Rita struck the gulf coast of the U.S. In a September interview on MSNBC . . . rewrite. Both burnable and on-burnable debris was present. Burnable debris included large trees that had to be put in grinder and then a chipper. The accumulated tree and vegetative material could then be stockpiled for disposal or burning. Non-burnable debris included various appliance and building accoutrements—some containing hazardous wastes (e.g., freon, mercury, lead)

ADDITIONAL READING

Analysis of the Clandestine CB Threat to USAF Strategic Forces (Unclassified), Volumes I and 11. Technical Considerations, Defense Technical Information Center (DTIC) Number AD379465, February 1967.

Antiterrorism Front-End Analysis (Unclassified), DTIC Number AD-C954865, June 1984.

Background Paper, Technologies Underlying Weapons of Mass Destruction, Office of Technology Assessment, U.S. Congress, OTA-BP-ISC-1 15.

Behind the first roar of machinery to drain the city, A tale of pluck and luck, John Schwartz, *New York Times*, September 8, 2005.

Biological Effects of Radiation, U.S. Nuclear Regulatory Commission, Office of Public Affairs.

Chemical/Biological Hazard Prediction Program, Technical Report, Defense Technical Information Center (DTIC) Number AD-BL 63245, 1991.

Chemical, Biological and Radiological Incident Handbook, Chemical, Biological and Radiological (CBRN) Subcommittee of the Interagency Intelligence Committee on Terrorism (IICT). October 1998.

Considerations of Flood Plains and Wetlands in Decision Making, Government Services Administration (GSA), GSA ADM 1095.2.

Corps closes landfill over asbestos fears EPA says grinding may release fibers, Mark Schleifstein, Staff writer, *New Orleans Times-Picayune*, February 17, 2006.

Effects of Terrorist Chemical Attack on Command, Control, Communications, and Intelligence (C^3I) Operations, Defense Technical Information Center (DTIC) Number AD-BL 65614, April 1992.

Facilities Standards for the Public Building Service, U.S. General Services Administration, Public Buildings Service, Office of the Chief Architect, Washington, DC 20405, March 2003.

Fundamentals of Protective Design for Conventional Weapons, TM 5-855-1. Washington, DC, Headquarters, U.S. Department of the Army, 1986.

Glossary of Terms in Nuclear Science and Technology, La Grange, IL, America Nuclear Society, 1986.

Guidance for Protecting Building Environments from Airborne Chemical, Biological, or Radiological Attacks, National Institute of Occupational Safety and Health (NIOSH), U.S. Department of Health and Human Services (DHHS) (NIOSH) Pub No. 2002-139, May 2002.

Hazardous Material Workshop for Law Enforcement, student manual from the Emergency Management Institute, Federal Emergency Management Agency.

HHS awards contract for radiation countermeasures, Global Security Newswire, February 14, 2006.

Manual for the Prediction of Blast and Fragment Loading on Structures, U.S. Department of Energy (DOE)/TIC 11268. Washington, DC, Headquarters, U.S. Department of Energy, 1992.

North American Emergency Response Manual, U.S. Department of Transportation, Transport Canada, Secretariat of Transport and Communications, 1996.

Nuclear Reactor Concepts, U.S. Nuclear Regulatory Commission, May 1993.

Nuclear Terms Handbook, U.S. Department of Energy, Office of Nonproliferation and National Security, 1996.

Proliferation of Weapons of Mass Destruction, Assessing the Risk, Office of Technology Assessment, U.S. Congress, OTA-ISC-559.

Protective Construction Design Manual, ESL-TR-87-57. Prepared for Engineering and Services Laboratory, Tyndall Air Force Base, FL. 1989.

Replacement pumps don't exist, officials say damaged equipment may take a week to dry out before it can be repaired. Peter Pae, *Los Angeles Times*, September 3, 2005.

Security Engineering, TM 5-853 and Air Force AFMAN 32-1071, Volumes 1, 2, 3, and 4. Washington, DC, Departments of the Army and Air Force, 1994.

Senate approves disease surveillance bill, Danielle Belopotosky, *National Journal's Technology Daily*, January 6, 2006.

Short supplies frustrate Floridians, South Floridians remained locked in long lines and frustration in their attempts to get ice, food and water from the government, Wanda J. Demarzo, Jacqueline Charles, Theresa Bradley, and Noah Bierman, *Miami Herald*, October 28, 2005.

Structures to Resist the Effects of Accidental Explosions, Army TM 5-1300, Navy NAVFAC P-397, AFR 88-2.Washington, DC, Departments of the Army, Navy and Air Force, 1990.

Water returned to lake contains toxic material, Sewell Chan and Andrew C. Revkin, *New York Times*, September 7, 2005.

What happens to all the Katrina debris? Vesay of Army Corps of Engineers explains volume, process of cleanup, MSNBC, September 12, 2005.

Weapons of Mass Destruction Terms Handbook, Defense Special Weapons Agency, DSWA-AR-40H, 1 June 1998.

Acronyms

μm	micrometer or micron, one-millionth of a meter
A	Acrit. Unknown probability that an aggressor intends to attack an asset.
AABC	Associated Air Balance Council
ACAMS	automatic continuous air monitoring system
ACGIH	American Conference of Governmental Industrial Hygienists
ADPI	air diffusion performance index
AEL	airborne exposure limits
AHU	air-handling units
ALARA	as low as is reasonably achievable
APRs	air-purifying pespirators
ARI	Air-Conditioning and Refrigeration Institute
ASC	allowable stack concentration
ASHRAE	American Society of Heating, Refrigerating, and Air-Conditioning Engineers
ASTM	American Society for Testing and Materials
ASZM-TEDA	Copper-silverzinc-molybdenum-triethylenediamine
ATSDR	Agency for Toxic Substances and Disease Registry
B	benefit of implementing a recommendation
BAS	building automation systems
BMS	balanced magnetic switches
°C	degrees Celsius
CAA	Clean Air Act
CAS	Chemical Abstracts Service
CBR	chemical, biological, or radiological
CCTV	Closed circuit television
CDC	Centers for Disease Control and Prevention
CEO	chief operating officer
CERCLA	Comprehensive Environmental Response, Compensation and Liability Act
CF	chemical facilities
CFATS	Chemical Facility Anti-Terrorism Standards
CFCs	chlorinated fluorocarbons
CFR	Code of Federal Regulations
cfm	cubic feet per minute
CG	phosgene; a choking agent*
Ci	curie
CIF	chemically impregnated fibers
CI/KR	critical infrastructure and key resources
CJD	Creutzfeldt-Jakob Disease
CMA	Chemical Materials Agency

COI	chemicals of interest
cpm	counts per minute
CWA	chemical warfare materiel
Da	expected damage level after implementing the recommendation
Db	expected damage level before implementing the recommendation
DCV	demand-controlled ventilation
DDC	direct digital control
DHHS	Department of Health and Human Services
DHS	Department of Homeland Security
DoD	Department of Defense
DoE	Department of Energy
DoJ	Department of Justice
DoT	Department of Transportation
dpm	disintegration rate
DSWA	Defense Special Weapons Agency
DTIC	Defense Technical Information Center
EMS	Energy management system
EPA	Environmental Protection Agency
EPCRA	Emergency Planning and Community Right-to-Know Act
FEMA	Federal Emergency Management Agency
FERC	Federal Electric Regulatory Commission
fpm	feet per minute
FRC	Federal Radiation Council
FRP	Facility Response Plan
FSA	Facility Security Assessments
FSP	Facility Security Plans
ft^2	square feet
GB	isopropyl methylphosphonofluoridate; a nerve agent (sarin)*
GC	gas chromatographic
GM	Geiger-Muller
GPL	general population limits
gsf	gross square feet
HAZMAT	hazardous materials
HD	mustard agent
HEPA	high-efficiency particulate air
HMR	hazardous materials regulations
HSPD	Homeland Security Presidential Directives
HVAC	heating, ventilating, and air-conditioning
ICRP	International Commission on Radiological Protection
IDLH	immediately dangerous to life and health
IICT	Interagency Intelligence Committee on Terrorism
in	inch
IND	improvised nuclear device
IRI	international risk insurance
IT	information technology

HP	horsepower
KeV	kiloelectron-volts
KI	potassium iodide
km/hr	kilometers per hour
kW/h	kilowatt/hours
L_A	likelihood of adversary attack
L_{AS}	likelihood of adversary success in causing a catastrophic event
LD_{50}	lethal dose; dose that would cause death in half of the exposed population
LEA	law enforcement agencies
LOTO	lockout or tagout
LPS	liters/second
m/s	meters per second
mCi	millicurie
MCRF	multi-cell radial flow
M^2	square meters
M^2/g	square meters per gram
M^3/min	cubic meters per minute
mR/hr	milliroentgen per hour
MERV	minimum efficiency reporting value
MINICAMS	miniature chemical agent monitoring system
mSv	milliSeverts = 100 mrem
mm	millimeters
mph	miles per hour
MPPS	most penetrating particle size
MTSA	Maritime Transportation Security Act
N95	95% efficient respirator filter for use in a non-oil mist environment
NAFA	National Air Filtration Association
NaI	sodium iodide
NBC	nuclear, biological, and chemical
NEBB	National Environmental Balance Bureau
NERC	North American Electric Reliability Council
NFPA	National Fire Protection Association
NIPP	National Infrastructure Protection Plan
NIOSH	National Institute for Occupational Safety and Health
NIST	National Institute of Standards and Technology
Nm	nanometers, one-billionth of a meter
NORM	naturally occurring radioactive materials
NPE	negative pressure enclosure
NPV	net present value
NRP	*National Response Plan*
NRT	near real time
NTP	normal temperature and pressure
OC	operating costs

OPT	optical microscope
OSHA	Occupational Safety and Health Administration
Pa	likelihood of preventing aggressor attempt after recommendations are implemented
P&A	precision and accuracy
Pb	likelihood of preventing aggressor attempt before recommendations are implemented
PCBs	polychlorinated biphenyls
PET	positron [β^+] emission tomography
P&ID	piping and instrument diagrams
PHAs	process hazard analyses
PLC	programmable logic controllers
PM	preventative maintenance
ppb	parts per billion
PPE	personal protective equipment
ppm	parts per million
PPS	protective physical system
PSE	particle size efficiency
PSM	process safety management
PVC	polyvinyl chloride
R	roentgen
rad	radiation absorbed dose
RDD	radiological dispersal device
rem	radiation equivalent in man
RFI	request for information
RMP	Risk Management Plan
ROE	rules of engagement
RTUs	remote terminal units
QIP	quench indicating parameter
Sa	likelihood of preventing aggressor success after recommendations are implemented
SAFETY ACT	Support Anti-Terrorism by Fostering Effective Technologies Act
Sb	likelihood of preventing aggressor success after recommendations are implemented
SCADA	supervisory control and data acquisition
SEL	source emission limit
SEM	scanning electron microscope
SIP	sheltering in place
SNAP	Significant New Alternatives Policy
SPCC	spill prevention controls and countermeasures
STEL	short-term exposure limits
STQ	screening threshold quantities
SV	severity value
SVA	security vulnerability assessment
TABB	Testing, Adjusting, and Balancing Bureau

TEDA	triethylene diamine
TIC	toxic industrial chemical
TIM	toxic industrial material
TDEs	transmissible degenerative encephalopathies
TENORM	technically enhanced (altered) normally occurring radioactive materials
TQ	threshold quantity
TSEs	transmissible spongiform encephalopathies
TSP	trisodium phosphate
TWA	time-weighted averages
UPs	Uninterrupted power service
USAF	U.S. Air Force
USCG	U.S. Coast Guard
VA	Vulnerability Assessment
VAM	Vulnerability Assessment Methodology
VAP	Vulnerability Assessment Program
VAV	variable air volume
VOC	volatile compounds
VX	nerve agent: phosphonothioic acid, methyl-S-(2-(bis(1-methylethyl)amino)ethyl)0-ethyl ester
WHO	World Health Organization
WPLs	worker population limits
XSD	Halogen-specific detector
yr	year(s)

Index

A

Absorbed dose, 178
Absorption, 138–139
Access control, 2
  airborne hazard protection, 197–198
  barriers, 59, 69, 122–125
  critical asset area entrances, 62–70
  detection methods, 48
  garages, 123–124
  information assurance procedures, 47
  locks/keys, 60
  mechanical rooms, 235
  physical security, 60, *See also* Physical
    security assessment
  request for information, 24
  security badges, 60
  security screening, 197–198
  service areas, 123–124
  vehicle gates, 66–69
  waiting areas, 124
Acoustical enclosures, 204
Acronyms, 395–399
Action plan development, 58, 204
Activated carbon filters, 217, 243–244, 246–247,
    250–251
  reactant impregnation, 217, 243, 247, 250–251
Activity, 175–176
Administrative support organizations, 71–72
Adsorption, 138–139, 243–244, *See also*
    Filtration systems; Sorbents
Adsorption capacity, 248–249
Adsorption/desiccant dehumidifiers, 326–327,
    362
Adversary characteristics, 13–14
Aerial access for fire protection, 391
Aerosols, 139, 242, 243, 256, 258
Agent HD, 141
Airborne contaminants, 242, *See also* Airborne
    hazard protection options; *specific*
    hazards or threats
  dilution rate, 139
  relative sizes, 242
  sampling methods, 144–149
Airborne exposure limits (AELs), 145–146
Airborne hazard protection options, 195–198,
    241–243, *See also* Biological threats;
    Chemical threats
  access control, 197–198
  airlocks, 236–238

collective protection filter units, 262–264
collective protection system testing and
    startup, 223
contaminant concentration buildup, 234
emergency air distribution shutoff,
    205–206
evacuation, 196, 207
filter and ventilation system, 205, 216–219,
    225, 261, *See also* Filtration systems
high-risk areas, 200–204
HVAC design approach, 219–223
HVAC inspection, 223–225
HVAC system performance, 197
mass notification system, 206
mitigation, 197
monitoring and detection, 272–273
outdoor air intakes, 198–200
pressure differentials, *See* Pressure
    differentials
prevention, 197
protection options, 198
protective equipment, 226
purging, 238
revolving doors, 200
sheltering in place, 207–216, *See also*
    Sheltering in place
smoke fans, 225
steady state and dilution ventilation control,
    232–234
ventilation protection strategies, 204–206
vestibules, 200
Air cleaning, 242, *See also* Filtration systems
  sorbent filters, 244–245, *See also* Sorbents
Air conditioner chillers, 302
Air conditioner de-energization, 206
Air conditioning units, computer room, 301
Air distribution systems design, 289–292, *See*
    also Air-handling unit; Variable air
    volume (VAV) system; Ventilation
    systems
  airside economizer cycle, 287
  insulation, 308
  underfloor system, 288
Air filtration, *See* Filtration systems
Airflow measuring devices, 268
Air-handling unit (AHU)
  access doors, 281
  air-delivery devices, 281
  baseline systems, 284
  controllers, 269

Printed and bound by CPI Group (UK) Ltd, Croydon, CR0 4YY

23/10/2024

01778262-0004